This important synthesis represents the first effort by present-day scholars to convey the variety of ways in which medieval scientists and natural philosophers used mathematics and mathematical modes of thought to describe natural phenomena.

Eleven distinguished historians of science have contributed original essays on the application of mathematics to natural philosophy, astronomy, cosmology, optics, and medicine. They interpret and describe a range of topics including how natural philosophers used mathematics to represent motion and the nature of a physical continuum; how diagrams and graphs were utilized in astronomy; how natural philosophers perceived eccentric and epicyclic orbs in the heavens; the manner in which Ptolemy and Alhazen employed mathematics to explain a visual anomaly in the sky; the process by which optics in the Latin Middle Ages was transformed from a nonmathematical to a mathematical discipline; the relationship in optics between mathematical theory and experiment; and how physicians applied quantitative methods to the determination of the "degree" of a compound medicine.

The book is a fitting tribute to Professor Marshall Clagett of The Institute for Advanced Study, Princeton, for his significant contributions to the history of medieval science.

Mathematics and its applications to science
and natural philosophy in the Middle Ages

Marshall Clagett

Mathematics and its applications to science and natural philosophy in the Middle Ages

Essays in honor of Marshall Clagett

Edited by
Edward Grant *Indiana University*
John E. Murdoch *Harvard University*

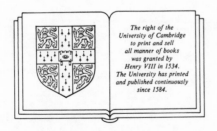

The right of the
University of Cambridge
to print and sell
all manner of books
was granted by
Henry VIII in 1534.
The University has printed
and published continuously
since 1584.

Cambridge University Press
Cambridge
London   New York   New Rochelle
Melbourne   Sydney

Published by the Press Syndicate of the University of Cambridge
The Pitt Building, Trumpington Street, Cambridge CB2 1RP
32 East 57th Street, New York, NY 10022, USA
10 Stamford Road, Oakleigh, Melbourne 3166, Australia

First published 1987

Printed in the United States of America

Library of Congress Cataloging-in-Publication Data
Mathematics and its applications to science and
natural philosophy in the Middle Ages.
Includes index.
1. Science, Medieval. 2. Mathematics – History.
3. Physics – History. I. Grant, Edward, 1926– .
II. Murdoch, John Emery, 1927– .
Q124.97.M38 1987 509'.02 86-23273

British Library Cataloguing in Publication Data
Mathematics and its applications to science
and natural philosophy in the Middle Ages.
1. Mathematics – History
I. Grant, Edward II. Murdoch, John E.
510'.9'02 QA23

ISBN 0 521 32260 X

CONTENTS

Contents

CONTRIBUTORS

BRUCE EASTWOOD Department of History, Patterson Office Tower, University of Kentucky, Lexington, Kentucky 40506

EDWARD GRANT Department of History and Philosophy of Science, Indiana University, Goodbody Hall 130, Bloomington, Indiana 47405

WILBUR KNORR Program in the History of Science, Building 200–031, Stanford University, Stanford, California 94305

DAVID C. LINDBERG Department of the History of Science, University of Wisconsin, 4139 Helen C. White Hall, Madison, Wisconsin 53706

MICHAEL R. MCVAUGH Department of History, University of North Carolina, Chapel Hill, North Carolina 27514

A. GEORGE MOLLAND Department of History and Philosophy of Science, King's College, Aberdeen AB9 2UB, United Kingdom

JOHN E. MURDOCH Department of the History of Science, Science Center 235, Harvard University, Cambridge, Massachusetts 02138

J. D. NORTH Filosofisch Instituut der Rijksuniversiteit, Westersingel 19, 9718 CA Groningen, The Netherlands

A. I. SABRA Department of the History of Science, Science Center 235, Harvard University, Cambridge, Massachusetts 02138

EDITH D. SYLLA Department of History, P.O. Box 8108, North Carolina State University, Raleigh, North Carolina 27695–8108

SABETAI UNGURU School of History, Tel Aviv University, Ramat-Aviv, 69 978 Tel Aviv, Israel

INTRODUCTION

Edward Grant and John E. Murdoch

Three names – Pierre Duhem, Anneliese Maier, and Marshall Clagett – have dominated scholarship on the history of medieval science and natural philosophy during the twentieth century. Of these, Duhem and Maier concentrated primarily on cosmological and physical conceptualizations. Marshall Clagett chose a different path and focused most of his enormous energies on the Latin text. As a medievalist doing research in the history of science, Clagett was determined to build his interpretations, conjectures, and syntheses on the rock of the meticulously established text. In a field where hundreds of works and thousands of Latin manuscripts are but little known and largely unedited, he chose to produce editions and translations so that in lieu of endless, vain speculation, the texts themselves would function as our guides.

Initially, Clagett's historical interests lay in Byzantium, specifically in the fifteenth-century Byzantine scholar Georgios Scholarios, or Gennadios II. But reproductions of the works of Giovanni Marliani, a fifteenth-century physician and natural philosopher, that were made available to him by Lynn Thorndike, his mentor at Columbia University, moved Clagett from the Greek into the Latin realm. With the publication in 1941 of his doctoral thesis, *Giovanni Marliani and Late Medieval Physics*, he began his lifelong preoccupation with medieval Latin science, applying his formidable paleographic skills to medieval physics and mathematics. His first book following the dissertation came in 1952, when, in collaboration with Ernest A. Moody, he published *The Medieval Science of Weights (Scientia de Ponderibus)*, a volume of eight texts on medieval statics, one of the most mathematized of medieval scientific disciplines.

This was but a prelude to *The Science of Mechanics in the Middle Ages* in 1959, a volume Clagett originally thought of entitling "Before Galileo," thereby indicating its relevance to the Scientific Revolution.

ix

Clagett sought to present "the substantial content and objectives of a few of the mechanical doctrines of the medieval period which were framed in mathematical terms or which had important consequences for a mathematical mechanics" (p. xxii). Subdividing the work into ten chapters organized under the three general rubrics of statics, kinematics, and dynamics, Clagett introduced the texts and translations of each chapter with superb summaries and analyses. Many of the major problems that had been identified in medieval physics – including the Merton mean speed theorem, the application of geometry to kinematics within the doctrine of the intension and remission of forms, Bradwardine's dynamic law of motion, impetus theory, the free fall of bodies, and so on – are included. More than any other volume, *The Science of Mechanics* made medieval physical thought intelligible to a wide audience. Along with Duhem's *Le système du monde,* it is undoubtedly the best-known and most widely quoted scholarly book on medieval science. In recognition of its significance and brilliance, the History of Science Society, in 1960, presented its prestigious Pfizer Award to Marshall Clagett.

Although the application of the doctrine of intension and remission of forms played a major role in *The Science of Mechanics*, Clagett devoted an entire volume to the subject in 1968 when he edited and translated Nicole Oresme's *Tractatus de configurationibus qualitatum et motuum* (*Nicole Oresme and the Medieval Geometry of Qualities and Motions*). Because Nicole Oresme was perhaps the most significant single author in the history of medieval science and the *Tractatus de configurationibus* easily ranks as one of the most innovative of his numerous treatises, Clagett's superb edition, translation, introduction, and analyses were of great moment for our understanding of medieval science.

To have published two such lengthy and momentous volumes as *The Science of Mechanics* and Oresme's *Tractatus de configurationibus* would have been more than sufficient achievement for most scholars. But Marshall Clagett's most notable accomplishment lies in his epochal five volumes (in ten tomes) on *Archimedes in the Middle Ages*. William of Moerbeke's translations of the works of Archimedes from Greek to Latin in 1269, which Clagett edited in the second volume, and their subsequent impact form the fundamental core of this monumental study. Published over a period of twenty years (1964–84), it not only describes the influx and impact of the works of Archimedes proper as well as Greco-Arabic and Latin works that reveal Archimedean influences, but it carries the story beyond the Middle Ages to the Renaissance, treating such fifteenth- and sixteenth-century figures as Nicholas of Cusa, Regiomontanus, Leonardo da Vinci, Niccolò Tartaglia, Francesco Maurolico, and Federigo Commandino. The number of Archimedean texts edited with full critical apparatus is truly staggering. All are thoroughly analyzed and most are also translated for the first time. For this magnificent and enduring contribution to the history of science, Marshall Clagett was awarded (in 1981) the Alexandre Koyré Medal of the International Academy of the History of Science (Paris).

Throughout his illustrious career Marshall Clagett has earned many awards and honors in recognition of his outstanding scholarship (for the precise extent of that scholarship, see the bibliography of his works at the end of this volume). In addition to the Pfizer Award and the Koyré Medal, Clagett was also awarded the Charles Homer Haskins Medal of the Medieval Academy of America (1969), the George Sarton Medal of the History of Science Society (1980) for outstanding scholarship, and the John Frederick Lewis Prize of the American Philosophical Society (1981). Other noteworthy honors are Clagett's election as President of the History of Science Society (1962–64) and his membership in a number of honorific societies such as Fellow in the Medieval Academy of America, Member of the American Philosophical Society (Vice-President 1969–72), Member of the Deutsche Gesellschaft für Geschichte der Medizin, Naturwissenschaft und Technik, and Member of the International Academy of the History of Science (Vice-President, 1968–71). Two major universities, George Washington University and the University of Wisconsin, have seen fit to award him honorary degrees.

Thus far we have focused exclusively on Marshall Clagett's scholarship – and for good reason. It is Marshall Clagett the extraordinary scholar whom we honor with this volume. But Marshall Clagett has functioned as more than a great scholar. He is one of a small group of devoted historians of science, who, immediately after World War II, were instrumental in transforming the History of Science Society into a modern organization capable of serving the ever changing needs of a growing society. Not only did he serve the History of Science Society as president, but he devoted time and energy to the advancement of his profession when he organized a significant meeting at the University of Wisconsin in September 1957 on the theme "Critical Problems in the History of Science." The meeting's goal was to stimulate the study of the history of science, and it may well have done so because a volume bearing that very title came out soon after.

As a professor of the History of Science (and later Vilas Research Professor) in the Department of the History of Science at the University of Wisconsin between 1947 and 1964, Clagett was instrumental in making Wisconsin an important center for the study of the history of science. Three contributors to this volume were privileged to have been his students during this period, and five others belong to the "Clagett tree" unto the second and third intellectual generations.

Since 1964, Clagett has been Professor of the History of Science in the School of Historical Studies of the Institute for Advanced Study in Princeton, New Jersey. As a professor of the Institute, Clagett played a unique and extraordinary role in the history of science. Through his generous auspices, quite a number of historians of science were privileged to spend at least one, and occasionally more than one, year as visiting members of the School of Historical Studies. Within the stimulating atmosphere of the Institute, numerous research projects were nurtured and ultimately brought to fruition. To all of his visiting colleagues, Marshall Clagett functioned as research resource, social companion, and, most important of all, as a true friend.

The articles in this volume have been brought together for publication during Marshall Clagett's seventieth year, the year of his retirement (1986) from the Institute for Advanced Study. The word "retirement," however, has no relevance to his life and plans. As if to herald this plain fact, Marshall Clagett quietly and unobtrusively began a new academic career approximately ten years ago when he undertook the study of Egyptian hieroglyphics, history, art, and culture. He has now become a professional Egyptologist with an excellent command of the language, able to read by sight inscriptions on tomb walls and sarcophagi. He is currently producing a multivolume source book in ancient Egyptian science, which, not surprisingly, is focused on texts.

Because mathematical texts and the uses of mathematics represent Clagett's basic research activities, it seems fitting that the theme of the essays in this honorific volume be concerned with both texts (Knorr's article) and the applications of mathematics to science and natural philosophy, applications that have been broadly conceived and include even the role of diagrams and graphs. Although no single volume could hope to embrace all aspects of the application of mathematics to science and natural philosophy in the Middle Ages, we hope that this first attempt to do so proves a significant contribution to the history of medieval science.

But we also hope that this volume conveys to Marshall Clagett our love and affection for him. It is our small expression of gratitude for his monumental scholarship and for the many kindnesses he has shown to all of us over the years.

Pure mathematics

1

The medieval tradition of Archimedes' Sphere and Cylinder

WILBUR KNORR

Marshall Clagett has devoted decades of patient labor to the textual tradition of Archimedes in the Middle Ages. This effort now amounts to five massive volumes documenting the Latin literature on Archimedes in Arabic-based and Greek-based works extending from the twelfth century to the Renaissance.[1] With access now to the critical texts prepared by Clagett, together with his translations and detailed commentaries, historians of medieval mathematics can begin to unravel the complexities of this central tradition. But Clagett's project of tracing the medieval transmission of the Archimedean corpus must also command the attention of those specialists like myself whose interests lie with the ancient field. I will here attempt to illustrate the complementary aspects of the ancient and medieval studies.

Within the Latin corpus of Archimedes no writing elicits more puzzles than the tract *De curvis superficiebus* (*CS*).[2] Its author, stated to be "Johannes de Tinemue" in most manuscripts, is otherwise unknown;[3] the geographical place name has not been identified for certain; and one cannot even be sure whether an ancient or a medieval scholar is intended. Compounding the confusion is that some manuscripts assign the writing to "Gervasius de Essexta," whom Clagett has tentatively identified as John Gervase of Exeter, a cleric with papal connections who died at Viterbo in 1268.[4] But the only secure indication of date appears to be the citation of *CS* as "De pyramidibus" by Gerard of Brussels sometime around the mid-thirteenth century.[5]

As indicated by the number of its extant manuscripts, *CS* was the principal source by which medieval Latin scholars had access to the

I wish to acknowledge the valuable assistance of my colleagues in the Stanford Classics Department, particularly Lionel Pearson, Edward Courtney, and Gregson Davis, in the effort to track down the literary allusions presented in sect 10 r I.

3

content of Archimedes' *Sphere and Cylinder*.[6] Even Maurolico in the sixteenth century exploited it as the basis for his own Archimedean editions, despite the appearance of complete Greek-based translations of *Sphere and Cylinder* by William of Moerbeke (1269) and Jacob of Cremona (c. 1453).[7] Yet *CS* is hardly a transcript of the Archimedean treatment: It condenses severely the exposition of Archimedes' main results on the surface area and volume of the sphere; it omits all his results on spherical sectors and segments, but includes other results not found in our Archimedean corpus; further, it frequently adopts forms of terminology and proof quite different from those in the Archimedean original. Thus, the relationship between the medieval tract and the Archimedean work is obscure. Older critics such as Heiberg supposed that *CS* was translated from an Arabic paraphrase of Archimedes; but Clagett, who subscribed to this view in his own first edition of *CS*, later argued its provenance as a translation from Greek.[8]

The issue is complex. Another adaptation of the Archimedean theorems on the sphere was produced in Arabic by the Banū Mūsā in the ninth century A.D. and became familiar in the West through its translation as the *Verba filiorum* (*VF*) by Gerard of Cremona in the twelfth century.[9] From our discussion below it will become clear that *VF* cannot be viewed as the source of *CS*. But if *VF* itself modified some Arabic rendition of these theorems, is it not possible that *CS* could be associated with the same? As I intend to show, stylistic considerations are inconclusive in deciding between the Arabic- and the Greek-source hypotheses. Only through a detailed comparison of related treatments of these materials in ancient and medieval works, as will be undertaken here, can a reasonably secure judgment be made.

That *CS* was produced as a translation, rather than as an original Latin work, is generally accepted. This view will also be reexamined here, but a preliminary remark on the nature of such translated works is appropriate. In general, the process of transmitting technical documents combines the efforts of authors, editors, translators, and copyists. The final product will bear vestiges of each kind of contribution in varying degrees. A simplified chronological scheme might distinguish four phases: (a) at the core, a technical composition framed in accordance with the mathematical style of a given tradition, for example, the rigorous format of Archimedean geometry; (b) secondary elaborations, attached as marginal comments or textual interpolations, to address the needs of later students; (c) changes due to the translator – these may be slight if he adheres to a policy of strict literalism, but substantial if his aim is toward use and readability; (d) another stratum of elaborations directed toward the new audience, whose needs and interests are likely to differ from those of the users of the original composition. The emphasis on logic within the medieval Latin curriculum, for instance, occasioned scholia on the logical nuances in proofs; such comments would, for the most part, be irrelevant to the concerns of the earlier technical tradition.

This general scheme would be expected to offer a guide for analyzing the transmission of any technical work passing through translation from antiquity to the Middle Ages. Specifically, *CS* has just such a layered character: It consists of a formal geometric nucleus punctuated by the kinds of comments and digressions one regularly encounters in ancient and medieval scholia. Disentangling these levels and inferring consequences for the genesis of the work is the project we now begin.

I

Readers familiar with the classical style of geometry, as they work through the tract *De curvis superficiebus*, will soon come to appreciate how firmly it adheres to this style. But the smooth flow of the deductive development is occasionally interrupted by remarks of a different sort, recognizable as secondary. These may be divided into three classes:

1. A few lemmas have been inserted, of a decidedly more elementary nature than the principal theorems. These include a proof relating to lines in the figure of proposition 5 (lines 19–27), an extension of prop. 5 for a modified configuration (lines 92–117), a lemma on the differences of cones in prop. 7 (lines 95–113), and a lemma on the products associated with certain inequalities of ratios in prop. 8 (lines 6–27). A few similar additions, such as an extension of prop. 7 analogous to that for prop. 5, are present in isolated manuscripts; their origin with Latin editors is clear.[10]

2. Citations of theorems in Euclid's *Elements* and of prior theorems in *CS* itself are found in every proposition. These are invariably short, for example, *per primam secundi, per IX duodecimi, per proximam huius*, and so on. Such cross-referencing is common in Latin mathematical works, but it is no less familiar among Arabic and Greek scholiasts; thus, no immediate conclusions can be drawn about provenance.[11] The great number of references made in prop. 7 to the ninth and eleventh propositions of *Elements* XII, however, suggests to me the relentless thoroughness of a medieval editor, for the ancient scholiasts tend to allow one such reference to suffice in a given theorem. (Remarkably, however, the scholiasts of *CS* never detect the more significant but less obvious assumption of Euclid's XII.16 made in half of the proofs.[12]) The four citations of Archimedes' *Dimension of the Circle* are likely to be part of the original (see section III of this chapter).

3. Interspersed are several longer comments, conceived in a self-consciously pedagogical tone strikingly different from the rest of the

work. Among these longer intrusions is the following from *CS*, prop. 1:

> Unde discolos et obiurgantes non cogit ad scientiam, cum et ipsi sapere non audeant, cum etiam laborent ut nesciant, cum et a pallade ultro vultus detegente oculos avertant. Hec igitur hiis, qui rerum subtilium fugas, quantitatum miracula, proportionum nexus, nature deposita rimantur, proposuit philosophus. (1.18–22)[13]

Just previously the writer has noted the requirement of a certain "hypothesis" to carry through the proof.[14] Presumably, then, it is this hypothesis, or the theorem itself with its need for it, which our writer claims "does not urge the peevish and chiding to knowledge." The writer's rhetorical flair is notable: for instance, *cum et . . . non audeant, cum etiam . . . nesciant, cum et . . . avertant* – these form three consciously balanced periods; again, *rerum subtilium fugas, . . . , nature deposita* – we read in succession four phrases of the same genitive-plus-accusative structure. In this florid language he thus berates the refractory students and encourages the diligent.

But we also detect here reminiscences of classical literature. The line about those "who do not dare to be wise" recalls a famous platitude from Horace:

> dimidium facti qui coepit habet: sapere aude, incipe.
> (*Epistles* 1.2.40: "the one who has started holds half the deed: dare to be wise, commence.")

The classical allusion gives an edge to the author's complaint: Not only do these recalcitrants ignore the proverbial wisdom; they even strive to achieve the opposite. The reminiscence of *sapere aude* could hardly reappear in *CS* through translation. It must result from a Latinist's imitating the ancient motto.

Similarly, the Latin poets are rich with accounts of Pallas (Minerva). In a grisly depiction of her entry into the cave of Envy, Ovid writes,

> ⟨Pallas⟩ videt intus . . . Invidiam visamque oculos avertit.
> (*Met.* 2.768–770: "Pallas sees within . . . the sight of Envy and averts her eyes.")

In another tale, Ovid relates how Pallas, disguised as an old woman, at last exhibits her true form, overawing all the nymphs but the reluctant Arachne (*Metamorphoses* 6.43–44).

No single narrative appears to provide an obvious model for the whole passage in *CS*. Our author seems to extemporize on familiar associations with the stories of Pallas, recalled in images such as her averting the eyes or transforming her countenance. Further, the closing admonition on the "flights of the subtle . . . miracles . . . intricacies . . . depositions of nature," while hardly a specific allusion, nevertheless does suggest a poetic reverence toward the secrets of wisdom.

CS concludes with a similarly remarkable passage:
> Sicque Tiphis noster portum tenet in quem iam dudum vela
> succinxerat. Iamque cum bibulis hereat harenis anchora
> Archimenidis remigii, Johannes navigationis grates ageret [!]
> summo creatori. (10.24–26)[15]

Vergil, Ovid, Statius, and Lucretius speak of "thirsty sands." In Ovid's
description of Scylla, for instance:
> bibula sine vestibus errat harena. (*Met.* 13.901: "without
> clothes she wanders along the thirsty sand.")

Again, he describes a beached dolphin
> quem . . . bibulis inlisit fluctus harenis. (*Heroides*
> 18[19].201: "which . . . the waves have dashed against the
> thirsty sands.")

Among the ancient poets, one of course encounters numerous references to the parts of ships. For instance, in Ovid,
> velo et remige portus . . . intrat. (*Met.* 6.445: "by sail and
> rower he enters the harbors.")[16]

Indeed, the phrase "by sail and oar" (*velis remisque*) is aphoristic for
haste.[17] Further, the metaphor of the writing as a voyage is also Ovidian. Closing the first book of *Ars amatoria*, he writes:
> Pars superat coepti, pars est exhausta laboris./ Hic teneat
> nostras ancora iacta rates. (1.771–772: "Part of the labor
> undertaken remains, part is complete./ Let the anchor
> outthrust here hold our ships.")

Moreover, the figure of Tiphis as helmsman – a commonplace among
the Latin poets, drawing from the legends of Jason's Argo – is invoked
by Ovid at the beginning of this same book:
> Tiphys in Haemonia puppe magister erat. (1.6: "Tiphys was
> master in the Haemonian ship.")

In much the same way, the author of *CS* conceives his own writing as
a voyage begun with Tiphys' preparations and concluded with the making fast of Archimedes' anchor.

These passages in *CS* do not mimic specific verses. The author appears to interweave freely a variety of familiar phrases in conscious
imitation of the style of classical Latin. As the ancient poets, Ovid in
particular, were standard in the academic curriculum of the twelfth and
thirteenth centuries,[18] our author would surely know them; and his
desire to impress his students through a display of such learning is
understandable, if not laudable.

The classical base of these passages has two important consequences. First, it identifies the man Johannes de Tinemue as the Latinist responsible for them. For the closing line cited above (10.25–26)
directly links the conclusion of the voyage (*cum bibulis . . . remigii*)
with the author's gratitude (*Johannes . . . creatori*). In effect, he is the

editor-commentator operative in phase (d) of our schema. We are surely warranted in further assigning to him several other textual intrusions of a similarly Latin cast,[19] and of course we do not wish to exclude the possibility that the same writer could have filled the role of translator in phase (c) as well. Clagett once conjectured that the enigmatic "Tinemue" might be Tynemouth.[20] Latin forms of the name Tynemouth, such as "Tinemutha,"[21] can be readily associated with the spellings in the manuscripts of *CS*. The latter reveal a clear pattern of scribal variation anyway, and the most likely form, "Tinemue," could well have already suffered from scribal error.[22] If Johannes still remains unknown, nevertheless we can terminate speculation on ancient commentators and places.

Second, the Grecisms in these Latin passages lose all bearing on the ultimate provenance of *CS*. Allusions to "Pallas" and "Tiphys," for instance, while clearly intended to evoke Greek antiquity, were easily possible for a medieval Latin scholar imitating classical Latin authors. Some works of obvious Greek origin, such as *philosophus* (1.22), *ypothesis* (1.15, 5.107), and *theorema* (1.16, 103; 2.40), and even *discolus* (1.18),[23] had become familiar in later Latin. To the contrary, it is remarkable how minimal the Grecisms are within the passages discernible as interpolated by Johannes.[24]

One ostensible Grecism is of particular interest: the term *falsigraphus*, which appears eleven times in *CS* to denote the hypothetical perpetrator of the false assumption introduced in an indirect proof.[25] As Clagett notes, the term may be associated with Greek *pseudographos* employed by Aristotle, for instance, in his criticism of the fallacious circle-quadratures of Bryson and Antiphon.[26] But Aristotle has in mind the arguer of fallacies, not the maker of indirect proofs, and in other authors *pseudographein* can mean simply "make a mistake."[27] Thus, the conception of *falsigraphus* in *CS* does not follow ancient usage. Its origin lies instead with the medieval commentators. For instance, the term is used in the same way as *CS* in the Corpus Christi manuscript of Archimedes' *De mensura circuli*.[28] Both this text and *CS*, in their turn, must owe their usage of *falsigraphus* to the Euclid recension called "Adelard III." For in the preface on terminology which heads this recension, the following account of the term is given:

> Instantie dissolutio est cum falsigraphus insistit non sic vel aliter accidere quam geometer affirmat. ("'Dissolution of the instance' is when the falsigrapher insists that what obtains is not as the geometer affirms, but otherwise.")[29]

Not only does *CS* agree with this use of the term *falsigraphus*, but also it adopts its *dissolutio* to signify indirect proof (*CS* 1.48). Indeed, the labeling of the parts of proofs adopted in *CS* conforms with that in

"Adelard III," namely, as *propositio, exemplum, dispositio, ratio, conclusio, ratiocinatio, et instantie dissolutio.*[30] All of these terms, with the exception only of *ratiocinatio*, are employed throughout *CS*.

CS employs a heavily Greek-based vocabulary in its technical body. In the first eight lines of prop. 1, for instance, one meets the terms *ypothenusa, cathetus, orthogonius, piramis*, and *diametros* (with the Greek termination for the nominative); other Greek cognates include *tetragonus* (2.2), *axis* and *basis* (2.3–4), *poligonium* (1.38), *trigonus* (4.13), *conica* (5.4), *spera* (6.1), *cubus* (10.1), *parallelogramum* (5.108). But as Clagett observes, all of these terms are standard within the medieval geometric tradition;[31] indeed, most of them appear in the Euclid recensions of Adelard, Hermann, and even Gerard, despite their use of Arabic sources.[32] It is clear, in fact, that the author of *CS* strongly prefers Latin forms over Greek when these are at hand. Thus, he uses *piramis rotunda* instead of *conus, columpna rotunda* instead of *cylindrus, circumferentia* instead of *periferia, equiangulus* instead of *isogonius*, and so on. Of all these Grecisms, only one (namely, the phrase with *tetragonus* in 2.2) appears to lie outside the terminology of the Euclid recensions.[33]

CS totally lacks those Arabicisms which mark many of the Arabic-based translations: there are no transliterations of Arabic terms, for instance, or peculiarities of diction or word order, such as frequently betray the Arabic sources of the Euclid translators.[34] But the absence of Arabicisms of itself need not exclude an Arabic source. Hermann of Carinthia's recension of Euclid's Book XII, for instance, adopts a Latin style similar to that of *CS*, also devoid of Arabicisms. Nevertheless, Hermann's editions of the other Euclidean books contain a few Arabicisms; moreover, the filiation between Hermann's version and the Adelard tradition is evident,[35] the latter being quite evidently based on Arabic sources. Thus, Hermann's recension of Book XII offers a clear instance of a thoroughly Latinized version purged of vestiges of the Arabic base.

As far as stylistic tests are concerned, then, we have reached an impasse. The absence of Arabicisms in *CS* does not betoken a Greek source to the exclusion of an Arabic source; the Grecisms in basic geometric terminology are all standard in the twelfth-century versions of Euclid, including those derived from Arabic; moreover, the most striking Grecisms, such as the allusions to "Pallas" and "Tiphys," all occur within interpolated remarks reminiscent of classical Latin authors. "Johannes de Tinemue" (presumably, John of Tynemouth), the Latinizing editor who inserted these remarks and, in general, set the work into its extant form, stands in a line with Hermann of Carinthia, the Latinizing Euclid editor. His effort falls within phase (d) of our general schema of transmission. The same Johannes may well also have translated *CS*. This seems probable, in view of the absence of an alternative un-Latinized tradition of the work.

As for the second contributor, "Gervasius de Essexta," we can do no other than assign to him those textual differences which separate the manuscripts naming him as author (Clagett's "Tradition II," in mss. A, E, and G) from the ones that name Johannes ("Tradition I"). As a perusal of the critical apparatus assembled by Clagett reveals, the discrepancies between these two lines of the text are minimal, merely affecting changes of word order, occasional minor additions or deletions, and the like.[36] A medieval editor might perhaps assume the honors of authorship on these grounds, but no modern writer would. Thus, whatever plausibility one assigns to the suggested identification with John Gervase of Exeter,[37] no essential question about *CS* is involved. For its date is fixed through other considerations, while the mathematical expertise of its editors is hardly reflected in these minor discrepancies among the manuscripts.

II

The hypothesis that *CS* is an original Latin composition crafted on Archimedean sources is difficult to maintain, since its author manifests, in Clagett's observation, "more detailed knowledge . . . of Archimedes than would have been possible for a Latin author of the early thirteenth century."[38] To this I would like to add some remarks on the Archimedean sources then available. In scrutinizing the hypothesis of original composition, we will begin to perceive the character of the technical source on which *CS* ultimately depends.

If *CS* were an original Latin writing, its principal source would have to be Book I of Archimedes' *Sphere and Cylinder* (*SC* I), for *CS* provides a reasonable rendition of its major results and methods. Because *SC* I became available in Latin only with Moerbeke's translation in 1269, the author of *CS* would have to consult Archimedes in the Greek. This would not be impossible, of course, for a scholar stationed at the Viterbo convent where Moerbeke made his translation.[39] Another Greek source to which *CS* betrays a certain affinity is the version of Archimedes' theorems on the sphere expounded by Pappus in Book V of the *Collection*.[40] Although Pappus was not widely known in the West before Commandinus' Latin translation in 1588,[41] the prototype of the extant Greek manuscripts may already have entered the papal collection in the thirteenth century; for some scholars seem to have consulted it from about that time.[42] A third source would be the *Verba filiorum* (*VF*) of the Banū Mūsā, available through the Latin translation by Gerard in the twelfth century.[43] Since *CS* betrays no affinities with the marked Arabo-Latin formulations adopted by Gerard, the supposition of its use by *CS* would entail that *CS* either reworked the Latin entirely or consulted it in Arabic.

Beyond these three sources I can name no others that a Latin writer of the thirteenth century might use. Conceivably, *CS* drew from a work no longer extant to us. But this hypothesis is effectively equivalent to supposing that *CS* is a translation from a lost prototype, since one cannot discriminate between the two hypotheses on the basis of extant evidence. We may thus eliminate this alternative here.

That *CS* does not use Pappus is clear. The technical methods of *CS* give no evidence of knowledge of the particular variants followed by Pappus. Particularly in the case of the measurement of the volumes of Archimedes' conical solids, formed by rotating regular polygons about their diameters, Archimedes takes the number of sides of the generating polygon to be divisible by four (cf. *SC* I.23). Pappus adopts a more general method (V, prop. 34), in which the sides of the generating polygon need not be equal and may be any number in multitude (V, prop. 34);[44] in Pappus' proof the generating semipolygon happens to have an odd number of sides. By contrast, in the proof of *CS*, prop. 7, the more restricted Archimedean method is followed, where the generating figure is a regular semipolygon having an even number of sides. But the theorem is *stated* without this qualification; thus, at the end of the proof the author calls attention to the fact that the odd-numbered case has not yet been dealt with and explains:

> quod quia auctor non proposuit, et nos illud investigare omittimus et diligenti relinquimus posteritati. (7.179–180: "Since the author has not proposed this, we too omit its investigation and leave it to diligent posterity.")

Since Pappus' method does cover this case, *CS* cannot be referring to him as the *auctor*. Even if we choose to assign this remark to the Latin interpolator, it is clear that Pappus' procedure has not been adopted by the composer of *CS*. For the generality of the method would have been evident, even if that were not explicitly posited in the construction and its proof.

It is further evident that neither *SC* I nor *VF* alone is sufficient as source for *CS*. For instance, *CS* measures the surfaces and volumes of the conical solids precisely in the manner of *SC* I. By contrast, in *VF* the volume theorem is assumed within that for the sphere (prop. 15), but is not actually proved; moreover, the surface of the conical solid is not worked out in a separate theorem, but is embedded within a lemma on inequalities (prop. 13) pertinent to the theorem on the sphere (prop. 14). In these cases, then, *CS* must rely on *SC* I.[45] Yet in the technique of convergence adopted in the measurement of the surface of the cone and the surface and volume of the sphere (*CS*, props. 1, 6, and 8, respectively), *CS* agrees with the approach in *VF*, distinctly different from that in *SC* I. Here *CS* would have to rely on *VF*. *CS* recognizes that the Archimedean theorems on circle measurement derive from a work different from that devoted to the theorems on the sphere; but *VF* presumes to originality for both blocks of material without distinction. Thus, the author of *CS* would require sources beyond *VF* to obtain a correct view of Archimedes' contribution. Such mixing of sources is difficult to account for, however, in an author assumed to have access to Archimedes' own treatment. It also conflicts with *CS*'s references to *philosophus* (1.22) and *auctor* (7.179) with their implication of dependence on a *single* reference source.

Beyond this, *CS* contains some results not present in either *SC* I or *VF*. For instance, *CS* expresses the volume of the sphere as equal to the cone whose height is the radius of the sphere and whose base equals its surface area (prop. 9). This is quite different from Archimedes' formulation in *SC* I 34.[46] *VF* provides yet another variant (*VF* 15: the sphere equals the product of its radius times one-third its surface area). As it happens, the formulation in *CS* agrees precisely with that given by Pappus (V, prop. 35), although their respective derivations differ.

Unlike both *VF* and *SC* I, *CS* provides a numerical estimate for the volume of the sphere, namely as 11/21 the cube of its diameter (prop. 10). Although no such result is proposed by Pappus, Archimedes (in *SC* I), or other writers in the formal geometric tradition, equivalents are regularly cited in the Heronian metrical literature.[47] Here, the author of *CS* must draw from works other than his primary Archimedean source.

CS thus betrays a pattern of freely mixing from a diverse range of Archimedean materials. This would be possible for an ancient editor who wished to assemble a new version on the basis of the several sources still circulating and who would have access to comparable adaptations where mixing had already occurred. But a medieval Arabic or Latin editor would be expected to adhere fairly closely to the one or two specific sources then available. This eclectic pattern in *CS* thus affirms the view that it was produced as a translation.

III

The hypothesis which remains to us, then, is that *CS* translates from a source freely conversant with the ancient Archimedean tradition. Noting specific parallels which link the ancient geometric works to *CS* will reveal the nature of the source on which *CS* directly depends.

As Clagett has emphasized, the formulation of the surface area of the cone in *CS* (prop. 1) imitates the Archimedean expression for the area of the circle in *Dimension of the Circle* (*DC*, prop. 1).[48] But the connection is explicit in the corollary which ends the same proposition of *CS*:

> [a] Archimenides enim in quadratura circuli ostendit
> circulum esse equalem triangulo orthogonio, cuius unum
> laterum rectum angulum continentium equatur
> circumferentie circuli, reliquum vero semidiametro. [b] Cum
> itaque ex ductu circumferentie in semidiametrum fiat
> superficies dupla circuli . . . (1.107–111)

The wording of the Archimedean reference happens to conform well
with Gerard's version of *DC*:

> Omnis circulus triangulo orthogonio est equalis, cuius unum
> duorum laterum rectum continentium angulum medietati
> diametri circuli equatur et alterum ipsorum linee circulum
> continenti.[49]

The discrepancies – *CS*'s omission of *duorum* and its substitutions of
medietati diametri by *semidiametro, alterum* by *reliquum,* and *linee
circulum continenti* by *circumferentie* – remove vestigial Arabicisms
in Gerard's Latin, as do other adaptations.[50] Part [b] of the statement
in *CS* parallels a corollary to Gerard's theorem, although the phrasing
is quite different.[51]

The wording of [a] in *CS* is much closer to Gerard's version than to
the extant Greek of *DC*.[52] But the medieval wording, where it differs
from the Greek, is supported by an ancient citation, that given by the
sixth-century A.D. commentator Eutocius.[53] Similarly, an anonymous
Greek tract *On Isoperimetric Figures,* which was translated into Latin
as *De figuris ysoperimetris* in the twelfth century,[54] cites the same
Archimedean proposition thus:

> [b] Quoniam vero quod sub ea que ex centro et perimetro
> circuli duplum circuli demonstratum est Archimenidi in
> mensuratione circuli. [a] Demonstravit enim quoniam omnis
> circulus equalis est trigono orthogonio cuius que ex centro
> equalis est uni earum que circa rectum, reliqua vero
> perimetro circuli.[55]

This passage from the anonymous isoperimetric writer (*AI*), like that
mentioned from Eutocius, indicates that the medieval recension of *DC*
continues a tradition of the Greek text alternative to that extant in our
codices of Archimedes.[56] Thus, although *CS* is likely to be influenced
by Gerard for its wording of [a], the basic formulation could still have
its place in *CS*'s source. This is affirmed by the appearance of [b],
stated in juxtaposition to [a] both in *CS* and *AI,* but assigned as a
corollary in Gerard's *DC*. Indeed, the product formulations for this and
other theorems, significant for *CS,* as also for *AI* and the ancient commentators, have been eliminated in *DC,* but for the extraneous corollary to its prop. 1 in the medieval recension. In *AI* the product formulation [b] is given priority, being attributed explicitly to Archimedes'
Dimension of the Circle.

This passage of *AI* typifies the severely literal Greek style adopted by some of the medieval translators.[57] Despite that, *AI* adopts the Arabo-Latin form of Archimedes' name ("Archimenides"); thus, the appearance of the same form in *CS* does not indicate its use of an Arabic source. On the other hand, *CS* is not executed in the same Graeco-Latin style as *AI*, but attempts a more natural Latin style. It is nevertheless difficult to view the present passage from *CS* as an editorial adaptation of *AI*, especially when the parallel from Gerard is taken into consideration. Their conformity seems due more to a dependence on similar source texts than to a direct connection.

CS and *AI* agree on their formulations of the volume of the sphere, as noted above, where *CS* is distinguished from *SC* I and *VF*. Moreover, *CS* and *AI* provide complete proofs in remarkably good technical agreement. I present their texts in parallel:

CS, prop. 9 (ed. Clagett):

[1] Omnis spera est equalis rotunde piramidi cuius basis equatur superficiei spere et altitudo semidiametro spere. . . .

[2] [a] GH columpna est duplo altior piramide E et sita est in equali basi cum E. Ergo columpna GH est sextupla ad E piramidem. . . . [b] Sed columpna GH sita est in basi FG circulo equali maximo circulo spere C et axis GH est equalis diametro spere C ex dispositione. Ergo columpna GH est sexquialtera ad speram C. . . .

[3] Ergo spera C est quadrupla ad E.

[4] [a] Item . . . curva superficies spere C . . . est quadruplus basi piramidis E. [b] Sed E et A piramides sunt eiusdem altitudinis. [c] Ergo A piramis est quadrupla E piramidi . . .

[5] Ergo A piramis est equalis C spere.

AI, prop. 7 (ed. Busard, lines 258–267):

[1] Quoniam vero conus basim habens circulum equalem superficiei spere altitudinemque equalem ei que ex centro spere equalis est spere colligitur ex eis que Archimenidis ita. [2] [b] Quoniam enim demonstravit quod cilindrus basim habens maximum circulum altitudinem dyametrum spere sesquialter est spere, [a] talis vero cilindrus sextuplus est coni basim quidem habentis eandem altitudinem vero eam que ex centro,

[3] quadrupla erit spera talis coni.

[4] [c] Est autem et eiusdem quadruplus et conus altitudinem quidem habens eandem, basim vero superficiei spere equalem; [b] sub eadem enim altitudine existentes ad invicem sunt sicut bases, [a] superficies autem spere quadrupla est maximi circuli.[58]
[5] Quare equalis spera dicto cono.

I have here given the whole text from *AI*, but only those portions of *CS* which have their analogue in *AI*. The only differences between them in *expository* order are the transpositions of steps 2 [a] and [b] and of steps 4 [a] and [c]; but the two proofs agree entirely on their *logical* order. The version in *CS* is ampler and more formal, explicitly labeling all the figures introduced; by contrast, *AI* gives a more concise proof, without *dispositio*. I have deleted from the quotation of *CS* its two citations of Euclid and one of Archimedes, its repetition of step 3 between steps 4 [c] and 5, and its derivation at step 4 [a] of the expression (merely asserted by *AI*) for the surface of the sphere. This derivation is necessary because the form of the surface theorem in *CS* (prop. 6) is different from the Archimedean expression stated in 4 [a].

AI is brief: This passage is merely an auxiliary lemma, explicating a step assumed without proof in its own source on isoperimetric figures. We might thus suppose that *AI* has abridged a version in the more detailed manner of *CS*. But *AI* explicitly cites "the things of Archimedes" as the basis of its proof, and so leaves the impression of improvising from basic Archimedean results, rather than adapting an extant proof. Indeed, its expression for the surface of the sphere (at *AI*:4 [a]) is precisely that adopted by Archimedes in *SC* I (prop. 33), not that used in *CS* (prop. 6). Similarly, a discrepancy relating to the generation of the conical solids aligns *AI* and *SC* I, rather than with *CS*.[59] Thus, despite their closeness in the handling of this lemma on the sphere, it is difficult to posit a direct textual link between *CS* and *AI*.

These may be compared with two other ancient treatments of this theorem. At the analogous place in his own rendition of the isoperimetric materials Theon of Alexandria proves the theorem somewhat more efficiently than *AI* by invoking directly Archimedes' formulation of the volume of the sphere in *SC* I.34 (namely, as four times the stated cone), rather than in the manner of *AI* (namely, as two-thirds the associated cylinder).[60] In this respect, Theon differs from both *AI* and *CS*. Theon gives an *ekthesis*, a feature comparable to the *dispositio* of *CS* but absent from *AI*. More generally, the differences between Theon and *AI* in their treatments of the isoperimetric figures are sufficient to set aside a direct textual connection between them; they appear to be drawing from the same or similar sources and explicating them in similar ways.[61] In the case of the lemma on the sphere, both writers apparently are innovating proofs on the basis of their Archimedean sources.

Another treatment of the lemma comes at the end of Pappus' account of Archimedes' *Sphere and Cylinder* in Book V of the *Collection* (prop. 37). Like *CS* and *AI*, Pappus proves the equivalence of the two formulations for the volume of the sphere: (i) as equal to the cone whose height equals the radius of the sphere and whose base equals its surface area, and (ii) as two-thirds the cylinder whose height equals the diameter of the sphere and whose base equals its great circle. In *AI* and *CS*, (i) is derived from (ii); but for Pappus, the opposite sequence is followed, since his principal theorem on the sphere is framed as in (i). The basic isoperimetric text also prefers form (i). By omitting an explanation it thus compels the commentators (Theon and *AI*), unfamiliar with this form, to bring it into agreement with what they know from Archimedes' *SC* I.[62]

In their treatment of this lemma on the sphere, then, the objective of Theon and *AI* is clear. As for *CS*, its cumbersome proof must employ the particular forms of the surface and volume theorems in its own earlier theorems (props. 6 and 8). But why should *CS* include this lemma (prop. 9) at all? The answer, I think, lies in the association with *AI*; for this has established a precedent for the interest of this form of the volume theorem. As with the citations of Archimedes' *DC*, discussed above, the conformity here of the Latin works, *CS* and *AI*, indicates an association between their underlying sources.

A numerical estimate for the volume of the sphere is given in *CS* 10, namely that the sphere has to the cube of its diameter the ratio of 11 to 21. Equivalent values appear in Hero's *Metrica* (II.11) and the Heronian metrical corpus; but the manner of derivation in *CS* is especially close to that given by Theon of Alexandria in his *Commentary* on Ptolemy. While the issue of sorting out the ancient origin and transmission of this numerical rule is vexed, I believe one can assign priority to Theon.[63] The rule is absent from other possible sources for *CS*, including both *SC* I and *VF*.

These comparisons, and others to be considered in the next section, reveal how the author of *CS* is a participant in the activity, represented by Pappus, Theon, the anonymous isoperimetric writer, and the Heronian editors, of commenting and improvising on Archimedes' *Sphere and Cylinder*. A medieval author, whether Arabic or Latin, would hardly have access to a spectrum of the ancient tradition wide enough to evoke so many parallels within an original writing. No such difficulty faces the hypothesis of translation: A technical work composed within the milieu of the ancient commentators could easily embrace the range of variants we have detected in *CS*. These would survive in a reasonably accurate translation, even if the translator had no direct knowledge of the works with which the source was once allied. The coincidences between *CS* and *AI* could arise through a process involving the independent translation of related sources, rather than indicate any direct links between the Latin versions. In what follows, then, I will assume the existence of a prototype underlying *CS* and seek to specify more precisely its nature and its place within the Archimedean tradition.

IV

Accepting *CS* as, at base, the translation of an ancient geometric tract has implications for one's view of another important work

in the Latin tradition of *SC* I, the *Verba filiorum* (or, to translate from
the Arabic title, the tract *On the Measurement of Plane and Spherical
Figures*) by the ninth-century geometers called the Banū Mūsā.[64] Although we have above emphasized the differences between *VF* and
CS, it is important now to consider the significance of their similarities.

Since both tracts undertake the project of providing geometric
proofs dealing with the surface and volume of curvilinear figures, limiting procedures are indispensable. Within the ancient Archimedean
tradition, two basic procedures are found: (1) The measurement of the
circle in *DC* 1 and of the parabolic segment in *Quadrature of the
Parabola*, prop. 24 both follow the Euclidean method used for the
circle, pyramid, cone and cylinder in *Elements* XII (props. 2, 5, 10–
11): The characteristic move is to introduce an inscribed (or circumscribed)[65] rectilinear figure whose measure differs by less than a preassigned magnitude from that of a specified curvilinear figure. (2) In *SC*
I, Archimedes elaborates a different procedure utilizing simultaneous
two-sided convergence: the curved figure is bounded above and below
by similar polygonal figures; as the number of sides is successively
doubled, the difference between the bounding figures can be diminished
to less than an arbitrary preassigned magnitude, so that the difference
between the intermediate curved figure and either bound must also be
less than that magnitude. Variants of this format are employed in all
of the limiting theorems in the Archimedean corpus, save those two
just cited which adopt the Euclidean format.[66]

CS and *VF* adopt neither of these Archimedean patterns. Instead,
they exploit a third method, exemplified by the Euclidean theorem on
the sphere (XII.18). The key step, as worked out by Euclid in two
preliminary constructions (XII. 16–17), is to inscribe in a given circle
(or sphere) a regular polygon (or a polyhedral solid) which does not
touch a second given circle (or sphere) concentric to the first and enclosed within it. The medieval tracts follow this model in their proofs
of the surface of the cone (*CS* 1; *VF* 9), the surface of the sphere (*CS*
6; *VF* 14) and the volume of the sphere (*CS* 8; *VF* 15); in addition, *CS*
employs this technique for its theorem on the proportionality of the
diameters and circumferences of circles (*CS* 3), and *VF* uses it for its
alternative proof of Archimedes' *DC* 1, on the area of the circle (*VF*
3–4).

This difference in method may be illustrated by the treatments of
the surface of the sphere in the three works.[67]

1. In *SC* 1.33, the surface is expressed as equal to an area four times
the greatest circle in the sphere. The proof is indirect: Supposing first
the surface (*S*) is greater than this area (*T*), *one introduces two lines
B, G such that* $1:1 < B:G < S:T$; then taking *D*, the mean proportional
between *B* and *G*, *one constructs two similar polygons, respectively
circumscribed and inscribed to a great circle of the sphere, such that*

the ratio of their sides is less than B:D. If the circle is now rotated about its diameter, it will generate the sphere and the polygons will generate two composite conical surfaces concentric with it. These surfaces will have the duplicate ratio of their sides, that is, less than $(B:D)^2$ = $B:G$, which is less than $S:T$. But the circumscribed surface is greater than S, the sphere, while the inscribed figure is less than T, so that the ratio of the conical surfaces must be greater than $S:T$. The alternative hypothesis, that $S < T$, similarly leads to contradiction.

2. In *CS* 6, the spherical surface is claimed to equal the product of the diameter and the circumference of a great circle. The proof is indirect: supposing inequality, one can first take the product to equal the surface of a sphere T concentric with the given sphere S but less than it; then *one inscribes in a great circle of S a regular polygon which does not touch the corresponding great circle in T*; the figure is rotated about a diameter to yield a composite conical surface (U) inscribed in (hence less than) sphere S but not touching (and thus greater than) sphere T. Now, the area of the conical surface U is known to be equal to the product of the circumference of the circle and a certain line less than its diameter; hence, the surface area of U is less than that of T, since the latter was supposed equal to the product of the circumference and the diameter; but the conical surface U contains sphere T, so that its surface must be greater. The alternative hypothesis is handled similarly.

3. In *VF* 14, the surface of the hemisphere is claimed to equal double the great circle. Again, the proof is indirect: supposing inequality, one can first set double the area of the circle equal to the surface of a hemisphere (T) less than that of the given hemisphere (S); *one then inscribes in S a composite conical surface (U) which does not touch T.* Now, the surface of U is less than the area double of the great circle of S and greater than the surface of hemisphere T (since it contains it); hence the double circle is less than the surface of T. This contradicts the hypothesis of their equality. Similarly, the alternative hypothesis leads to contradiction.

As indicated in the lines emphasized in (1), the critical move for convergence is the introduction of the circumscribed and inscribed polygons whose ratio falls within the specified range; these constructions have been given in *SC* I.1–6. But in (2), convergence depends on the inscription of the polygon within the circle of S so that its sides do not touch the inner circle of T; this construction is given in *Elements* XII.16. The key step in (3) is the inscription of the conical surface which does not touch the inner sphere; this is the same procedure as in (2), save that *VF* tacitly assumes the preliminary construction of the inscribed polygon which in *CS* is explicitly stated. Interestingly, neither *CS* nor *VF* justifies the introduction of such polygons, whether by a construction or by a citation of *Elements* XII.16, wherever that assumption is made.[68]

Thus, *CS* and *VF*, different from Archimedes, employ the alternative convergence method in the manner of Euclid's XII.16–18, although neither gives any explicit acknowledgment of this dependence. They further assume without justification the existence of curvilinear figures equal to given magnitudes (e.g., that there is a sphere whose surface is equal to the stated area); and their assumptions about the relative magnitudes of containing and contained figures (e.g., that the surface of the conical solid is greater than the spherical surface contained within it) are neither proved nor covered in explicit postulates.[69] Within the ancient tradition, the theoretical apparatus developed in Archimedes' *SC* I (for instance, the postulates on the relative magnitudes of convex figures) would be known. However, the omissions are striking in medieval authors, whose access to such reference works was limited.

These shared aspects of the technical execution in *CS* and *VF* are unaccountable as mere coincidence. But we have seen in section II that no direct use of *VF* by *CS* is plausible. Their clear connections, as in the content of their theorems and the technique of convergence in their proofs, must follow from textual connections between their respective source documents.

CS and *VF* both prove the theorem that the circumferences of circles are proportional to their diameters. In *CS* (prop. 3) an indirect proof is used, eliciting a contradiction through the introduction of an inscribed regular polygon. The convergence technique follows the alternative Euclidean pattern via XII.16, rather than the more familiar method of XII.2.[70] The proof of the proportionality theorem in *VF* (prop. 5) is also indirect; but it adopts a different strategy based on the Archimedean rule for the area of the circle (*DC* 1), which *VF* (prop. 4) has enunciated in an alternative form: that the product of half the diameter times half the circumference equals the area of the circle.[71]

A proof of the same theorem is given by Pappus in *Collection* V, prop. 11.[72] His procedure is direct, but in its appeal to the Archimedean circle theorem (*DC* 1) is effectively equivalent to that in *VF*. At the end of the proof, Pappus remarks:

> This is also proved without supposing that the product of
> the diameter of the circle and the circumference is
> quadruple of the circle. For the similar polygons inscribed to
> the circles, or circumscribed, have perimeters which have to
> each other the same ratio as the (lines) from the centers
> [namely, the radii]; so that also the circumferences are to
> each other as the diameters.[73]

The assumption, here stated by Pappus, on the perimeters of similar polygons does in fact enter into the proof in *CS* (e.g., 3.13–14). It is thus clear that proofs of the sort suitable as models for that in *CS* 3 can be assigned to the ancient tradition. At the end of *CS* 3, a similar note appears:

> The same could also be demonstrated via the third (proposition) of Archimedes' *Dimension of the Circle*. But it is not (the case) as satisfactorily [*sed non est adeo sufficienter*]. (3.34–35)[74]

There is a puzzle here: In *DC* 3, where Archimedes computes bounds on π (e.g., that π is less than $3\frac{1}{7}$), he must *assume* as already proved that the ratio of the circumference to the diameter of any circle is a constant. It would thus be a clumsy logical error to use this theorem in a proof of the constant proportionality. Presumably, then, *CS* intends the *first* proposition of *DC*, and if so, would envisage an alternative proof in the manner of Pappus or *VF*. Why such a proof should be judged unsatisfactory is not clear. Certainly, Pappus' direct proof would be seen as entirely unobjectionable. Perhaps a specific, somehow defective, version of the proof instigated this comment. Since notions of mathematical elegance are subjective to a large degree, the remark should be taken as the expression of a personal preference and nothing more.

To whom is the remark due? If the Latin translator-editor has inserted it, then we would infer that he had two forms of the proof before him and by this comment was explaining his omission of one of them. But the insertion of such technical explanations was a familiar practice among the ancient commentators, as the parallel in Pappus illustrates.[75] It thus seems more likely that the comment was in the base source. If, further, we suppose the sources of *VF* and *CS* to be identical, then it would follow that *VF* has innovated its proof at the suggestion of this comment. That would help account for *VF*'s failure to follow Pappus' model, but instead to adopt the more cumbersome form with its unnecessary use of the indirect argument.

The coincidences of content and technique here noted indicate a textual link between *VF* and *CS*. From the preceding section, we have been able to place *CS* in the context of the ancient geometric tradition, and the parallels noted in the present section likewise have been seen to have ancient precedents. Since *CS* depends ultimately on an ancient source, it follows that *VF* does too. We may now consider what the relation between their sources is. The salient fact is that *CS* generally adheres more closely to the procedures in Archimedes' *SC* I than *VF* does. To establish this point, we may consider their respective treatments of two results: the area and volume measurements of the auxiliary conic solids.

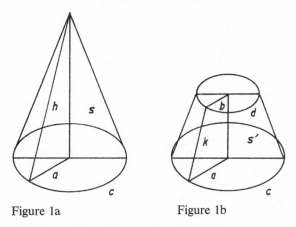

Figure 1a   Figure 1b

All three versions set out preliminaries on the surface area of conical segments. In *SC* I.14, Archimedes uses his standard limiting technique to express the surface (s) of the cone as a circle whose radius (r) "equals in square" (*dynatai*) the product of its slant height (h) and the radius of its base (a); that is, $s = \pi r^2$, for $r^2 = ha$ (Figure 1a). In *CS* 1 the same surface s is equated with a right triangle whose legs equal the slant height (h) and the circumference (c) of the base. In a corollary, this is recast as half the "product" (*ductus*) of h and c, en route to showing that the ratio of the conical surface to its base is the ratio of h to a; Archimedes establishes the latter result in *SC* I.15. In *VF* 9, as in *CS*'s corollary, the conical surface is given as the "product" (*multiplicatio*) of h by $c/2$. As noted above, *CS* and *VF* also agree on their use of the alternative limiting technique, modelled after *Elements* XII.16. Unlike *CS*, however, *VF* has no parallel to the Archimedean ratio theorem of *SC* I.15.

To evaluate the surface (s') of the truncated cone, that is, the difference between two conical surfaces, Archimedes figures it as the difference between the two corresponding circles (*SC* I.16). It follows that s' itself can be expressed as a circle whose radius "equals in square" the product of the slant height (k) and the sum of the radii (a and b) of the bases; that is, $s' = \pi \kappa(a + b)$ (Figure 1b). In *CS* 4 the same result is obtained by representing the difference as the trapezoidal area between the two triangles corresponding to the cones; hence, $s' = \kappa(c/2 + d/2)$, for c and d denoting the circumferences of the lower and upper circular bases. *VF* 11 expresses s' in precisely the same way as *CS*, but its proof depends on algebraic manipulations of the product forms for the cones taken from *VF* 9.

On the basis of these results, one can evaluate the composite surfaces, made up of truncated cones, to approximate the surface of the

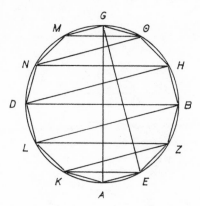

Figure 2

sphere. In *SC* I.21, Archimedes inscribes in a given circle a regular polygon the number of whose sides is a multiple of 4; he then shows via similar triangles that $EK + ZL + BD + HN + \Theta M : AG = GE : AE$ (Figure 2).[76] In *SC* I.23 the polygon is rotated about its diameter AG so that its sides form the surface of a composite conical solid, while at the same time the circle traces out a sphere; the surface S of the conical solid is shown to be less than the surface of the sphere. In *SC* I.24, S is shown to equal a circle whose radius "equals in power" the product of AE by $EK + ZL + BD + \cdots$; that is, $S = \pi AE \cdot (EK + ZL + \cdots)$. This utilizes the results on conical and truncated conical surfaces from *SC* I.14–16. In *SC* I.25 he shows that $S = \pi AG \cdot GE$, hence $S < \pi AG^2$, the latter being 4 times the great circle in the sphere. In *SC* I.28–30 the analogous results are established for the circumscribed surfaces; in *SC* I.33 these are applied in an indirect proof to show that the surface of the sphere equals 4 times its great circle.

The *CS* follows a closely similar procedure. In *CS* 5 it is shown first, on the basis of the results on truncated conical surfaces in *CS* 4, that $S = LD (\pi EK + \pi ZL + \pi BD + \pi HN + \pi \theta M)$ (Figure 2).[77] Next, by similar triangles, one shows that $AX : XE (= LD : BL) = AG : EK + ZL + \cdots$; thus, $S = BL \cdot \pi AG$. In *CS* 6 the result $S < AG \cdot \pi AG$ is applied in an indirect proof to establish that the spherical surface equals the product of the diameter (AG) and circumference (πAG) of a great circle. In a corollary, this result is expressed first as 4 times the great circle, then as two-thirds the closed cylindrical surface containing the sphere.

Thus *CS* adopts for the surface of the inscribed solid the same procedure used by Archimedes in *SC* I. In *CS* 5, after the solid is formed by rotating the regular polygon (as in CS I.21), its surface is expressed

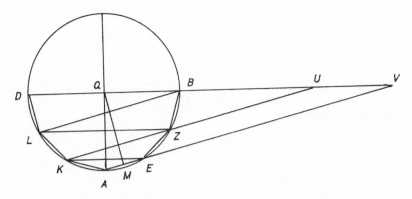

Figure 3

as the product of one of its sides by the sum of the circumferences of the circles whose diameters are, respectively, the chords joining vertices of the polygon disposed symmetrically relative to the diameter (as in *SC* I.24). This is reexpressed as the product of chord *LB* by the circumference of the circle (as in *SC* I.21). *CS* 5 notably simplifies Archimedes' treatment by eliminating his *dynamis* terminology.[78] Further, *CS*'s form for the second product renders obvious the inequality established separately in *SC* I.25. This product also motivates directly the expression for the spherical surface in *CS* 6, namely as the product of the diameter times the circumference of the great circle of the sphere: for the circumference is one of the factors in both, while the chord in *CS* 5 corresponds in the limit to the diameter in *CS* 6.

Thus, but for the order of steps, *CS* adheres to the technical model of *SC* I. It condenses the Archimedean treatment considerably by merging *SC* I.21 and 24 into one theorem, *CS* 5, by assuming without proof the inequality of *SC* I.23, and by embedding *SC* I.25 within *CS* 6. Further, the alternative convergence technique used in *CS* 6 makes dispensable the equivalent of Archimedes' treatment of the circumscribed cases (*SC* I.28–30).

By comparison with the treatments in *SC* I and CS, that in *VF* is considerably compressed. *VF* 12 proves two preliminary results relating to parallel chords in the semicircle *BAD* (Figure 3):[79] $QV = EK + ZL + QB$ (via parallelograms *EKUV, ZLBU*) and $QM^2 < MA \cdot QV < QB^2$.[80] Then, *VF* 13 shows that the surface S' of the conical solid in the hemisphere, generated by rotating the semipolygon about its altitude *QA*, is less than twice the circle of its base (of radius *QB*) but greater than twice the circle of the hemisphere inscribed within the solid (i.e., of radius *QM*).[81] That is, deducing from the results on truncated conical surfaces in *VF* 11 that $S' = AE\,(\pi EK + \pi ZL + \pi QB)$, one finds that $2\pi QM^2 < S' < 2\pi QB^2$. These inequalities are then applied within the indirect proof in VF 14 to show that the surface of the hemisphere equals twice its base circle.

VF 13 can be viewed as a reorganization of CS 5. The inequalities proven for the conical surface in VF 13 are actually the initial steps in the two parts of the indirect proof in CS 6.[82] Although the exact expression for the surface of the conical solid is not the stated objective of VF 13, it is developed in the course of the proof, by effectively duplicating the argument of CS 5. Like CS 5, VF here circumvents the Archimedean *dynamis* terminology, but by an alternative strategy in which the constant proportionality of circles and their diameters (VF 5; cf. CS 3) is framed as a quasi-numerical multiplier, the "quantity into which when the diameter is multiplied it results in the circumference" (cf. 13.77–79). Thus, by multiplying the terms of the inequalities in VF 12 one obtains the inequalities sought in VF 13. In this, VF is closest in concept to the modern transcription via the symbol π.

Although one could try to explicate the treatment in VF with reference to SC I directly, rather than through the mediation of CS, that would not account well for the alternative convergence method which VF shares with CS. But for the convergence technique, CS follows the model of SC I closely, and this could not be explained by positing VF as its source. Only on one point does VF appear closer to SC I than is CS, namely, in its expression for the surface of the sphere. For in equating it with 4 times the great circle, VF (14, corollary) seems to conform with SC I.33, and not with CS, where the spherical surface equals the product of the diameter and the circumference of its great circle. But the choice of phrasing in VF 14 probably derived from the inequalities around which it has organized the technical argument in VF 12 and 13; that is, since the stated product is bound above and below by the squares of specified lines, the spherical surface will accordingly be bound by the circles of which these lines are the diameters. Moreover, in the corollary to CS 6 (lines 42–43) the surface of the sphere is expressed as 4 times its great circle, just as in SC I.33. Thus, VF does not require direct access to SC I for this form.

A similar pattern applies for the respective treatments of the volume of the conical solids. In SC I.26, the volume of the solid inscribed in the sphere is shown to be equal to the cone whose base is the surface of the solid and whose height is the line drawn from the center perpendicular to a side of the generating polygon. The proof applies theorems on the volume of the cone already shown (in SC I.17–20). Thus, in Figure 4 the volume of the "solid rhombus" consisting of the two cones generated by rotating the triangles NAZ, NZX equals a single cone whose base is the conical surface formed by NAZ and whose height is the perpendicular from X to AZ (SC I.18); similarly, the solid

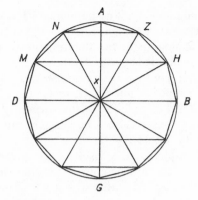

Figure 4

bounded by the three conical surfaces *NZHM, XNZ,* and *XMH* is equal
to the cone whose base is the conical surface *NZHM* and whose height
is the perpendicular from *X* to *ZH* (SC I.20);[83] and so on. Since the
heights of the cones are equal and their surfaces sum to the surface of
the solid, the whole solid equals the cone of the stated base and height.
By combining this result with *SC* I.25, it follows (*SC* I.27) that the
solid is less than 4 times the cone (*K*) whose base equals the great
circle of the sphere and whose height equals its radius. Analogous
results are established for the circumscribed solids in *SC* I.31. Finally,
in *SC* I.34, Archimedes proves that the volume of the sphere is 4 times
the cone *K*.

    CS 7 condenses into one theorem the entire Archimedean treatment
of the volume of the inscribed compound solid, including the preliminary results on the cones, solid rhombi, and related figures (i.e., *SC*
I.17–20, 26). Corresponding to Archimedes' "solid rhombus," *CS* introduces the "double pyramid" formed by rotating two isosceles triangles of common base. This, and related solids formed as the "difference" of two such double pyramids,[84] are found to equal a cone
whose height is the line drawn from a given center perpendicular to
the side of the generating triangle, and whose base is the surface of
the solid. One can then infer that the whole solid equals the cone of
corresponding height and base. As for the surface theorem, the use of
the special convergence technique in *CS* 8 makes unnecessary a separate treatment of the circumscribed solids (in contrast with *SC* I.31–
32).

    VF 15 determines the volume of the sphere as the product of its
semidiameter by one-third its surface area. The result is proved through
an indirect argument which ostensibly depends only on the result for

the surface of the sphere proved in *VF* 14. Nowhere does *VF* establish lemmas on the volumes of cones or an explicit form for the volume of the composite conical solids, analogous to the stereometric results in *SC* I and *CS*. Yet in *VF* 15 a critical role is played by these very results:

> it will be necessary *from what we have prefaced* (*ex eo quod premisimus*) that the product of line SB [namely, the radius of the sphere] by one-third of the area of the surface of the solid which contains the sphere ABGD be greater than the sphere *ABGD*. (15.16–19; cf. 15.29–31)

Despite the claim here set in emphasis, the result is not covered by any of the previous theorems. One can, of course, assume that the solid is greater than the sphere enclosed within it; but it still must be proved that the solid equals the stated product. An author working directly with the Archimedean sources is not likely to have omitted such a result, since its treatment extends over a series of theorems interwoven into the structure of *SC* I (namely, I.17–20, 26–27, 31). By contrast, *CS* consolidates all this material into a single proposition (*CS* 7), so that a process of scribal error could well have resulted in its omission.

Since Gerard's text is here supported by the later Arabic recension by al-Ṭūsī, it is evident that the original Arabic text of the Banū Mūsā never held the missing steps; for it is unlikely that both of the later editions should have been based on documents scribally corrupted in the same way. On the other hand, *VF* includes a cross-reference to "what has been prefaced." It would be remarkable that the authors failed to recognize correctly the content of their own previous theorems. But conceivably they came upon some such words in their source – for instance, a Greek phrase like *katha prodedeiktai*, commonly used by ancient technical authors – and then transmitted them in their own version without realizing that the materials so designated were lacking. In this event, the omission would be due to their use of a defective source.

From our comparisons of the volume theorems, one sees that the source of *VF* agreed with *CS* in its basic organization, in particular in its incorporating into a single proposition all the items bearing on the volume measurement of the composite solids. This affirms the findings based on our survey of the surface theorems and their preliminaries. In both respects, the closeness of *CS* to the technical model of *SC* I rules out the hypothesis that *CS* has worked from *VF* or a text resembling it. But the general technical affinities between *CS* and *VF* (particularly in their method of convergence) indicate further that *VF* draws from a source comparable to *CS*.

V

To account for the data presented in the preceding sections, the simplest thesis would be (a) that the technical core of *CS* is a Latin translation, adhering closely to a Greek prototype, and that *VF* is a freely edited rendition of the same prototype into Arabic. Before explicating this thesis, we may briefly consider two related alternatives: (b) that the common prototype was an Arabic work; and (c) that *CS* and *VF* depended on two different sources.

Hypothesis (b) is effectively equivalent to (a), for it would still require a Greek prototype for the assumed Arabic source; otherwise, the Archimedean parallels, especially strong in *CS*, could not be accounted for. Further, we would have to suppose in either event that *CS*, as a close translation, provides better witness to the source text (whether Arabic or Greek) than *VF* does.

Nevertheless, hypothesis (b) encounters difficulty, since it requires that the presumed Arabic source would be available to a Latin translator three or four centuries after its composition. For a work of the necessary description could not ever have circulated widely; otherwise, the Banū Mūsā could hardly have perpetrated the fraud of presenting *VF* as entirely the product of their own researches. In their preface, the authors explain that they will not treat of things which are well known through works in circulation (*quorum scientia est publicata*), but will expound less familiar results with due notice of the ancient writers (pref., lines. 17–26). Even so, only two items are so designated: the treatment of Archimedes' approximations for π in *VF* 6, and Archytas' solution for the cube duplication (which they assign to Menelaus) in *VF* 16. In their epilogue, they confidently affirm that all else is their own (19.63–70) and explicitly claim originality for the theorems on the sphere (19.74–78).[85] From our own knowledge of the ancient sources, however, we can recognize their claims to be false: Their theorems on the area of the circle (*VF* 1–5) and on the surface and volume of the sphere (*VF* 9–15) are Archimedean, while the rule for measuring triangles (*VF* 7) is given by Hero, and may also be due to Archimedes.[86] On the kindest reading, we could assign to the Banū Mūsā responsibility for the particular alternative proofs presented in *VF*. But it is clear that they were working from ancient materials.

In connection with hypothesis (c), that *CS* and *VF* worked from different prototypes, we observe first that the strength of the parallels between *CS* and Archimedes' *SC* I indicates that *CS* has rendered its source closely, whereas the editorial intervention separating *VF* from *SC* I is considerably greater. Certainly, many of these differences are due to the Arabic authors; but on the present hypothesis, some of them must lie in their source. To warrant the view of multiple prototypes, we must be able to identify features of *VF* which cannot be accounted for by reference to *CS*, for example, places where *VF* is closer to *SC* I than *CS* is. I have been able to detect only one such place, however – *VF* 14 expresses the surface of the sphere as in *SC* I.33, but differently from *CS* 6. Yet, as we have seen, even this can be accounted for, since

CS 6 produces the Archimedean form in its corollary. In sum, nothing in *VF* requires a source other than *CS*, and nothing in *CS* requires a source other than *SC* I. The evidence thus seems not to warrant the two-source hypothesis.

For critical purposes, thesis (a) has the advantage that all the discrepancies between *CS* and *VF* can be assigned to the Arabic editors. These would include the framing of the alternative proofs of Archimedes' circle theorem (*VF* 3), the alternative form of the summation in *VF* 12, and the consolidation of the sequence of theorems on the surface of the conical solids into a single proposition on the inequalities between their surface and the associated circular areas (*VF* 13). The omission of the volume theorem for conical solids, however, may be explained through their use of a defective copy of the source.

In now considering thesis (a) more closely, let us designate the proposed Greek source as *CS**. Placing *CS** is possible by considering once again the parallels associating *CS* with *DC* and *AI*. As we have seen, *CS* knows Archimedes' *Dimension of the Circle* in the same form extant to the medieval translators, which but for minor discrepancies is the same as the Greek of *DC* extant today. *CS* states *DC* 1 much as we have it in Greek, formulating the area of the circle as equal to a right triangle whose legs equal its radius and circumference; moreover, *CS* imitates this same formulation for other results (cf. *CS* 1). Yet the formulation of *DC* 1 in terms of such a right triangle is not found in ancient authors before the fifth century A.D.[87] The theorem is known to earlier writers, such as Pappus and Theon in the fourth century A.D, Hero in the first century A.D. and Zenodorus in the third or second century B.C.; but they always adopt the alternative product format: such as, that twice the circle equals the product of its radius and its circumference.[88] Comparable product formulations are prominent throughout *CS* and *VF;* for instance, *CS* knows the product form for the circle (cf. 1.110–111) and formulates the surface of the sphere as equal to the product of the diameter and circumference of a great circle in it (*CS* 6). In this respect, there is a parallel with *AI*, where both forms for the circle are stated, but where the product formulation is assigned explicitly to Archimedes' *DC*.[89]

Further, *CS* applies *DC* 2 in this form:

> proportio GF quadrati an O circulum est sicut XIIII ad XI,
> per secundam Archimenidis de quadratura circuli. (10.12–13)

This closely parallels the form extant to us in Greek:

> The circle has to the (square) on the diameter the ratio that
> 11 (has) to 14.[90]

But these expressions are markedly different from that stated by Hero in *Metrica* I.26:

Archimedes proves in the Measurement of the Circle that 11
squares on the diameter of the circle become equal, very
nearly, to 14 circles.[91]
In contrast with the wording in Hero, CS and DC both fail to distinguish
between approximate and exact parameters. A further weakness of DC
2 is that its proof violates the proper deductive ordering of theorems
by assuming the result of DC 3. These lapses have become standard
criticisms of DC. As early as the thirteenth century, for instance, al-
Ṭūsī sought to restore the correct logical sequence by transposing DC
2 and 3 in his edition.[92] Modern critics are in agreement that on such
points the extant DC cannot accurately represent what Archimedes
actually wrote.[93] Thus, CS must have been composed when the modified edition of DC had taken hold as the familiar Archimedean
reference.

In these respects, then, CS* is associable with the extant DC. Since
this edition of DC can be placed in the early part of the fifth century,
the same becomes a *terminus post quem* for CS*.

That CS employs both the triangle and the product forms of DC 1
suggests an association with AI, which also uses both. This connection
is further indicated, as we have seen, by their very similar proofs of
the product expression for the volume of the sphere. Since AI too can
be assigned to the early fifth century A.D. (that is, not later than mid-
fifth century),[94] the dating of CS* to this same period is again indicated.

Other links between AI and CS* can be detected. In the surface and
volume measurements of the conical solids (CS 5 and 7, respectively),
the proofs tacitly assume that the number of sides of the generating
semipolygon is even, although this restriction is not actually stated in
the enunciations. In SC I the analogous restriction is explicit (cf. I.23,
25–26); but it presents no difficulty there, since these theorems are
merely ancillary to the theorems on the sphere (I.33–34). To the author
of CS*, however, there is a logical gap which must be filled by the
proofs of the omitted odd cases. He provides this for the surface theorem (5.92–117), but apparently is unable to do so for the volume theorem. Thus, he comments:

> since the author has not proposed this, we too omit its
> investigation and leave it to diligent posterity. (7.179–180)

The remark is reminiscent of the lines which close AI, relating to a gap
in its own argument on isoperimetric solids:

> For the rest, although it is necessary to prove (this) also . . .
> our philosopher has added nothing, but has left off, . . .
> having entrusted it to us to seek the proof agreeable to
> geometers. And this has not yet been provided by us, but
> we will accord thanks of benefit to the one who has found
> it. (1164.15–21)

Further, *AI* and *CS* share certain aspects of technical style. Within its proofs, *CS* occasionally elaborates its figures in specific terms, even though the construction is intended to be general. For example, a given sum consists of the "five circumferences" of specified circles in *CS* 5.80; "three perpendiculars" are drawn to the "three sides" of the polygon in *CS* 7.13–14; a single pyramid is taken to equal the "six pyramids," which together equal a given composite solid, in *CS* 7.160–170. In formal writing, one would usually speak of "each" or "all." But *CS* here seems to echo *AI* where it refers directly to the "hexagon" and "pentagon" in its theorem on isoperimetric polygons.[95] Moreover, *CS* frequently refers to angles by a single letter: for example, "the angle D" (1.85) means "the angle at the point D" (so also *passim* in props. 4, 5, and 7), even where the reference is ambiguous. *AI* does the same in its prop. 1, whereas in *CS* more than one angle could so be designated.[96] While *AI* and *CS* are hardly unique in adopting these mannerisms, they hereby become associated with a style of technical writing different from the classical treatises of the formal tradition.

Since *AI* follows the forms for the surfaces of the conical solid and the sphere adopted in Archimedes' *SC* I, rather than the variants appearing in *CS*, we cannot view *AI* as dependent on *CS**. Both works must be put in a direct relation with *SC* I. But on several points, *AI* and *CS* adopt like terms in divergence from Archimedes. For instance, in its volume theorem (prop. 7), where *CS* writes of "every solid of conic surfaces inscribable and circumscribable in a sphere" and takes the height of the corresponding conical volume not as the perpendicular but as "the semidiameter of the sphere inscribed in the solid," it conforms with usage in *AI* (prop. 10, 1160.13–17). By contrast, Archimedes in *SC* (e.g., I.26) consistently expresses that height as the perpendicular drawn to a side of the generating polygon. Moreover, in emphasizing the relation of the conical solids and the associated circumscribed and inscribed spheres (prop. 10), *AI* provides a plausible motive for the alternative convergence method in *CS**, where the critical move involves introducing spheres equal to volumes supposed to be greater or less than stated magnitudes.

These comparisons place *AI* in an intermediate position between *SC* I and *CS**. But the reminiscences of *AI* in *CS* (and hence in *CS**) are of a subtle kind, the signs of a deep familiarity rather than an overt exploitation of a source text; indeed, it is difficult to perceive why the author of *CS** would use *AI* at all, in the presence of *SC* I. It appears, then, that *CS** was composed within the same scholarly group, and conceivably by the same individual, responsible for the production of *AI* and the extant recension of Archimedes' *DC*. We could well suppose that *CS** was intended as a new version of Archimedes' theorems on the sphere, complementing the applications of these theorems in *AI*

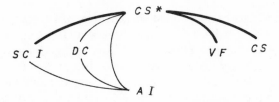

Figure 5.   Sources and affiliates of *De curvis superficiebus*.

and extending the format used for the circle theorems in *DC*. Since *AI* and *DC* betray influence from the editions and commentaries of Theon of Alexandria, we can assign them, with *CS*, to the early part of the fifth century A.D. Our information about Theon's disciples is limited; indeed, we know the name of only one – his daughter Hypatia.[97] The possibility that she was responsible for the editions of these works is certainly tantalizing, but making a specific attribution overreaches our present evidence.[98]

These findings can be summarized schematically (Figure 5), where *CS* is a translation of the prototype *CS**, a technical tract in the tradition of Archimedes' *SC* I. But the Latin work is a composite: The core is amplified by technical interpolations, some doubtless due to ancient scholiasts, and by pedagogical comments, assignable to medieval editors explicating aspects of the logical connections of steps and encouraging students to diligence. The whole is executed in a firmly Latin style that all but obliterates vestiges of the Greek language of the source. One might suppose that the two roles of translator and commentator were served by separate individuals; for the process of Latinizing the style would be expected to depend on an earlier literal rendering. Such appears to characterize the Euclid editions of Hermann and Campanus, for instance, and to obtain in the cases of other technical works, like the Euclidean *Catoptrics*.[99] But for *CS*, I think it more likely that the roles were conflated; for we lack evidence of a primitive form of the text, despite the relatively wide circulation of the work. Adopting this view, we can assign the work to the early thirteenth century, since the editorial interpolations reflect familiarity with the Euclid recension "Adelard III" (middle or late twelfth century), while it is cited in the *De motu* of Gerard of Brussels (mid-thirteenth century or earlier).[100] We may identify the translator-editor as Johannes de Tinemue (presumably, John of Tynemouth), of whom we know nothing more. The other figure associated with *CS*, Gervase of Essex, is of little interest; as the one responsible for the minor editorial changes held in a few of the manuscripts of *CS*, his mathematical contribution is negligible.

The present analysis has sustained the position of Clagett, that *De curvis superficiebus* is at base the translation of a Greek source. But supporting this view has proved to be far from straightforward. Linguistic considerations, for instance, fail to establish provenance, since the style has been thoroughly Latinized, removing virtually all traces of a non-Latin source. Turning instead to *CS*'s affiliations with ancient geometric works of related content and technique, I have elaborated the connections with the anonymous tract *On Isoperimetric Figures* and with the extant version of Archimedes' *Dimension of the Circle,* and I have argued that the Greek prototype was composed within the same ancient group that produced them, and possibly by the same scholar. A central aspect of my analysis is recognizing the composite nature of *CS*, which retains vestiges of the contributions of editor, copyist, and translator superimposed over the original writing. The resultant view is perhaps more complex than one usually maintains for ancient or medieval technical works, but, if anything, less complex than the truth of the matter is likely to be. We have no alternative, however, but to seek the simplest view that organizes the extant evidence in a manner consistent with our general knowledge of the Greek, Arabic, and Latin traditions of scholarship. If that falls short of factual certainty, it nevertheless can provide a useful position from which to explore these traditions further.

Notes

1 *Archimedes in the Middle Ages* I: *The Arabo-Latin Tradition* (Madison, 1964), contains the documents of principal interest for the present study. William of Moerbeke's translation of Archimedes from the Greek is presented in vol. II (Philadelphia, 1976), and the Archimedean tradition through the sixteenth century is documented in vol. III (Philadelphia, 1978). Two sequel volumes collect texts on conic sections (IV, 1980) and Archimedes-based studies, such as Gerard of Brussels' *De motu* (V, 1984). In the notes, I will designate these as *AMA* followed by the volume number. To these volumes should be added the collection of mechanical texts edited by Clagett with E. A. Moody, *The Medieval Science of Weights* (Madison, Wis., 1952), for these writings ultimately depend on an Archimedean technical base. (For a discussion, see my *Ancient Sources of the Medieval Tradition of Mechanics* [Florence, 1982]).

2 The critical text is edited with translation and commentary by Clagett, *AMA* I, chap. 6.

3 *AMA* I, p. 440. Alternative forms for *Tinemue* include *Tinennie, Chinemue, Tin', Tiñ, Thiñ, Thin 9, Tlñ* (cf. pp. 450n., 506n.).

4 Ibid., p. 443.

5 Ibid., p. 440, 444; cf. also *AMA* V, p. 7. Gerard cites *CS*, prop. 1, twice, while two scholia cite its prop. 4. Agreement in phrasing and content also indicate Gerard's dependence on *CS* (*AMA* V, 14–15). Dating Gerard to ca. 1250 rests on a citation of his work in Richard of Fournival's bibliography, assignable to before 1260; see *AMA* V, p. 6.

6 Clagett cites twelve manuscripts ranging from the thirteenth century to the sixteenth century (*AMA* I, pp. 446–449).

7 On Moerbeke, see *AMA* II, pt. 1, chap. 1; on Jacob of Cremona, see *AMA* III, pt. 3, chap. 2; on Maurolico's use of *CS*, see *AMA* III, pp. 798–808.

8 Cf. the view of J. L. Heiberg in *Archimedis Opera,* 2nd ed., III (Leipzig, 1915), p. xcviii. Clagett's own position, with notice of the change from his earlier view, is stated in *AMA* I, pp. 441–442.

9 The text is edited by Clagett in *AMA* I, chap. 4; see the discussions in sections II and IV of this chapter.

10 Clagett includes these additions with discussion of their provenance in his edition of *CS* (*AMA* I, chap. 6, sec. 3–5).

11 As Clagett observes (*AMA* I, p. 44ln.), the citations of *Elements* XII.9 conform with the Arabo-Latin versions, rather than with the standard Greek text (where the same proposition is numbered XII.10). This cannot be viewed as decisive for an Arabic or Latin provenance for *CS*, since there exists a Greek manuscript that holds a text of Book XII agreeing with the Arabic and Latin versions (see note 35 below). Nevertheless, it seems to me unlikely that this single manuscript alone represents the prevalent tradition of the *Elements* in late antiquity, against the testimony of all the other extant manuscripts. Hence, citations by ancient scholiasts would be expected to conform to the latter. Thus, either the bulk of the Euclidean references in *CS* were due to a Latin editor, or they were harmonized with the edition familiar to such an editor.

12 See note 68 below.

13 Here and throughout, I adopt the text in *AMA* I; the translations are my own, in part modifying those given by Clagett:
    Whence it [namely, the theorem or the hypothesis] does not urge the peevish and chiding to knowledge, since they are the very ones who do not dare to be wise, since indeed they strive to be ignorant, since even from Pallas, revealing her countenance of her own accord, they avert their eyes. These things, then, the philosopher has propounded to those who search into the flights of subtle things, the miracles of quantities, the intricacies of proportions, the depositions of nature.

14 *CS* 1.15–17: "The hypothesis of the present demonstration, and the entire series of following theorems, postulates for itself that there be admitted a straight line equal to a curved and a straight surface equal to a curved." The same hypothesis is noted by al-Ṭūsī in his edition of the Banū Mūsā (quoted in *AMA* I, p. 254n.) and by Maurolico in his expansion of *CS* (cf. *AMA* III, p. 799; cf. also note 69 below).

15 "And thus does our Tiphis hold the port toward which he had so long ago made ready the sails. And now, since the anchor of Archimedes' boat clings to the thirsty sands, let Johannes heap up thanks of navigation to the Supreme Creator." By "heap up" I render *ageret* as if a corruption of *aggeret* (present subjunctive of *aggerare*); for the imperfect subjunctive (*ageret,* from *agere*), usually connoting a counterfactual, is here impossible. Forms of *agere* with words of thanks are common, and could easily induce a scribal alteration of the less familiar form from *aggerare*. (I am indebted to Lionel Pearson for this suggestion.) Further, by "made ready" I render *succinxerat*. The verb *succingere* usually means "gird," whence "prepare." It is odd with *vela* ("sails"), however, for that ought to imply "furling the sails," that is, taking them *in* at the end of a voyage, not letting them out at its start. The writer might mean just this: that the voyage is nearly over, so

we take in the sails and enter harbor on power of oars. But perhaps he has merely confused the assorted navigational idioms.

16 Cf. also *Ars amatoria* 1.368; *Heroides* 13.101; *Remedia amoris* 790; *Tristia* 5.14.44.

17 Cf. Cicero, *Tusculan Dissertations* 3.11.25; also Plautus, *Asinaria* 157.

18 On the use of classical Latin authors in the twelfth century, see C. H. Haskins, *The Renaissance of the Twelfth Century* (Cambridge, Mass., 1927; repr. 1957), chap. 4.

19 In addition to 1.19–22 and 10.24–27, already discussed, these suspect passages include 1.28–32, 1.78, 1.102–104 (a technical writer would not describe his own theorem and its corollary as "elegant"), 2.5–9 (an overelaborate formulation of the hypothesis of indirect argument), 2.21–22, 2.39–40, 3.33, 5.16–18 (a note of encouragement for "even the less diligent reader"). 5.66–67, 5.90–91, 5.117, 6.24–25, 7.24–26, 7.95, 7.113, 7.172–173 (on the end of the "hunt"), 8.6–8, 8.26–27, 8.62–63, 8.66–67, 8.86–87 (that we are not "defrauded" of our claim), 9.26–27, 10.4–7. At several places second-person forms are used (1.54–55, 5.59, 5.63–64, 7.155); the practice is met rarely in ancient scholia, but never in the body of formal technical works. Where parallels to the *content* of any of these passages can be cited in scholia to ancient technical writings, it is possible that the Latin writer is offering his own florid rendition of a comment in his source. It is nevertheless noteworthy that none of these passages contains words of Greek derivation. The invocation of the *summus creator* at 10.26 raises the possibility that biblical passages also influenced the style of the author. One such reminiscence might occur at 5.16–17: *ne . . . lector scrupulum quo progrediens pedem offendat possit reperire,* which may recall the famous line from Ps 90.12 (in the Septuagint-based Latin version): *ne forte offendas ad lapidem pedem tuum* (cf. Mt 4.6, Lk 4.11, Rom 9.32). Similarly, when the author concludes at 8.86–87: *sic ergo proposito non sumus defraudati,* he may be echoing Sir 37.23–24: *qui sofistice loquitur odibilis est, omni re defraudabitur, . . . omni enim sapientia defraudatus est.*

20 See "The *De curvis superficiebus Archimenidis,*" *Osiris* 11 (1954): 299. The suggestion is not repeated in Clagett's later edition of *CS* (cf. *AMA* I, pp. 440, 442–443).

21 Other variants are *Thinemutha, Tinomoutum;* cf. C. T. Martin, *The Record Interpreter,* 2nd ed. (London, 1910), p. 413.

22 Compare the variant forms mentioned in note 3 above. Since the letters *t, c,* and *e* are easily confused in Latin orthography, it seems possible that an abbreviated form *Tinemut'* could be altered to *Tinemue* by a scribe expecting an ablative form after *de*.

23 Medieval Latin sources employ the Greek derivative *discolus* (or *dyscolus*), to mean "peevish," "ill-tempered," "surly," and the like; cf. *Mediae Latinitatis Lexicon Minus,* ed. J. F. Niermeyer (Leiden, 1976), p. 338; and *Revised Medieval Latin Word-List,* ed. R. E. Latham, (London, 1965), p. 159.

24 See note 19 for a list of passages.

25 The term appears at 1.33, 46, 72, 76, 80, 84, 94; 3.13; 6.39; 8.65, 80.

26 *AMA* I, pp. 169, 442. Cf. *Sophistical Refutations* 171b34–2a7; and *Physics* 185a15–18.

27 Aristotle discusses *pseudographêmata* in *Topics* I. 1.101a5–17; see T. L. Heath, *Mathematics in Aristotle* (Oxford, 1949), pp. 76–78. Pappus uses the cognate verb to designate errors committed by his colleagues and predecessors; cf. ed. F. Hultsch, *Pappi Collectio* (3 vols; Berlin, 1876–1878), I, p. 40; II, pp. 474, 530.

28 The text is edited by Clagett in *AMA* I, chap. 3, sec. 5; the term *falsigraphus* appears at line 218.

29 Text from Clagett, "King Alfred and the Elements of Euclid," *Isis* 45 (1954), p. 274, lines 49–50 (cf. also *AMA* I, p. 444n.). As Clagett notes, other versions of the *Elements* adopt a different terminology for indirect proofs ("King Alfred," p. 273).

30 "King Alfred," p. 274, lines 43–44; for comparison with terms in other versions, see ibid, p. 273.

31 *AMA* I, p. 442.

32 See the comparative list of terms collected by H. L. L. Busard in "Addendum 2" of his edition of "Adelard I" (*Studies and Texts* 64 [1983], Toronto), pp. 397–399. Note that even the Greek nominative form *diametros* is found in the Arabic-based versions.

33 *CS* (2.2) uses *tetragonus* to mean "rectangle," not "square." This anomaly is found also in *Verba filiorum*, prop. 5 (lines 15, 16, 21, 26, 30), where the term *quadratum* ("square") is used to signify a *rectangle* (cf. also *VF*, props. 7, 12). Although the error in *VF* may be due to the Latin translator, Gerard (so Clagett, *AMA* I, pp. 233–234), the Arabic recension used by Gerard is not extant (our text is in the recension by al-Ṭūsī, so that direct confirmation is impossible. The Arabic for "square" (*murabbaᶜ*) sometimes signifies the more general "quadrilateral" (cf. my discussion of the term in *Ancient Sources of the Medieval Tradition of Mechanics*, pp. 83, 201). The failure by the Arabic writers to separate the terms for "square" and "rectangle" might possibly be due to misreadings of Greek sources, where the terms are typically given in the elliptical forms *to apo* . . . and *to hypo* . . . , respectively. But this cannot explain the case of *CS* 2. For there, the whole phrase "tetragono qui continetur sub lineis equalibus axi columpne et circumferentie" (2.2–4) makes clear that the writer understands by *qui continetur sub* . . . the precise equivalent of the Greek phrase *to hypo* . . . *periechomenon* (*orthogônion*). With the full Greek expression in mind, one can readily explain the presence of *tetragonus* as the Latin writer's incorrect fill-in for the missing substantive, *orthogônion*. This passage is of particular interest in the present context, since it contains a distinct Grecism not explicable by the standard technical usage.

34 Busard compiles a list of the Arabicisms in the medieval Euclid versions in "Addendum 2," *Studies and Texts*, pp. 391–396.

35 Clagett has compared the principal Euclid versions in "The Medieval Latin Translations from the Arabic of the *Elements* of Euclid," *Isis* 44 (1953);16–42; he argues that although Hermann may have consulted "Adelard I," he unquestionably did work with "Adelard II" (ibid., pp. 26–27), and Clagett illustrates this convincingly in his excerpts from the texts (pp. 30–42). Unlike "Adel. I," which provides complete proofs, "Adel. II" usually gives only outlines and cross-references. As one can see from Busard's editions of Hermann (Leiden, 1968; Amsterdam, 1977) and "Adel. I" (Toronto, 1983), Hermann agrees with the organization of proofs in "Adel. I." For Book XII this is particularly interesting, since the medieval translations follow a line of the Greek different from the majority of the extant Greek manuscripts and represented now only by ms. Bonon. 18–19 (for text, see J. L. Heiberg, *Euclidis Opera* IV, Appendix II); Heiberg discusses this manuscript and its relation to the Arabic recensions in "Die arabische Tradition der Elemente Euklid's," *Zeitschrift für Mathematik und Physik*, hist.-lit. Abth., 29 (1884), pp. 6–14. Moreover, special coincidences

have been noted strengthening the link between Hermann and "Adel. I" (see J. E. Murdoch, "Euclid: Transmission of the *Elements, Dictionary of Scientific Biography* IV, p. 447). A possible account would have Hermann exploit the abbreviated "Adel. II" to assist his retranslation of the Arabic, in the same recension of al-Ḥajjāj as that used for "Adel. I." Interestingly, *CS* shares with "Adel. III" both its basic terminology and such stylistic idiosyncracies as Ovidian allusions (cf. Murdoch, "Euclid-Transmission," p. 446).

36 Among the discrepancies which are not merely scribal variations (e.g., transpositions, spelling variants and errors, additions or deletions of single words, etc.) are these: 1.18–22 (*unde . . . philosophus*) is deleted; 1.27–32 is replaced merely by *et cetera* (cf. 1.57); 2.14 amplifies *eidem linee* to *uni et eidem linee;* 2.39–40 alters *suo junctum theoremati* to *premissi theoreumatis;* 3.14–16 is omitted (but likely by error); 3.33 omits *inconcussum;* 3.35 omits *non* (deliberately?); 3.38 adds *scilicet prioris* for clarification; 5.65 omits a phrase beginning with *scilicet;* 7.37, and so on, alter the correct references to theorem "IX" into "XI;" 7.78 alters the correct *columpna* into *conica* (cf. 7.81); 7.114–115 amplifies the wording of the other manuscripts; 8.62 alters the literary *teramus* to the more familiar *teneamus;* 8.86–87 alters the literary *sic ergo proposito non sumus defraudati* to the more standard *et sic constat propositum;* and, of course, in the closing lines *Johannes . . . creatori* is removed (10.26) and the name *Johannes de Tinemue* is replaced by *Gervasius de Essexta* (10.27). This is not a complete listing, but it is representative.

37 *AMA* I, p. 443. The identification requires a scribal change of *Exonia* ("Exeter") into *Essexta* ("Essex"); if correct, the confusion would doubtless have occurred between the English names rather than between the Latin. What makes the suggestion attractive is the possible connection with the translators at the Viterbo court in the 1260s (see note 4 above).

38 *AMA* I, p. 441.

39 Ibid., p. 443n., citing the monograph on Moerbeke by M. Grabmann.

40 Props. 20–37; *Collectio*, ed. F. Hultsch, I, pp. 362–410.

41 Ibid., I, pp. xvii–xix.

42 S. Unguru in "Pappus in the Thirteenth Century in the Latin West" (*Archive for History of Exact Sciences* 13 [1974]: 307–324) proposes that Witelo consulted Pappus' Book VI in the 1270s; he also reviews the evidence for the history of the Pappus ms. Vat. gr. 218.

43 The text is edited with translation and commentary by Clagett in *AMA* I, chap. 4.

44 Pappus' procedure is based on the rotation of any triangle about an axis, where the whole triangle lies on one side of the axis save for one of its vertices; he proves that the volume of the solid of revolution so generated equals the cone whose base equals the conical surface traced by the side opposite that vertex and whose height equals the perpendicular drawn from the vertex to the same side. This result provides an alternative treatment of Archimedes' results in *SC* I.17–20.

45 We will explore these technical correspondences more fully in section IV.

46 But Archimedes adopts the analogous form for spherical sectors in *SC* I.44, as also for the conical solids in *SC* I.26 and 31. In *SC* I.34 the sphere is measured as four times the cone whose base equals the great circle of the sphere and whose height equals its radius. The principal form of the volume

theorem in *CS* (prop. 8) is as two-thirds the cylinder whose base equals the great circle and whose height equals the diameter; for *SC* I.34, this is also derived, but only as a corollary.

47 Derivations or applications appear in Hero, *Metrica,* ed. H. Schöne, III.11; *Stereometrica* (*Heronis Opera,* ed. J. L. Heiberg, V), chaps. 2, 3, 7–8, 65–66, 68–71; Theon of Alexandria, *Commentary on Ptolemy's Book I* (ed. A. Rome [Vatican City, 1936], pp. 395–396); and elsewhere. (Cf. note 63 below.)

48 *AMA* I, pp. 443–444.

49 Ibid., p. 40. One may note the marked difference from the phrasing employed in the translation attributed to Plato of Tivoli: "omnis circulus triangulo ortogonio, cuius unum ex duobus lateribus recto adiacentibus angulo dimidio diametri eiusdem circuli, alterum vero latus linee circumferenti equatur, existit equalis" (ibid., p. 20).

50 Cf., for instance, the version in the Florence ms. conv. soppr. J. V. 30, ed. Clagett, ibid., p. 106 (lines 1–4): "quod omnis triangulus orthogonius, cuius unum latus equatur circumferentie, reliquum latus semidyametro, equalis est ipsi circulo."

51 Ibid., p. 46 (lines 62–64): "tunc quod fit ex multiplicatione eius [sc. medietatis diametri circuli] in medietatem sectionis [!] circumferentie est area figure accepta equalis aree trianguli *E.*" Here, the triangular area *E* is the right triangle that the proposition has shown equal to the circle itself.

52 For comparison, I present a translation from Heiberg's text of *DC* 1 (*Archimedis Opera* I, p. 232): "Every circle is equal to the right-angled triangle of which the ⟨line⟩ from the center [namely, the radius] ⟨is⟩ equal to one of the ⟨sides⟩ about the right ⟨angle⟩, and the perimeter ⟨is equal⟩ to the base."

53 Quoted in ibid., p. 233n.; cf. also III, p. 230 (lines 8–10): "Having set out a right-angled triangle he says, let it have the one of the ⟨sides⟩ about the right ⟨angle⟩ equal to the ⟨line⟩ from the center, the remaining ⟨side equal⟩ to the perimeter."

54 A Greek text of the tract *On Isoperimetric Figures* has been edited with Latin translation and notes by Hultsch in *Pappi Collectio* III, pp. 1138–1165. It is based on only one manuscript, however, and no critical edition of the work has yet been made. A critical edition of the medieval Latin translation (Latin title, *De figuris ysoperimetris*) has been presented by Busard in "Der Traktat *De isoperimetris* . . . ," *Mediaeval Studies* 42 (1980):61–88.

55 *De figuris ysoperimetris,* ed. Busard, lines 239–242.

56 A discussion of the discrepancies between the Latin and Greek versions of *DC* 1, with suggestions on the history of the text, is given by T. Sato, "Archimedes' *On the Measurement of a Circle,* Proposition 1: An Attempt at a Reconstruction," *Japanese Studies in the History of Science* 18 (1979):83–99. I modify and extend some of his findings as a consequence of my examination of the Greek evidence in "Archimedes' *Dimension of the Circle:* A View of the Genesis of the Extant Text," *Archive for History of Exact Sciences* 35 (1986), 281–324. A survey of the medieval Arabic, Hebrew, and Latin versions of *DC* will be included in a textual study currently in preparation as a sequel to my *Ancient Tradition of Geometric Problems* (Boston, 1986).

57 An important group of Greco-Latin tracts was produced in Sicily in the latter part of the twelfth century. It includes Ptolemy's *Almagest,* Euclid's

Elements, Data, Optics, and *Catoptrics,* the anonymous *De ysoperimetris,* and others. On these translations, see C. H. Haskins, *Studies in the History of Mediaeval Science* (Cambridge, Mass., 1924; repr. 1960, 1967), chap. 9. The translation of the *Elements* is examined by Murdoch in "Euclides Graeco-Latinus," *Harvard Studies* 71 (1966):249–302. A text of the *Data* with English translation and commentary has been prepared by S. Ito (Tokyo, 1980). Heiberg transcribed one manuscript of the Latin version of the *Optics* and set it in parallel with his critical text of the Greek (cf. *Euclidis Opera* VII); a critical edition of the Latin has since been prepared by W. Theisen in *Mediaeval Studies* 41 (1979):44–105. For the Latin *De ysoperimetris,* see the edition by Busard cited in note 54 above.

58 I here modify the punctuation in Busard's text to conform with the Greek (ed. Hultsch, *Pappi Collectio,* p. 1162). The rendering of *de* as *autem* in 4[c] is a subtle mistranslation; for it thus sets 4[c] as a premise toward step 5. Rendering instead by *vero* would adhere better to the Greek, setting 4[c] in parallel with 4[b] as a premise toward 4[a]. The logical ordering in *CS* is correct.

59 In *SC* I (e.g., prop. 23) the generating figure is a regular polygon the number of whose sides *is measured by four;* the same manner is specified in *AI* (prop. 7 *ad init.*). But in *CS* 5 and 7 the qualification is omitted. (See section IV of this chapter for additional comment.)

60 Cf. Theon's *Commentary on Ptolemy's Book I,* ed. A. Rome (*Studi e Testi* 72, p. 377. In taking the volume of the sphere to be two-thirds the containing cylinder, Theon follows the manner in Archimedes' corollary to *SC* I.34; but the same form is adopted by Archimedes as its principal form in the statements in his prefaces to *SC* I, *SC* II. In the *Method* (prop. 2), as in *SC* I.34, the first form proved is the sphere-cone relation, while the sphere-cylinder appears as a corollary.

61 I am currently preparing a study of this issue.

62 Where the assumption of form (i) occurs in Pappus' version of the isoperimetric theorems, Pappus gives the following source reference: "This is manifest from the things proved by Archimedes in the ⟨book⟩ On the Sphere and Cylinder, and from the other lemmas set down by us" (ed. Hultsch, I, p. 360, lines 18–21). The Archimedean reference is to *SC* I.34; Pappus' "other lemma" appears in *Collectio* V, prop. 35.

63 I discuss the origin of this rule and its appearances in the ancient metrical corpus in "On Two Archimedean Rules for the Circle and the Sphere," *Bollettino di Storia delle Scienze Matematiche* (forthcoming).

64 The Latin text has been edited with translation and notes by Clagett in *AMA* I, chap. 4. The Arabic text, in the recension by Naṣīr al-Dīn al-Ṭūsī (d. 1274) has been published as no. 1 of vol. 2 in the Osmania University edition of al-Ṭūsī's tracts (2 vols.; Hyderabad, 1939–40); an examination of the work, with partial translation into German, appears in H. Suter, "Ueber die Geometrie der Söhne des Mūsā ben Schākir," *Bibliotheca Mathematica,* 3rd ser., 3 (1902), pp. 259–272. Clagett cites the Arabic, where it diverges from the Latin, in his notes to *VF.*

65 Actually, Euclid dismisses the circumscribed cases by reducing them to the inscribed case via inverting ratios. It appears, however, that the Eudoxean sources available to Archimedes included both inscribed and circumscribed constructions, separately converging to the given curvilinear figure via successive bisection. For details of the argument, see my "Archimedes' *Dimension of the Circle*" (cited in note 56 above).

66 In Pappus' account of Archimedes' theorems on the sphere, for instance, a method of type (1) is followed (*Collectio* V, ed. Hultsch, props. 28, 35); his accounts of the Archimedean and spherical spirals appear to envisage an argument of type (2) (*Collectio* IV, ed. Hultsch, props. 21, 30). On the latter, see my "Archimedes and the Spirals: The Heuristic Background," *Historia Mathematica* 5 (1978):43–75.

67 In the paraphrases given here, the steps of the convergence argument are set in emphasis. The lettering of the original treatments has been altered to facilitate comparison. The three versions of the argument are reproduced in detail by T. Sato, "Quadrature of the Surface Area of a Sphere in the Early Middle Ages – Johannes de Tinemue and Banū Mūsā," *Historia Scientiarum* 28 (1985):61–90.

68 The same construction is assumed in *CS* 1, 2, 3, 6, 8, and *VF* 5, 14, 15; in all cases the Euclidean citation is absent. *VF* establishes an associated construction in a separate lemma (prop. 3): that given any line less than the circumference of a circle, one can inscribe in the circle a regular polygon of perimeter greater than the line; and analogously for the circumscribed polygon. But the proof, having introduced auxiliary concentric circles of circumference equal to the given line, then assumes, without explanation or citation of Euclid, the polygon constructed in *Elements* XII.16.

69 As quoted in section I above (cf. note 14), the Latin editor of *CS* singles out as a critical "hypothesis" the assumption that there exist straight figures equal to given curved figures (*CS* 1.1.15–17). The same is noted by al-Ṭūsī in his edition of the Banū Mūsā (see the line quoted by Clagett in *AMA* I, p. 254n.). These may be compared with an observation by the commentator Eutocius, who, in defense of Archimedes' assumption of this same hypothesis, separates the actual construction of such a line from the issue of its existence (cf. *Commentary on Dimension of the Circle,* ed. Heiberg, III, p. 230, lines 14–20):

> it is somehow clear to anyone, I think, that the perimeter of the circle is some magnitude, and that this is among the things extended in one ⟨dimension⟩, while the straight line is also of the same kind. Even if it has in no way been made clear that it is possible to produce a straight line equal to the perimeter of the circle, nevertheless that there is some straight line by nature equal ⟨to it⟩ is an object of research for no one.

In the older Aristotelian literature, however, the utter incommensurability of curved and straight was dogma, and used to argue the corresponding incommensurability of circular and linear motion (cf. *Physics* VII.4). For related reasons, some raised objections to proposed circle quadratures which appealed to geometric motions; cf. Pappus' account from Sporus in *Collectio* IV, chap. 31 (ed. Hultsch, I, pp. 252–254).

70 The Euclidean convergence technique is also followed in Archimedes' *DC* 1. For a discussion, see my "Archimedes and the *Elements,*" *Archive for History of Exact Sciences* 19 (1978):211–290.

71 Note that *VF* 4 provides an alternative proof of the circle theorem (*DC* 1), exploiting the convergence procedure of *Elements* XII.16.

72 The text is virtually identical with that in Book VIII, prop. 22.

73 *Collectio,* ed. Hultsch, I, p. 336.

74 The text may be corrupt, for the adverb (*sufficienter*) reads strangely with *est*. This could be avoided by taking *et* instead of *est*. The phrase *sed non et* ("but not actually") has classical precedents; cf. Vergil, *Aeneid* 6.86, and ps.–Vergil, *Copa* 24. (I am endebted to E. Courtney for this suggestion.)

75 The ancient commentators frequently interpose their editorial assessments. For instance, Pappus notes his own omission of analytic treatments and his preference for the comparative conciseness of syntheses in his preface to the theorems on isoperimetric regular solids (*Collectio* V, ed. Hultsch, pp. 410–412). In the *Mechanics,* Hero presents one method for solving the cube duplication as "the one best suited for practical application" (cited by Pappus, *Collectio* III, chap. 9, ed. Hultsch, I, p. 62). The ps.-Diophantine author of the *Epipedometrica* presents the method of measuring the sphere via the 3:2 ratio as a "better way" than that via the cylinder (cf. *Diophanti Alexandrini Opera,* ed. P. Tannery [2 vols; Leipzig, 1893–5] II, p. 28, line 26). In general, one of the major functions the commentators take on is making judgments on the relative merits of different forms of construction and proof.

76 I here adopt the lettering of the diagram in *SC* I.21 and maintain it throughout this discussion of Archimedes' treatment. In subsequent propositions Archimedes himself introduces modified figures with different letterings.

77 I continue to follow the lettering of Archimedes' figure in *SC* I.21. Although *CS* 5 adopts essentially the same diagram, it employs its own quite different lettering. Note that my term πEK abbreviates what *CS* would express as "the circumference of the circle of diameter *EK*."

78 Archimedes typically expresses a curvilinear area as equal to a circle whose radius *dynatai* (namely, "equals in square") the product of two given lines. *CS* avoids this by substituting circumferences for their diameters, in accordance with the proportionality proved in *CS* 3 (cf. the discussion of this theorem above). The *dynamis* terminology is prominent in *SC* I and among writers in its tradition (e.g., Pappus in *Collectio* V, ed. Hultsch, props. 23–25, 28–30, 32, 35). Euclid retains it in his studies of irrationals and regular solids (Books X and XIII, respectively), but otherwise replaces it by formulations with *tetragônon.* For other references, see my "Archimedes and the *Elements,*" p. 264.

79 For convenience of comparison, I have modified the lettering of the diagram in *VF* 12 to conform to that of *SC* I.21 (as in Figure 1).

80 Because, by similar triangles, $QA^2 (= QB^2) = MA \cdot AV$ for $AV > QV$; and also $QM^2 = AM \cdot MV$ for $MV < QV$. A somewhat more direct method would note, from similar triangles, that $MA:QM = QA:QV$, so that $MA \cdot QV = QM \cdot QA$. Since $QM < QA$, $QM^2 < MA \cdot QV < QA^2$.

81 Note that the use of the radius of the inscribed sphere for measuring the solid is important for *AI* and *CS* (prop. 7); this will be discussed further below.

82 Cf. *CS* 6.16–9, 35–7.

83 The result is obtained by forming the solid as the "remainder" (*perileimma*) between two solid rhombi.

84 The term *differentia* used here by *CS* corresponds to *perileimma* in *SC* I.20, 26 (cf. the preceding note).

85 The closing section (19.71–79) is held in Gerard's version of *VF* but is absent from the extant Arabic recension. As Clagett observes (*AMA* I, p. 354n.), the meaning of the Latin is not fully clear.

86 The Heronian rule for triangles (*Metrica* I.8) is explicitly assigned to Archimedes by al-Bīrūnī on the authority of al-Shannī; cf. H. Suter, "Das Buch der Auffindung der Sehnen im Kreise . . . ," *Bibliotheca Mathematica,* 3rd ser., 11 (1910–11), p. 39.

87 The earliest datable citation is with Proclus (*In Euclidem,* ed. G. Friedlein, Leipzig, 1873, p. 423), that is, ca. A.D. 450. The agreement with the enunciation of the extant *DC* 1 is precise. For other references, see my "Archimedes' *Dimension of the Circle*" (cited in note 56).

88 For references, see Heiberg's note in *Archimedis Opera* I, p. 233n., and my article on *DC* (see note 56).

89 The passage is reproduced in section III of this chapter.

90 *Archimedis Opera,* ed. Heiberg, I, p. 234.

91 *Heronis Opera,* ed. Schöne, III, p. 66.

92 Cf. the Hyderabad edition of al-Ṭūsī's tracts (II, no. 5, pp. 129, 133).

93 Such reservations are expressed by E. J. Dijksterhuis in his *Archimedes* (Copenhagen, 1956; New York, 1957), p. 222; cf. also *Archimedis Opera,* ed. Heiberg, I, p. 233n.

94 As extant in Greek, *AI* forms a section of the anonymous "Introduction to the Almagest." J. Mogenet has argued that the "Introduction" was assembled by Eutocius, hence early in the sixth century; cf. *L'introduction à l'Almageste* (Brussels, 1956), chap. 2. The "Introduction" is a composite work, however, so that its editor need not actually have composed each of its constituent parts. In the particular case of *AI,* certain affinities with passages in the commentators suggest a date of composition around A.D. 400. I defer a full discussion of this question to a study now in preparation, which is also cited in note 61.

95 *AI* prop. 1, ed. Hultsch, *Collectio,* pp. 1138–1140; the analogous treatments in Pappus (*Collectio* V, prop. 1) and Theon (ed. Rome, pp. 356–357) do not particularize the figure in this way. *AI* likewise speaks of a specific "hexagon," "pentagon" and "quadrilateral" in its prop. 8 (ed. Hultsch, pp. 1154–1156). While a similar approach is adopted in the correlative places in Pappus (prop. 10) and Theon (pp. 372–374), the agreement between *AI* and *CS* remains.

96 *AI,* ed. Hultsch, *Collectio* pp. 1140–1142. Similar phraseology appears in the analogous places in Theon (ed. Rome, p. 357) and Pappus (V, prop. 1).

97 See E. Kramer, "Hypatia," in *Dictionary of Scientific Biography* VI (New York, 1972), pp. 615–616. Hypatia (d. A.D. 415), a prominent teacher of philosophy and science at Alexandria, composed a commentary on Diophantus and assisted her father, Theon of Alexandria, in the completion of his Ptolemy commentaries. None of her writings is extant, but she is addressed with respect and admiration in the correspondence of her disciple Synesius.

98 This suggestion could be explored by determining whether linguistic parallels exist among the several works with which Hypatia is or may be associated. That is, if *CS, AI,* and *DC* happened to share stylistic features found in Theon's commentaries and Diophantus, the coincidences would result from the editor, since no other link among these works is plausible. A full examination is necessary, however, before conclusions can be drawn.

99 Two traditions of the Latin *Catoptrics* agree precisely on the enunciations of the theorems; but one adheres verbatim to the Greek text in its proofs also, while the other presents entirely different proofs. (I have examined these versions in manuscript; to my knowledge, no published edition is yet available.) A similar relation links the Euclid versions of "Adelard III" and Campanus to "Adelard II," in that both take over the statements of propositions in the third version but rework its proofs; cf. Murdoch, "Euclid: Transmission," p. 446, and Clagett, "Medieval Latin Translations," pp. 23–25, 29–30.

100 If "Adelard III" is indeed by Adelard of Bath, it must antedate the mid-twelfth century; if it is not by him, the existence of a twelfth-century manuscript of the work nevertheless assures its date before 1200 (see Busard's edition of "Adelard I," pp. 5–6). On the dating of Gerard of Brussels, see note 5 above.

Applied mathematics

2

Colonizing the world for mathematics: the diversity of medieval strategies

A. GEORGE MOLLAND

There is nothing new about trying to make the world mathematical, and the work of Marshall Clagett in particular has shown that numerous such efforts were made in the Middle Ages. In this chapter, I shall try to get some sense of the diversity of medieval strategies by making use of the metaphors of begetting (usually parthenogenetically)* and colonizing. History cannot be forced into a straitjacket, but the rough outlines of the story are that long ago arithmetic went colonizing and begat geometry and music to rule over the new territories. Geometry later declared a considerable degree of independence from arithmetic, and also colonized on its own and begat astronomy, optics, and mechanics. In the Middle Ages all six of these tried to colonize, but little successful begetting occurred. Logic also appears as a character, but in a less clear-cut way. A recurrent theme will be the tension between holism and atomism, of the extent to which a whole may be understood in terms of a set of determinate parts.

Arithmetic I
Arithmetic must be accounted the oldest mathematical science, and it is hard to conceive of human society without an ability to count and perform simple operations of addition and subtraction, and even of multiplication and division. Again, at an early stage there must have been a need for rudimentary measurements of lengths, heights, and areas, that is to say, the expression of the size of these in numerical terms. Thus, arithmetic may be said to have given rise to geometry. By the time of the philosophical Greeks, arithmetic (in its pure state)

---

* Begetting is usually regarded as a male affair, while the mathematical sciences were traditionally female. In this chapter no sexist overtones are intended: The metaphors are loose and mixed.

was conceived of as dealing with discrete collections of abstract units. (This had the effect of meaning that fractions were not properly numbers.) From this more formal standpoint, arithmetic among the Pythagoreans was able to colonize part of the realm of sound, and so engender the science of music, because of the association of the principal musical consonances with simple whole number ratios.

Geometry I

Geometry may originally have conformed nearly to its etymological meaning of the measurement of fields, but among the Greeks a particularly pure form was developed in which it was conceived as dealing generally and abstractly with continuous quantity. With the discovery of incommensurability, it was found to be not so subservient to arithmetic as was once thought, and henceforth there was a considerable chasm between the continuous and the discrete, and geometry had to develop many of its own special methods. Nevertheless numbers could not be completely banished from geometry, and it was usually allowed that arithmetic retained a certain priority. For instance, at the beginning of his *Geometria speculativa*, Thomas Bradwardine wrote that "Geometry is subsequent to arithmetic in a certain way, for it is of posterior order, and the properties of numbers are of service in magnitudes."[1]

In Greek pure geometry the range of figures that could appear was determined by the permissible constructions, which were sometimes, as in Euclid's *Elements*, enumerated in the postulates. In the plane geometrical part of that work the only lines to appear are either straight ones or circular arcs. In works by Archimedes, Apollonius, and others the range was widened but still nowhere nearly approached the extent of variation that could be seen (at least apparently) in nature or in freehand drawing. In fact, most of the figures of Greek geometry may be regarded as built up from a series of rotations about axes.[2] To Plato it seemed that the consequent "purity" of geometry signified that its objects were not sensibly based but existed in a separate realm of mathematicals. Aristotle on the other hand insisted that the objects of geometry had no separate existence but were derived by a process of abstraction from sensible things: for him, too, the geometer was not speaking of his diagrams but of what they symbolized.[3] The separation, whether real or mental, of the objects of geometry meant that there was no guarantee that nature herself could be captured in the limited net provided by human mathematics, even in its most sophisticated forms.

Nevertheless it was part of the Greek genius to attempt the feat, and so there arose what Aristotle called "the more physical of the branches of mathematics,"[4] which in the Middle Ages were often called "middle sciences." Plato is reputed to have posed the problem: "By the assumption of what uniform and ordered motions can the apparent motions of the planets be accounted for?"[5] Whatever the truth of this may be, the question was answered by Eudoxus in terms of a series of uniformly rotating spheres. Later astronomical theories were more complicated, in order to achieve a better fit with the observed data, but in all there was the implicit assumption that the heavenly bodies behaved in ways similar to what a geometer was doing when he performed his constructions, and certainly not in a capricious or wholly unpredictable fashion. Likewise in optics, visual rays were proposed and postulated to travel in straight lines, except in certain circumstances that had to be investigated. Once again a quite good fit with observation was obtained, but there had to be the initial confidence that nature, at least in some of her parts, would succumb to mathematics. Mechanics took its starting point more in the realm of human constructs, and so needed less initial confidence. We may, however, note that whereas Aristotle said that only "lines, surfaces, bodies, and also, besides these, time and place" were, strictly speaking, continuous quantities,[6] statics also treated weight as one, and so for that matter did Aristotle himself![7] Again, although music was traditionally regarded as a child of arithmetic, it still made the crucial assumption (remarked upon below) that pitch could be treated as a continuous quantity.

Optics

In both the Neoplatonic and the Christian traditions light had for long been a potent symbol of divine causal action, but for us it is convenient to start with Robert Grosseteste's metaphysics or philosophy of light. As David C. Lindberg has pointed out, this had at least four components: epistemological; metaphysical or cosmogonical; etiological or physical; theological.[8] Only the second and third need concern us here, and they may be seen as pulling in somewhat different directions. Let us first consider Grosseteste's cosmogony. The principal source is the short treatise *De luce*, which he began as follows:

> First corporeal form, which some call corporeity, I judge to
> be light (lux), for light of itself diffuses itself in every
> direction, so that from a point of light a sphere of light as
> great as is possible (*quamvis magna*) is suddenly generated,
> unless something shady comes in the way. Corporeity is
> what the extension of matter in three dimensions necessarily
> follows, but, since each, namely corporeity and matter, is
> itself a simple substance lacking any dimension, it was
> impossible for form that was in itself simple and lacking
> dimension to induce dimension in every direction into matter
> that was similarly simple and lacking dimension, except by

> multiplying itself and diffusing itself suddenly in every
> direction, and in its diffusion extending matter.[9]

In this process of multiplication the matter farther from the center became more rarefied, and multiplication was brought to a halt with a still finite sphere as a result of matter's limited capacity for rarefaction. The extremity of this sphere was the firmament, and Grosseteste then proceeded to discuss the formation of the other spheres, but we may ignore that here.

Grosseteste's scheme asserts that extension itself is a product of the self-multiplication of light. Although the sphere produced was finite, the multiplication had to be infinite, "for a simple thing finitely replicated does not generate a quantity (*quantum*), as Aristotle shows in *De caelo et mundo*, but it is necessary that it be infinitely multiplied to generate a finite quantity."[10]

Although Grosseteste invokes the authority of Aristotle, he is in fact adopting a profoundly un-Aristotelian position, for Aristotle was adamant that continua were not composed of indivisibles, but could only be divided into other continua which were then susceptible of similar division in a never-ending process. The idea that continua have ultimate components may be seen as holding out an earnest that wholes may be fully understood in terms of their parts, an attitude more characteristic of seventeenth-century mechanism than of Aristotelian holism.[11]

Grosseteste was by no means unique in the Middle Ages in holding a compositionist view of continua, but his type of stance became far more widespread in the seventeenth century. Grosseteste showed some awareness of the type of tradition he was in when he said, with regard to his production of quantity from the multiplication of light, that "This, as I believe, was the understanding of the philosophers who posited all things to be composed from atoms, and said that bodies were composed from surfaces, and surfaces from lines, and lines from points".[12] It would be absurd to try to make Grossesteste into either an ancient or a seventeenth-century mechanist; but his conception does at least make the physical universe at heart more like a collection of undifferentiated discrete objects, and so places a quasi-numerical structure at the heart of physical reality. It also makes geometry once again more subservient to arithemetic, and fittingly the treatise ends with a piece of numerology, although this probably does not derive in any direct way from his attitude toward continuity.

The physical side of Grosseteste's philosophy of light was more conventionally geometrical, and attempted the assimilation of a whole range of natural effects to the laws of geometrical optics.

> All causes of natural effects have to be given by lines,
> angles and figures for otherwise it is impossible for the
> reason why (*propter quid*) in them to be known. This is

manifest thus. A natural agent multiplies its virtue from itself right into the patient, whether it acts on sense or on matter. This virtue is sometimes called species, sometimes similitude, and it is the same whichever way it is called, and the agent will send the same into sense as into matter or a contrary, in the way it sends the same hot (*calidum*) into touch as into cold, for it does not act by deliberation or choice, and so it acts in one way whatever it meets, whether sense or something else, whether animate or inanimate, but the effects are diversified on account of the diversity of the patient.[13]

Grosseteste then proceeds to discuss in a rather sketchy way how different strengths of action result from the different orientations of the lines along which the virtues travel, and also the circumstances in which their rays are reflected or refracted.

In all this Grosseteste is not completely reducing physics to mathematics, but he does continually assert the superiority of the latter, and gives the impression that the physicist is a mere underlaborer to the mathematician, to whom belongs the privilege of providing the best *propter quid* explanations. Thus at the beginning of his treatise on the rainbow he wrote, "Consideration of the rainbow is for both the perspectivist and the physicist, but it is for the physicist to know the fact (*quid*), the perspectivist the reason why (*propter quid*), on account of which Aristotle in the book of *Meterorologica* did not set forth the reason why, which is for the perspectivist, but compressed the fact about the rainbow, which is for the physicist, into a brief discourse."[14] However, as Lindberg has emphasized, Grosseteste's actual optical knowledge was very limited,[15] and so, to an even greater extent than with many other medieval attempts at mathematization, his schemes remained merely programmatic.

Grosseteste's most important follower in essaying a generalized optics was Roger Bacon, whose faith in the potency of mathematics paralleled Grosseteste's own. "Demonstration by cause is necessarily far more powerful than demonstration by effect, and Aristole holds this in the book of *Posterior Analytics*. Therefore, since in natural things demonstration by cause is had along mathematical paths, and demonstration by effect is had along natural [philosophical] paths, the mathematician is far more powerful in relation to natural things than the natural philosopher himself."[16] But Bacon did not follow Grosseteste's philosophy of light in all its ramifications, and, in particular, omitted the cosmogonical aspect. Instead, he insisted that light was an accident and not prior to quantity.[17]

Where Bacon most enthusiastically followed Grosseteste was in the doctrine of the multiplication of species, and this he developed in far greater detail. He was particularly aided in his enterprise by having assimilated more thoroughly than had previous Latin scholars the optical theories of Alhazen, which were to dominate subsequent optics until the time of Kepler.[18] Most of Bacon's discussion of the geometry of the multiplication of species is based on optical examples, but he is insistent that the doctrine is of far wider application and essential for a proper understanding of natural effects.

> These species bring about every alteration in the world and in our bodies and souls. But, because this multiplication of species is unknown to the crowd of students, and only to three or four Latins, and that [only] in optics (*in perspectivis*), that is in the multiplication of light and colour for vision, therefore we do not perceive [as such] the wonderful actions of nature that are all day brought about in us and in things in front of our eyes, but we judge them to be brought about either by a special divine operation, or by angels, or by demons, or by chance and fortune. But it is not so, except insofar as every operation of a creature is in some way from God. But this does not preclude the operations being brought about by natural reasons, because nature is an instrument of divine operation.[19]

The heavens and the human rational soul are particularly important sources of species. The species of the former account for astrological influences, while those of the latter, especially when conjoined with appropriate heavenly influences, can produce all sorts of marvelous effects. "By this power bodies are healed, venomous animals are put to flight, brutes of all kinds are called to hand, both serpents from caves and fish from the depths of the waters."[20] In this way the doctrine of multiplication became a cornerstone in Bacon's program of naturalizing phenomena that were often accounted magical.[21]

In his major treatise on the subject, the *De multiplicatione specierum*, Bacon gives quite a detailed account of how species are actually multiplied.[22] Species of corporeal objects are of a bodily nature but are not themselves individual bodies. Instead, the individual parts of the medium are successively assimilated to the nature of the emitting object. Each part of the medium acts upon the next, in order to draw out the relevant potentiality, so that the species are transmitted rectilinearly (unless impeded) in a stepwise fashion. This has often reminded scholars of later wave theories of light, and certainly Bacon's scheme merits at least a superficial comparison with how Huygens has pulses of light transmitted from particle to particle. Also, by concentrating on individual parts of the medium and analyzing the motion into a series of discrete units, Bacon suggests how wholes may be understood in terms of their parts, and thus veers in this instance toward an atomistic stance.

Bacon's account also imbeds the mathematics of transmission firmly in the nature of physical things. This was not to be the case with a later compatriot, John Dumbleton, who pursued a long, rambling inquiry into the status of the lines, angles, and figures employed by the perspectivists, and came to the conclusion that these were mere fictions useful for predicting where an image seen, for example, by reflection, would appear.[23] Dumbleton's case may have been extreme, but Bacon's vision of mathematizing nature on the analogy of optics seems to have attracted few significant followers until the time of John Dee.[24]

Astronomy

Being made of the fifth element, the heavens behaved on radically different principles from the sublunary regions, and so astronomy had less scope for colonization than, say, optics or music. Nevertheless, it did have some influence on the ways in which a more general mathematical science could be conceived. One locus for this was the question of the status of the epicycles and eccentrics used in mathematical astronomy, a topic considered by Edward Grant in Chapter 7. Were they mere mathematical fictions useful for saving the phenomena, or did they correspond to actual orbs in the sky? These questions raised in explicit form the problem of the extent to which one was justified in holding that one's mathematical imaginings represented the actual structure of nature, and gave much scope for conflict between what I have elsewhere called realist and conceptualist attitudes on the question of the relation of mathematics to nature.[25] It is interesting to note that in this instance Roger Bacon, usually so firmly realist, seems to have wavered toward the conceptualist pole,[26] even though he held astronomy to be one of the principal reasons for the formation of geometry: "The authors of perspective show us that lines and figures declare the whole operation of nature, its principles and effects, and this is similarly evident through celestial things, which are considered by both natural philosophy and astronomy."[27]

One writer who emphatically asserted the real existence of epicycles was Nicole Oresme, who went on to suggest that angels or intelligences may not be needed for moving the heavens, and in doing so, invoked a memorable mechanical analogy:

> Perhaps, when God created [the heavens], he placed in them motive qualities and virtues, just as he placed weight in terrestrial things, and placed in them resistances against these motive virtues. And these virtues and resistances are of another nature and matter from any sensible thing or quality that is here below. And the virtues are moderated, tempered and harmonised against the resistances, so that the movements are made without violence. And, except for the

violence, this is in some way similar to when a man has
made a clock, and leaves it to go and be moved by itself; so
did God leave the heavens to be moved continually
according to the ratios that the motive virtues have to the
resistances, and according to the established decree.[28]

This image can suggest that similar mathematical laws may apply both
to the celestial regions and to sublunary machines. This could in turn
open the prospect of mathematizing the sublunary world by treating it
mechanically.[29] Also, John Buridan's suggested extension of impetus
to the heavens[30] could invite a closer connection between celestial and
sublunary motion, and hold out the promise of making the latter more
precisely mathematical.

Mechanics

For the Middle Ages we must understand this as confined essentially to statics and hydrostatics, and not yet as extending to considerations of motion. Although there was a vigorous tradition, deriving
largely from treatises associated with the name of the enigmatic[31] Jordanus de Nemore, this science effected little colonizing in the Middle
Ages, despite Nicolaus Cusanus's enthusiastic encomia of weight as a
means of getting experimental and quantitative knowledge of a whole
range of phenomena.[32] Much of the trouble arose from the fact that in
the Aristotelian tradition, weight was a concept lacking in independence and almost always had to be considered in conjunction with its
contrary lightness. Pure elemental fire was absolutely light, and pure
elemental earth was absolutely heavy, but in all sublunary bodies there
was a mixture of elements, and so a warring combination of weight
and lightness. That this did not preclude mathematization is shown by
quite a striking passage in which Bradwardine applied his law of motion
to the conception.

All mixed bodies of similar composition will be moved with
equal speed in a vacuum, for in all such the movers are
proportional to the resistances. Therefore by the first
conclusion of this chapter, all such will be moved equally
swiftly. From this also you will know that, if two unequal
mixed heavy bodies of similar composition are suspended in
a balance in a vacuum, the heavier will go down, for let A
and B be two such heavy bodies, A greater and B less, and
let C be the heaviness (*gravitas*) of A, and D similarly the
lightness of the same. Let E be the heaviness of B, and F
the lightness of the same. Then C, D, E, F are four
proportionals, and C is the greatest, F the least. Therefore,
by the eighth supposition of the first chapter, C and F
gathered together exceed D and E brought together. And C

> and *F* strive to raise B, and only D and E resist. Therefore,
> by the second part of the ninth conclusion, B will ascend,
> and A descend.[33]

The second part of this passage makes it clear that, while a mathematical statics on these assumptions is possible, it will be considerably
more complicated than ordinary statics.

Bradwardine's earlier conclusion, that bodies of similar composition
would fall equally swiftly in a vacuum, can look notably "modern,"
but its very modernity made it unacceptable to even such a figure as
Nicole Oresme.

> I suppose that the speed of a motion is according to the
> ratio of the motive virtue to the resistance, and hence, if
> these ratios are equal in two moved bodies, they will be
> moved with equal speeds. Then I take, for example, two
> bodies of fine silver of which one is quadruple the other in
> size (*qualité*), and I posit that in the smaller one the heavy
> (*pesans*) elements with regard to the light ones are as 6 to 1,
> and then in the big one, which is of a similar mixture, they
> will be as 24 to 4, which is such a ratio as 6 to 1. And then,
> by the supposition made above, the small and the big will be
> moved downwards with the same speed, if other things are
> equal, and consequently the two bodies weigh equally, and
> this is manifestly false. Therefore elements are not in their
> proper form in mixed bodies.[34]

Oresme's conclusion, that elements change their forms when entering
into mixtures, effectively rules out the possibility of mathematizing
heaviness and lightness. Ironically, among the culprits in backing up
the Aristotelian association of weight with speed of descent was the
Jordanus statical tradition, where we meet the postulate "That which
is heavier descends more rapidly," and the conclusion "Among any
heavy bodies, the ratio of speed in descending and weight, taken in
the same order, is the same . . . ," although Bradwardine held that the
latter was being misinterpreted.[35]

Music

According to Aristotle, the Pythagorean association of number
with music was a potent factor in making them extrapolate and hold
"the whole heaven to be a musical scale and a number."[36] In so doing
they introduced the seductive doctrine of the music of the spheres.
Aristotle admitted the seductiveness, but rejected the theory. Boethius,
on the other hand, regarded this type of *musica mundana* as very important, and there had also been hints of it in Plato's *Timaeus*.[37] Thus,
in the twelfth century, before the translation of Aristotle's *De caelo*,
it had reasonable authority, although it was by no means clear what

the actual quantities were that the musical ratios were between. Even later, Aristotle's criticisms did not completely destroy the doctrine, for it could be held that the music was not perceptible to the senses but of an intellectual nature. This was basically the strategy of Nicole Oresme, who, after providing a substantial discussion of what sorts of ratios might be involved and how, went on to affirm a close relationship between human, celestial, and divine music.[38] But, even though he went into some detail, Oresme did little to try and tie the doctrine in with the theories of mathematical astronomy. Kepler was probably the first to take that task seriously.

Plato, in the *Timaeus*, provided further intimations of musical structure in the world in his accounts of the bonding of elements and of the formation of the World Soul,[39] but his expression was laconic in the extreme and very difficult to interpret precisely, and in any case the influence of the *Timaeus* declined dramatically after the twelfth century. I therefore move to a less obvious but arguably more significant example of musical influence to be found in a fourteenth-century work, the *Tractatus de proportionibus* of Thomas Bradwardine.[40]

This work contains his by now famous proposal for a quantitative law of motion, which, translated literally, reads: "The ratio of speeds in motion follows the ratio of the power of the mover to the power of the thing moved."[41] In seeking to establish his own position, Bradwardine examined and rejected four other positions, the last of which is the following: "There is no ratio nor any excess of motive power to resistive power, and so the ratio of speeds in motions does not follow any ratio or excess of motive power to power of the mobile, but a certain dominion and natural relation of mover to moved."[42] The nub of this is to assert that powers, not being quantities, cannot properly be compared quantitatively, and so the search for any mathematical relation between them is otiose. Bradwardine's opening objection to the position is the following:

> If there were no ratio between powers on account of their
> not being quantities, there would not be a ratio between
> sounds, and then the whole modulation of music would
> perish, for the *epogdous* or tone consists in a sesquioctaval
> ratio, the *diatesseron* in a sesquitertial, the *diapente* in a
> sesquialteral, the *diapason*, which is composed from
> *diatesseron* and *diapente*, in a double, *diapson cum diapente*
> in a triple, and *bis diapason* in a quadruple ratio.[43]

Although music was regarded as subordinate to arithmetic, Bradwardine's argument points out that it also includes an implicit quantification of pitch in the manner of a continuous quantity. And if pitch, then why not powers? The very existence of music as a mathematical science thus gave confidence in the possibility of extending mathematics' domain.

But there was also, I would maintain, a linguistic influence from music that allowed Bradwardine to formulate the law in the way he did. The law, as quoted above, appears very simple in expression, but modern writers have often spoken of its logarithmic or exponential nature. The paradox depends on a shift in mathematical language. In the ancient Greek and medieval mathematical traditions there is quite frequent talk of compounding ratios together.[44] If A is greater than B and B greater than C, the result of compounding the ratio of A to B with that of B to C was the ratio of A to C. Our natural tendency is to interpret this as multiplication, and this was sometimes done in the Middle Ages. However, in musical works, such as that by Boethius, it was regularly regarded as addition. This was fairly natural, since ratios corresponded to intervals, and compounding ratios corresponded to combining intervals, or adding them together.

For long this interpretation of compounding ratios remained unobtrusive and innocent, but Bradwardine and later Nicole Oresme developed its consequences in an extensive and systematic manner. For instance, doubling a ratio now corresponded to what we would call squaring a fraction, and halving it to taking a square root. If we symbolize Bradwardine's law itself by saying that it asserts that V (speed) is proportional to the ratio of F (power of the mover) to R (power of the thing moved), then in modern terms this translates into V being proportional to the logarithm of F/R. This translation in itself suggests what a potent weapon was available for creating a rich mathematical account of the natural world, and this promise was well borne out in the work of such figures as Richard Swineshead, even though the mathematical physics thus created had very vague empirical reference.

Arithmetic II

The one area of direct medieval colonizing by arithmetic that I shall consider is numerology. To us this can seem one of the most baseless of traditional superstitions, but when the world was conceived to contain determinate numbers of things of the same kind (four elements, seven planets, etc.) and was also thought of as the product of design, it was perfectly reasonable to seek a rationale for the particular numbers chosen. As Boethius said,

> All things whatsoever that were fabricated from the first age
> of nature seem to have been formed on the rationale of
> numbers, for this was the principal exemplar in the mind of
> the framer, for hence is obtained the number of the four

elements, hence the succession of times, hence the motion of the stars and revolution of the heavens.[45]

Aristotle attacked the arithmeticism of the Pythagoreans, but still gave a numerological explanation for there being just three dimensions:

> For, as the Pythagoreans say, the world and all that is it is determined by the number three, since beginning and middle and end give the number of an 'all,' and the number they give is the triad. And so, having taken these things from nature as (so to speak) laws of it, we make further use of the number three in the worship of the Gods.[46]

The peculiar position of this number was further enhanced with the advent of the Christian doctrine of the Trinity, but, despite the abundance of numerology in medieval theological and devotional literature, it made curiously little impact on the major works of natural philosophy.[47] That was to be far more a Renaissance trait, exemplified in profusion in works such as Henry Cornelius Agrippa's *De occulta philosophia.*[48]

Geometry II

An obvious starting point for seeking direct geometrical colonizing would seem to be Plato's theory of geometrical atomism.[49] In this, four of the five regular solids gave their shapes to the small particles of the four elements, while the fifth, the dodecahedron, was in a vaguer way assigned to the "whole." If accepted, Plato's theory would surely have given great confidence in the possibility of mathematical science, for it promised that from the smallest parts outward, nature was mathematically designed. But this was not to be. Aristotle launched a vigorous attack on the theory,[50] and Plato's own account was not included in Chalcidius's incomplete Latin translation of the *Timaeus.* Thus the main source of medieval knowledge of the theory was Aristotle's criticism, and this was universally or almost universally accepted, either tacitly or, in such writers as Albertus Magnus, with delight.[51] More effort was spent in considering a mathematical difficulty that arose from Aristotle's criticisms, about filling space with regular polyhedra.[52]

Far more mileage can be obtained by considering the geometrization of qualities, a subject that has received great attention in recent years.[53] For Aristotle, "A quantity does not seem to admit of a more and a less. Four-foot for example: one thing is not more four-foot than another."[54] Moreover, unlike the term "four-foot," " 'large' or 'small' does not signify a quantity but rather a relative."[55] On the other hand, "Qualifications [but, it turns out, not all of them] admit of a more and a less; for one thing is called more pale or less pale than another, and more just than another."[56] In all this Aristotle seems to be making a

big gap between quantities and qualities, but in fact it was the ability
of qualities to admit a more and a less, to be intended and remitted,
that opened the way for their quantification, for there is a strong temptation to ask, "How much more?" or "How much less?" This happened at an early stage in the pharmacological tradition, and Roger
Bacon explicitly used a straight line to represent the range of intensities
of, say, heat in a drug.[57]

In the natural philosophical tradition the ontological problems associated with intension and remission had received considerable attention from thirteenth-century thinkers, but in the fourteenth century
there was a tendency to emphasize the logical and quantitative aspects
of the question,[58] and with this we regularly find ranges of intensity
being assimilated to straight lines. This tendency reached its zenith in
Nicole Oresme's *De configurationibus qualitatum et motuum*, which
attempts a thoroughgoing mathematization of the world. In Oresme's
conception the degree of intensity of a quality at each point of a subject
was represented by a perpendicular proportional in length to the degree
of intensity at that point. In the case of a linear subject this resulted
in a sort of graph, and in a two-dimensional subject, a relief map. For
a body one should by analogy have a four-dimensional figure. This
option was not available to Oresme, and so he settled for, as it were,
superimposed relief maps. In this way he was postulating how the entire
qualitative makeup of a body may be mathematically represented. But
he did not stop with mere representations: His configurations were
meant to have explanatory power in a way similar to ancient atomism.

> It is manifest that bodies have different powers (*diversimode
> potest*) in their actions according to the difference of their
> figures, on account of which the ancients who maintained
> that bodies were composed of atoms said that the atoms of
> fire were pyramidal on account of its vigorous activity.
> Thus, according to the difference in the pyramids, bodies
> can pierce more or less, and, according to this or that
> sharpness, it is certain that they can cut more or less
> strongly, and so for other actions and figures. And, since it
> is thus for the figures of bodies, it seems reasonable for it to
> be able to be spoken conformably about the aforesaid
> figurations of qualities, so that there is a quality whose
> particles are proportional in intensity to small pyramids, and
> on account of this it is more active, other things being equal,
> than an equal quality that was either simply uniform or
> proportional to another, not so penetrating, figure. Or, if
> there were two qualities of which the one's particles were
> proportional to sharper pyramids than the other's particles,
> the quality that corresponded to the sharper pyramids

would, other things being equal, be more active, and
similarly for other figures.[59]

In this Oresme is appealing to mathematics to provide *propter quid*
explanations, and proposing the formation of a science of qualities that
would be subalternate to geometry. Although he did not develop it
much, a particularly important area of application of the doctrine was
in the explanation of occult virtues,[60] those properties of things that
did not follow directly from their elemental constitution, and hence
often seemed brute facts, incapable of being rendered fully intelligible.
Oresme's suggestion was that, if the configurations of the qualities be
taken into consideration as well as the ratios, then the virtues succumb
to a mathematical rationale. In this he may be compared directly with
Descartes, who held it to be a major benefit of his mechanical philosophy that it showed the way to discovering the causes of the "won-
derful effects that are usually referred to occult qualities."[61] A major
difference between the two is that, although Descartes's mechanistic
explanations are often fantastical, he stuck to his program tenaciously,
whereas Oresme showed far less confidence in the possibility of
grounding his doctrine empirically, and may ultimately have abandoned
it.[62] The urge to beget a new middle science was strong, but the territory
did not seem sufficiently subdued.

We may also use Oresme as preeminently representative of the general geometrization of motion.[63] Oresme treated motions analogously
with qualities, and held that "intensity of speed" was to be graphed
over both time and the parts of the moving body. This and the work
of related writers, notably the Mertonians, can suggest that there were
being developed sophisticated techniques for dealing with "instantaneous velocity," thus putting the fourteenth century well on the way
to the seventeenth. I have argued elsewhere that there is a very real
sense in which these thinkers can be seen as occupying a halfway house
between Aristotle and the likes of Galileo,[64] but it is important to recognize how different was their approach from that of their seventeenth-
century successors, a difference that largely arose from their holistic
attitudes. In the seventeenth century there was a strong tendency to
regard motions as being built up from the instantaneous velocities of
individual mass points, but in the Middle Ages the primary focus was
on the whole motion of the whole body, from the beginning to the end.
This resulted in treating speed as something like a five-dimensional
object, as is at least apparent from an otherwise not very clear passage
from Oresme:

> There is a twofold difformity of speed, subjective and
> temporal, which, as regards the purpose in hand, differ in
> this, that subjective punctual speed, just like punctual
> quality is to be imagined by a straight line, and linear

subjective speed is to be imagined by a surface or superficial
figure, but the speed of a surface is to be imagined by a
body, and similarly the speed of a body by a body, entirely
in the same way as was said about the figuration of
qualities. . . . But instantaneous punctual speed is imagined
by a straight line, and temporal punctual speed is to be
imagined by a surface, but the temporal speed of a line is
imagined by a body, and similarly superficial and corporeal
[speeds] are imagined by a body because there is not
available (*non contingit dare*) a fourth dimension, as was
said [above].[65]

This conception can easily cause problems in comparing motions together. The so-called Merton Rule, which equates a uniformly difform
motion with a uniform motion, follows from a simple equation of areas,
but a similar equation leads to Oresme's assertion that "A uniform
speed that lasts for three days is equal (*est equalis*) to a speed three
times as intense that lasts for one day,"[66] contrary to one's intuitive
understanding of what it is to be speedy.

Moreover, even the concept of intensity of speed is not as unambiguously quantified as one might have expected. Oresme held that
intensity of speed was to be measured by the amount of the particular
"perfection" that was being gained or lost as a result of the motion.
"I say therefore that universally that degree of speed is unqualifiedly
more intense or greater at which in an equal time there is gained or
lost more of that perfection following (*secundum*) which the motion is
made."[67] But this means that intensity of speed can depend on how
the perfection is described. For instance, in a rectilinear motion of
descent:

The speed of motion is measured by the space traversed,
but the speed of descent by nearness to the centre. Thus it
is possible that A and B be moved equally swiftly but not
descend equally swiftly, in that A is moved along a direct
line to the centre and B along an oblique line, and so A
descends more swiftly than B, and yet B is moved equally
swiftly. Similarly, since descent is measured by the ratio of
nearness to the centre, it happens that what is moved
uniformly or regularly along a direct line from the centre
descends difformly, because it nears the centre more swiftly
when near than when far away, but with there always
remaining an equal speed of motion.[68]

In similar circumstances Richard Swineshead would try to deploy the
resources of rational argument in order to determine which was the
correct quantification, but Oresme was more inclined to revel in the
multiplicity of possibilities. "According to the multiple denominations,

speed is multiply varied or denominated."[69] In this way it can often seem that it is language rather than nature that is being quantified, and so despite the intensity of his geometrical vision, mathematics can for Oresme remain very much on the surface of things.

Oresme spoke of his graphical representations as imaginations. This was appropriate, for in the Middle Ages there was a close association between geometry and the imagination, since it was in the imaginative faculty that geometrical objects were "viewed." Moreover, the imagination was often associated with the positing, contrary to Aristotle, of an infinite space outside the heavens, and such a space was often referred to as imaginary, not in order to deny its existence but to assert that it was pictured in the imagination.[70] The two themes were brought together by Henry of Ghent, who distrusted the imagination, and described some of his opponents as

> those of whom the Commentator says that in them the imaginative virtue dominates over the cognitive virtue, and so, as he says, they do not believe demonstrations unless the imagination accompanies them, for they cannot believe that there is neither plenum nor vacuum nor time outside the world, nor can they believe that there are here non-corporeal beings that are neither in place nor in time. They cannot believe the first because their imagination does not stop in finite quantity, and so mathematical imaginations and what is outside the heaven seem to them infinite. But in this it is not right to believe the imagination. . . . Therefore such people are melancholy and make the best mathematicians, but the worst metaphysicians, because they cannot extend their understanding beyond site and magnitude, on which mathematicals are founded . . . , and they make inept natural philosophers.[71]

With this connection it should not surprise us that two of the most notable fourteenth-century proponents of an infinite space were the mathematicians Thomas Bradwardine and Nicole Oresme; and I have suggested elsewhere that geometrical exigencies were a major factor in bringing Bradwardine to this view, even if theological considerations have to be given at least equal weight.[72] In both the doctrine of configurations and in the assertion of infinite space we seem to have clear examples of how imaginative representations of the world can affect one's view of the actual constitution of the world – in these instances casting it in a geometrical mode.

Logic
This section needs to be included, but it must either be very long or very short. The former option would involve a disproportionate

amount of space and lengthy time for research and reflection, and so it has to be the latter. The reason logic has to be considered is that in important fourteenth-century traditions of natural philosophy the urge to mathematize was inextricably bound up with what later ages would regard as the most futile of logic-chopping exercises – *quisquiliae* in the typical humanist view of Richard Swineshead. The context must be seen as the scholastic disputation, where one had to use logical, and especially dialectical, techniques to examine an opponent's arguments. At first there was little mathematics involved, but with the fourteenth-century tendency to move from ontological issues to descriptive ones, which were usually concerned with things that were in principle more accessible to the senses, the question "How much?" more and more frequently raised its head. Not that the Schoolmen went out and empirically measured things, but they did give much attention to the theory of how things that were not directly measurable, such as speeds or intensities of qualities (cf. above, under "Geometry II") should be quantified and "denominated." Such contexts could often give rise to quite complicated logico-mathematical arguments,[73] but if one tries to make diagrams of their forms, one often finds the need to see certain propositions as reinforcing the arrows that represent the passage from one assertion to another.[74] Dialectic has infiltrated into the demonstrative arguments that are normally characteristic of mathematics, and this is one reason for approving of Leibniz's description of the tradition as "semi-mathematical."[75] In the work of a Swineshead or an Oresme one can feel that knowledge is very much the aim, but in a text such as Themo Judei's "Question on the Motion of the Moon," one can easily wonder whether one is not descending simply into sophisticated sophistry.[76]

Afterword

The seventeenth century saw the world mathematized in a far more effective way than had the Middle Ages. The process involved many people, but among the most outstanding were Galileo and Descartes. From his Pisan days onward, Galileo had been strongly influenced by "the superhuman Archimedes, whose name I never mention without a feeling of awe."[77] The researches of Marshall Clagett have amply demonstrated the extent of Archimedes' penetration into the Middle Ages, but they also serve to highlight how in the sixteenth century there arose a far greater tendency not only to admire Archimedes and invoke his authority, but to emulate him and advance beyond what he had achieved.[78] It was the tradition of mechanics, with its statical and hydrostatical works, that was most influential on Galileo. However, as we have seen, mechanics did little colonizing in the Middle Ages. This helps to explain how it was that in the seventeenth century the mathematization of the world picture was intimately bound up with its mechanization.[79]

Complementary to the Archimedean influence was the emphasis on experimentation. This is not the place to investigate the precise role that experiments played in Galileo's achievement, but only to note how Galileo was wont to present his quantities in such a way that one could see how at least in principle they were empirically measurable. Cannonballs may not have followed his theoretical parabolic trajectory exactly, but at least this curve could be compared with their actual paths in a more direct and precise way than could, for example, Bradwardine's law be tested against nature.[80]

Descartes wrote that Galileo "only sought the reasons of certain particular effects without having considered the first causes of nature, and so built without foundation."[81] There is some justice in this stricture (if so it should be described), and at the ontological level, at least, Descartes was the more radical thinker. His program involved seeing extension as the only essential property of matter, and so virtually identifying matter and space. He was thus able to claim to Mersenne that his physics was nothing but geometry,[82] although he did notably little technical geometry within the context of his natural philosophy. Descartes's strategy is basically a wholesale and direct colonization by geometry, but even here mechanics played a substantial part, for the imagination of mechanical instruments formed an essential part of Descartes's conception of geometry, and his mathematization of natural philosophy was abetted by comparisons of animals, and so on, with mechanical automata.

In the Middle Ages optics and music made considerable efforts to increase their territory, but in the seventeenth century they were far overtaken by mechanics, although Kepler allowed them to flex their muscles in his celestial physics, and even Newton contemplated a musical theory of colors.[83] Nevertheless, in order to see them make significant and lasting inroads, we probably have to wait for the theory of electromagnetic waves and for wave mechanics.

Notes

1 A. G. Molland, "The Geometria Speculativa of Thomas Bradwardine: Text with Critical Discussion" (Ph.D. dissertation, Cambridge University, 1967), 59.

2 Cf. A. G. Molland, "Shifting the Foundations: Descartes's Transformation of Ancient Geometry," *Historia Mathematica* 3 (1976):25–27.

3 *Anal. Post.* I.10.76b40–77a3.

4 Aristotle, *Physics* II.2.194a7–8.

5 M. R. Cohen and I. E. Drabkin, *A Source Book in Greek Science* (Cambridge, Mass.: Harvard University Press, 1948), 97.

6 Aristotle, *Categories* 6. 4b22–23, 5a38–39.
7 For example, in *Phys.* IV.8.
8 D. C. Lindberg, *Theories of Vision from al-Kindi to Kepler* (Chicago: University of Chicago Press), 1976. Cf. Lindberg, "On the Applicability of Mathematics to Nature: Roger Bacon and his Predecessors," *British Journal for the History of Science* 15 (1982):12.
9 *Die philosophischen Werke des Robert Grosseteste, Bischofs von Lincoln*, ed. L. Baur. Beiträge zur Geschichte der Philosophie des Mittelalters, IX (Münster: Aschendorff, 1912), 51.
10 Ibid., 52.
11 Cf. A. G. Molland, "The Atomisation of Motion: A Facet of the Scientific Revolution," *Studies in History and Philosophy of Science* 13 (1982):31–54.
12 Grosseteste, *Philosophischen Werke*, ed. Baur, 53–54.
13 Ibid., 60.
14 Ibid., 72.
15 Lindberg, *Theories of Vision*, 94–102.
16 *The "Opus Maius" of Roger Bacon*, ed. J. H. Bridges (Oxford, 1897–1900; repr. Frankfurt, 1964), I, 188–189.
17 *Roger Bacon's Philosophy of Nature: A Critical Edition, with English Translation, Introduction, and Notes, of De multiplicatione specierum and De speculis comburentibus*, ed. D. C. Lindberg (Oxford: Clarendon Press, 1983), 14; *Opera hactenus inedita Rogeri Baconi*, ed. R. Steele et al. (Oxford: Clarendon Press, 1905–40), XVI, 62–63.
18 Cf. Lindberg, *Theories of Vision*, 107–116.
19 *Fr. Rogeri Bacon Opera quaedam hactenus inedita*, ed. J. S. Brewer (London, 1859; repr. Kraus Reprint, 1965), I, 99–100.
20 Bacon, *The "Opus Maius,"* I, 395.
21 Cf. A. G. Molland, "Roger Bacon: Magic and the Multiplication of Species," *Paideia*, forthcoming.
22 Bacon, *De multiplicatione*, ed. Lindberg, passim.
23 A. G. Molland, "John Dumbleton and the Status of Geometrical Optics," *Actes du XIIIe Congrès International d'Histoire des Sciences* (Moscow: Nauka, 1974), III–IV, 125–130.
24 *John Dee on Astronomy: Propaedeumata Aphoristica (1558 and 1568)*, ed. and tr. Wayne Shumaker, int. J. L. Heilbron (Berkeley: University of California Press, 1978); N. H. Clulee, "Astrology, Magic, and Optics: Facets of John Dee's Early Natural Philosophy," *Renaissance Quarterly* 30 (1977):632–680.
25 A. G. Molland, "An Examination of Bradwardine's Geometry," *Archive for History of Exact Sciences* 19 (1978):131, and "Mathematics in the Thought of Albertus Magnus," in *Albertus Magnus and the Sciences*, ed. James A. Weisheipl (Toronto: Pontifical Institute of Mediaeval Studies, 1980), 467.
26 Cf. P. Duhem, *To Save the Phenomena*, tr. E. Doland and C. Maschler (Chicago: University of Chicago Press, 1969), 38–40, and Grant's Chapter 7 in this volume.
27 MS Oxford, Bodleian, Digby 76, fol. 78ar: "Auctores enim perspective nobis ostendunt quod linee et figure declarant nobis totam operationem nature, et principia et effectus, et similiter patet per celestia, de quibus naturalis et astronomia communicant." Fols. 69r–79r of this manuscript contain some geometrical writing following immediately after a version of Bacon's *Communia mathematica* (edited by Steele in Bacon, *Opera*

hactenus inedita, XVI). I have little doubt about Bacon's authorship of these unpublished passages, and hope soon to publish an edition of them.

28 Nicole Oresme, *Le livre du ciel et du monde*, ed. A. D. Menut and A. J. Denomy (Madison: University of Wisconsin Press, 1968), 288. The clock image is not original to Oresme, but goes back at least to Cicero *De natura deorum* 2.34–38; cf. Oresme, *Livre du ciel*, 282.

29 For a suggestive study of the relationship between clocks and other automata on the one hand and mechanical philosophy on the other, see D. J. de Solla Price, "Automata and the Origins of Mechanism and Mechanistic Philosophy," *Technology and Culture* 5 (1964):9–23.

30 Iohannes Buridanus, *Quaestiones super libris quattuor de caelo et mundo*, ed. E. A. Moody (Cambridge, Mass.: The Mediaeval Academy of America, 1942), 180–181; cf. M. Clagett, *The Science of Mechanics in the Middle Ages* (Madison: University of Wisconsin Press, 1959), 524–525.

31 Cf. A. G. Molland, "Ancestors of Physics," *History of Science* 14 (1976):64–67, and M. Clagett, *Archimedes in the Middle Ages* (Madison: University of Wisconsin Press, 1964; Philadelphia: American Philosophical Society, 1976–1984), V, 145–146.

32 *Idiota de staticis experimentis*, in Nicolaus de Cusa, *Opera omnia* V, (Hamburg: Felix Meiner, 1983), 219–241.

33 Thomas Bradwardine, *Tractatus de proportionibus*, ed. H. L. Crosby (Madison: University of Wisconsin Press, 1955), 116.

34 Oresme, *Livre du ciel*, 670.

35 E. A. Moody and M. Clagett, *The Medieval Science of Weights* (Madison: University of Wisconsin Press, 1952), 128, 154, 174. Cf. Bradwardine, *Tractatus de proportionibus*, 100–104.

36 *Metaph*. A5. 986a2–3; cf. *De caelo* II.9.290b12–291a27.

37 Boethius, *De institutione arithmetica libri duo. De institutione musica libri quinque* . . . , ed. G. Friedlein (Leipzig, 1867; repr. Frankfurt: Minerva, 1966), 187–188; *Timaeus* 36D.

38 Oresme, *Livre du ciel*, 476–486.

39 *Timaeus* 31b–32C, 35B–36B.

40 On what follows, cf. Molland, "Ancestors of Physics," 67–70.

41 Bradwardine, *Tractatus de proportionibus*, 110.

42 Ibid., 104.

43 Ibid., 106.

44 On all this, cf. Molland, "Examination of Bradwardine's Geometry," 150–160, and E. D. Sylla, "Compounding Ratios: Bradwardine, Oresme and the First Edition of Newton's *Principia*," *Transformation and Tradition in the Sciences*, ed. E. Mendelsohn (Cambridge: Cambridge University Press, 1984), 11–43.

45 Boethius, *De institutione arithmetica*, p. 12.

46 *De caelo* I.1.268a11–16.

47 Cf. V. F. Hopper, *Mediaeval Number Symbolism* (New York: Columbia University Press, 1938; repr. New York: Cooper Square, 1969).

48 Cf. A. G. Molland, "Cornelius Agrippa's Mathematical Magic," forthcoming in a volume edited by C. Hay.

49 *Timaeus* 53C–56C.

50 *De caelo* III.8.306b3–307b23.

51 Cf. Molland, "Mathematics in the Thought of Albertus Magnus," 474–475.

52 Molland, "Examination of Bradwardine's Geometry," 170–174.

53 See, for instance, A. Maier, *An der Grenze von Scholastik und*

Naturwissenschaft (Rome, 1952), 255–384; *Nicole Oresme and the Medieval Geometry of Qualities and Motions. A Treatise on the Uniformity and Difformity of Intensities, known as Tractatus de configurationibus qualitatum et motuum*, ed. M. Clagett (Madison: University of Wisconsin Press, 1968); E. Sylla, "Medieval Quantification of Qualities: The Merton School," *Archive for History of Exact Sciences* 8 (1971):9–39.

54 *Cat.* 6.6a19–20.

55 *Cat.* 6.5b27–28.

56 *Cat.* 8.10b26–27.

57 *Opera hactenus inedita*, IX, 144–149. Clagett, *The Science of Mechanics*, 334, doubts the authenticity of the ascription of this work to Bacon.

58 Cf. A. Maier, *Zwei Grundprobleme der scholastischen Naturphilosophie*, 3rd ed. (Rome: Edizioni di Storia e Letteratura, 1968), 74–109.

59 Oresme, *De configurationibus*, 226.

60 Ibid., 234–238. Cf. A. G. Molland, "The Oresmian Style: Semi-Mathematical but also Semi-Holistic," *Université de Nice, Cahiers du Séminaire d'Epistémologie et d'Histoire des Sciences* 18 (1985):7–12. A fuller version of this paper is scheduled to appear in a volume edited by P. Souffrin.

61 *Principia philosophiae* IV.187, in *Oeuvres de Descartes*, ed. C. Adam and P. Tannery (Paris, 1897–1913), vol. 8, pt. 1, p. 314.

62 Cf. A. G. Molland, "Nicole Oresme and Scientific Progress," *Miscellanea Mediaevalia* 9 (1974):213.

63 On what follows, cf. Molland, "Atomisation of Motion."

64 Molland, "The Oresmian Style."

65 Oresme, *De configurationibus*, 292.

66 Ibid., 406.

67 Ibid., 276.

68 Ibid., 278.

69 Ibid., 280.

70 On medieval theories of infinite space, see especially E. Grant, *Much Ado about Nothing: Theories of Space and Vacuum from the Middle Ages to the Scientific Revolution* (Cambridge: Cambridge University Press, 1981), chap. 6.

71 Quodlibet II, questio 9, in *Quodlibeta Magistri Henrici Goethals a Gandavo Doctoris Solemnis* (Paris, 1518), fol. 36r.

72 Molland, "Examination of Bradwardine's Geometry," 132–136.

73 Cf. A. G. Molland, "Richard Swineshead and Continuously Varying Quantities," *Actes du XIIe Congrès International d' Histoire des Sciences* (Paris, 1971), 127–130.

74 Cf. Molland, "Examination of Bradwardine's Geometry," 147–148.

75 *Leibnizens mathematische Schriften*, ed. C. I. Gerhardt (Halle, 1849–63). Erste Abtheilung, IV, 13–14.

76 H. Hugonnard-Roche, *L'Oeuvre astronomique de Themon Juif* (Geneva: Minard, 1973), 251–411. Cf. my review in *Journal for the History of Astronomy* 7 (1976):68–69, where I remark on Themo's delight in displaying intellectual pyrotechnics, to the possible detriment of a disinterested search for knowledge.

77 *Le opere di Galileo Galilei. Nuova ristampa della edizione nazionale* (Florence: Barbera, 1968), I, 300; Galileo Galilei, *On Motion and on Mechanics*, tr. I. E. Drabkin and S. Drake (Madison: University of Wisconsin Press, 1960), 67.

78 Cf. A. G. Molland, "Archimedean Fortunes," *History of Science* 19 (1981):143–147.
79 The phraseology is from the title of E. J. Dijksterhuis, *The Mechanization of the World Picture* (Oxford: Clarendon Press, 1961).
80 Cf. A. R. Hall, *Ballistics in the Seventeenth Century* (Cambridge: Cambridge University Press, 1952), 79–101, and M. Segre, "Torricelli's Correspondence on Ballistics," *Annals of Science* 40 (1983):489–499.
81 *Oeuvres de Descartes*, II, 380.
82 Ibid., II, 268.
83 *The Optical Papers of Isaac Newton*, ed. A. E. Shapiro (Cambridge: Cambridge University Press, 1984–), I, 544–547; I. Newton, *Opticks* (New York: Dover, 1952), 154, 225.

PART I. NATURAL PHILOSOPHY

3

Mathematical physics and imagination in the work of the Oxford Calculators: Roger Swineshead's On Natural Motions

EDITH D. SYLLA

This chapter is part of a larger project to understand the scientific work of the so-called Oxford Calculators, how it arose at Oxford in the first half of the fourteenth century, and what its influence or fortuna was in later periods and places. In earlier papers I have tried to show how the nature of the Oxford Calculators' work was influenced by the importance of undergraduate logical disputations at Oxford and particularly by the prominence of disputations *de sophismatibus*.[1] Looking at Richard Swineshead's *Book of Calculations* in light of William Heytesbury's *Rules for Solving Sophismata* and *Sophismata*, for instance, one can see how mathematics such as the summing of infinite series might have been valued for its use in setting up and unraveling complex sophismatical cases and not only for solving natural physical problems.

In this chapter, I want to examine an early-fourteenth-century Oxford physical work that seems to exist on a midground between previous Oxford physical works, such as those of Walter Burley, and the developed calculatory traditions. The work I intend to discuss is one variously entitled *Descriptions of Motions, On the Prime Mover* (from its incipit), or *On Natural Motions*. I will call it *On Natural Motions*.[2] The work is now, following James Weisheipl, generally ascribed not to the Richard Swineshead, fellow of Merton College, who wrote the *Book of Calculations*, but to one Roger Swineshead, not known to have been associated with Merton College, who later became a Master in Sacred Theology, and who was at the time of his death ca. 1365 a Benedictine monk of Glastonbury.[3]

On Natural Motions was written after Thomas Bradwardine's *On the Ratios of Velocities in Motions* of 1328 to which it refers, but before about 1337 or 1338, when it was copied in Erfurt Amplonian MS F135, the only complete extant copy. Roger Swineshead may also have been the author of logical treatises on obligations and insolubles recently studied by Paul Spade among others,[4] but although there are some reverberations (which I will mention again below) between Roger Swineshead's *On Natural Motions* and Heytesbury's *Rules for Solving Sophismata*, written in 1335, the *On Natural Motions* itself does not have the extensive treatment of logical topics that Heytesbury's *Rules* has. Roger himself, near the end of his discussion of the proper definition of motion, says that he will not prolong his discussion of this topic, explaining, "Since, however, the present subject matter is not real, but more nearly logical, nor will I in this business apply my effort in the way of logic to notions and not to things, I intend to end the aforesaid subject in a lively but not extended manner."[5]

In my recent paper examining the role of mathematical physics in John Dumbleton's *Summa of Logic and Natural Philosophy*, probably written in the 1340s,[6] I concluded that Dumbleton's mathematical physics could be characterized as an Aristotelian mathematical physics because its mathematical descriptions were assumed to arise by abstraction of quantities from real physical objects and processes. Dumbleton was not Platonistic in his use of mathematics in that he did not operate as if mathematics was higher than or metaphysically prior to physical objects, but neither was he trying to make a mathematical model fitting the data within an acceptable margin of error. Dumbleton's goal was not to create a sublunar analogue to a Ptolemaic system of epicycles and eccentrics saving the phenomena. Rather, the mathematics was assumed to be a quantitative description of real physical objects and processes.

Like Dumbleton's *Summa*, Roger Swineshead's *On Natural Motions* is a work of natural philosophy, but it is more focused than the *Summa*, limiting itself quite closely to problems of motion. Its subject matter includes the definitions of motion and time (Part 2) and the four kinds of motion – generation (Part 3), alteration (Part 4), augmentation and diminution (Part 5), and local motion (Part 6). It ends with discussions of the relation of motion to its causes (Part 7) and of maxima and minima in active and passive powers or elsewhere (Part 8).[7] In the discussions of alteration, augmentation and diminution, and local motion, a general pattern emerges of first treating the subject physically, refuting erroneous opinions and establishing the true one, and then considering the proper measure of velocity in motions of that type.

The work is not, however, as a whole a work of applied or mixed mathematics, as Thomas Bradwardine's *On the Ratios of Velocities in Motions* or Albert of Saxony's similar work could be labeled. There is considerable detail about various natural phenomena. So, for instance, Part 3, on generation, mentions climates of various regions of the world, burning mirrors, the production of tides by the moon, the production of a spark by striking iron on a flint stone, lightning, comets, rain, and snow, the evaporation of water into air, the mixture of milk and water, the production of butter and whey from milk, the sugar in an apple, the drying up of mud, the generation of frogs in the region of the clouds, the production of compound forms by the heavens, the generation of flies and worms from putrifying material or in the viscera of men, that if an eel is cut in half either half can move away, that when an animal's heart is violently cut out it continues to beat while the rest of the body may also move or jump about, and that, unlike parts of plants which when cut off grow back into whole plants, parts of animals do not grow into whole animals.[8] Doubtless, many if not all of these observations are also to be found reported in the various works of Aristotle or other previous authors. Nevertheless they are repeated at some length in the *On Natural Motions*, giving it a naturalistic as well as a mathematical or quantitative emphasis.

The relation of the *On Natural Motions* to previous work

In view of the extensive naturalistic material in the work, one might ask, in fact, why the *On Natural Motions* contains as much as it does about the measures of motion in the various categories. Thomas Bradwardine's *On the Ratios of Velocities in Motions* is obviously the major model and precedent for Part 7.[9] Indeed, when Swineshead comes to presenting his own view in Chapter 2 of Part 7 on the relations of velocities to movers and resistances, he omits any mathematical preliminaries, saying that "since the difference of ratios is sufficiently extracted from arithmetic and unified in a certain treatise entitled *On Ratios*, and since despite the falsity of the third chapter of that work [where Bradwardine expounds his function relating forces, resistances, and velocities], the truth of the first chapter remains, I will, for the sake of brevity, presuppose the first chapter of that treatise in the present work."[10] A precedent for the mathematical part of Swineshead's Chapter 6 also can be found in Bradwardine's Chapter 4, where he argues in favor of measuring the velocity of circularly moved bodies by the distance covered by the fastest moving point.

Elsewhere there are looser precedents for Swineshead's use of latitudes in measuring qualitative differences in Part 4 in Walter Burley's earlier treatises on alteration[11] and parallels to Swineshead's discussions of all three major categories of motion in William Heytesbury's *Rules for Solving Sophismata*.[12] Although the relative dating of the *On Natural Motions* and *Rules for Solving Sophismata* is not firmly established, a comparison of the discussions of augmentation in the two works has led me to the conclusion that Roger Swineshead's discussion occurred after Heytesbury's (written in 1335), to which it apparently refers. I will describe some of the evidence for this conclusion below. Heytesbury's *Rules*, then, may have provided a model for the mathe-

matical parts of Swineshead's *On Natural Motions* that are not obviously related to Bradwardine's work.

In sum, it is not self-evident from Swineshead's work why his particular combination of naturalistic and mathematical treatments of motion would be thought useful or relevant to some further end, and he
does not explain why he has made such a naturalistic-mathematical
combination. What he does is perhaps understandable, however, if he
is taken to be incorporating ideas patterned after the recent successful
mathematical work of Bradwardine and Heytesbury into a more traditional Aristotelian naturalistic context. John Dumbleton's *Summa*
appears likewise to try to incorporate recent mathematical ideas like
those of Bradwardine and Heytesbury into a larger natural philosophical context.[13]

Distinctive characteristics of the work

When I began my reexamination of the *On Natural Motions*,
then, with an eye to characterizing the nature of the mathematical
physics it includes, I expected to find a situation more or less analogous
to the one I previously found in Dumbleton's *Summa of Logic and
Natural Philosophy*, with what I have characterized as its Aristotelian
attitude toward the role of mathematics in physics. In considering
Roger Swineshead's *On Natural Motions* in some detail, however, I
found several interrelated ways in which Swineshead's mathematical
physics is unlike Dumbleton's.

First of all, Swineshead differs from Dumbleton in favoring idiosyncratic or paradoxical-sounding conclusions, saying, for instance, that
some altered body will be altered for an hour with infinite velocity.[14]
Swineshead's work, like Heytesbury's, seems to reveal more traces of
the influence of the wide logical-disputational context at Oxford than
does Dumbleton's. As a work written at Oxford for the use of students,
as it describes itself at the start,[15] the *On Natural Motions* seems to
presuppose, without saying so explicitly, an educational context in
which students are expected to correlate and use together in disputations knowledge and skills gained in the various disciplines of the
liberal arts and philosophies. Further evidence relevant to this supposition will be discussed later in this chapter.

But an additional factor differentiates Roger Swineshead's work from
Dumbleton's. Dumbleton, like most of the later Calculators, has an
ontology similar to that of William of Ockham, restricting real entities
to substances and qualities. Roger Swineshead, in contrast, admits
three categories of things, adding quantity to substance and quality,
and, moreover, referring to the other categories as *modes* of things
(*modi rerum*).[16] Although they are not things, these modes seem to
have a certain ontological status in Swineshead's view. Swineshead

considers motion, his topic in the present work, to be not a thing, but an action or passion, that is, a mode of things.[17] On the other hand, although he includes quantities among the things (*res*), he does not seem to push this farther in a Platonistic direction. Time, often described as the measure or quantity of motion, he defines by saying that temporality or time "is the temporal being of a mutable thing."[18] I will return below to a discussion of the status of modes of things.

On first inspection, then, Roger Swineshead's mathematical physics is dissimilar from Dumbleton's, in favoring more idiosyncratic or paradoxical sounding conclusions and in adopting a less Ockhamist or minimalist ontological position. To provide a more extensive body of evidence to indicate the nature of the work, I have made a preliminary edition of a section of the *On Natural Motions* in an appendix to this chapter. But Roger Swineshead is not only idiosyncratic. As I will argue below, he evinces an attitude toward the nature of mathematical physics which can be compared favorably to that of the more successful Nicole Oresme and which may well have been of importance in the later development of a more thoroughly mathematical physics.

Roger Swineshead's use of *imaginatio*

But before I turn to a comparison of Swineshead's procedure to that of Oresme, I want to provide the basis for a more thorough understanding of the characteristics of Roger Swineshead's mathematical physics by means of an examination of his diverse uses of the term *imaginatio*, or imagination. Fourteenth-century authors have often been characterized as relying heavily on arguments *secundum imaginationem*.[19] Ockham, Buridan, and many others often clarify issues by supposing in imagination that God by his absolute power creates things other than they naturally are and then investigating what the consequences would be for our understanding of the necessary distinctions between various concepts or entities.

Roger Swineshead does not use *secundum imaginationem* cases in quite this way, but he makes fairly frequent use of "imagination" as a synonym for "reason" and as opposed to "real." Thus, he contrasts things that are really distinct with things that are distinct only rationally or in imagination. He says, for example, that there are two magnitudes, namely quality and the latitude of quality, that are different rationally although not really or as things.[20] Then, to understand motion of alteration more fully he posits three distinct latitudes, each of which he distinguishes into two latitudes different not really or as things but rationally.[21] The first latitude is the latitude of quality, the second is the latitude of motion of alteration, and the third is the latitude of the latitude of motion. After speaking in more detail of the first two latitudes, each being distinguished in reason into two others, he speaks of the third being distinguished *ymaginarie* into two others,[22] using that word as a substitute for *secundum rationem*. Swineshead is thus here not using *imaginatio* in any technical sense that would distinguish it from reason in general.

An objection to the positing of these three latitudes raises overtones of something different from Dumbleton's Aristotelian-realist attitude toward scientific theory, saying, "Especially against the latitude of the latitude [of motion], it is argued as follows. If with respect to the latitude of motion a second, wholly distinct latitude were posited, for a similar reason, of the second latitude a latitude would be posited in the same way, and so forth. But infinitely many latitudes distinct from the latitude of motion are superfluous. Therefore the latitude of the latitude should not be posited."[23] Swineshead replies to this by denying the inference and by saying that the latitude of the latitude of motion is not some (new) thing to be acquired by motion. Rather it and its two sublatitudes are acquired along with the latitude of motion and vice versa.[24] In other words, I understand him to be saying that the latitude of the latitude is a distinct concept, but it is not a separate real entity. This reply thus seems to reveal an attitude toward theory somewhat different from that of Dumbleton, who tightly linked physics and the correlated mathematical theory. Unlike the Ockhamists, Roger is not concentrating on minimizing the number of entities posited. He writes very blandly of things that are distinct rationally although not *in re*. He is not taking an explicitly conventionalist or instrumentalist attitude toward theoretical entities, and yet the link between real things and the quantitative concepts of the theory is not so tight as it was in Dumbleton. The effect is that there is more room for theoretical creativity or the invention of new and useful concepts not necessarily corresponding on a one-to-one basis with real physical things.

Swineshead's conclusions concerning alteration further exemplify the uses he is making of imagination. According to his first conclusion, only the four primary tangible qualities hot, cold, wet, and dry are in a proper sense intensible and remissible,[25] where the proper sense of intension and remission is defined as the strengthening or weakening of a quality solely by natural action of a quality similar or contrary in species or effect without something coming from outside.[26] According to Swineshead's second conclusion, however, heaviness and lightness, taste and smell, can intend and remit in a more general sense,[27] where this more general sense is defined as a quality changing either really or in imagination, so that modes of things and not only things may intend or remit.[28] True intension in the general sense occurs when a quality is made stronger by a quality that has a similar effect but is dissimilar in species, without anything added from outside.[29] So heaviness and lightness may change in the broader sense by the increase of heat or cold. Imaginary intension in the general sense, on the other hand, occurs when a thing is more strongly transformed by a mode and so the mode of the thing is said to be increased.[30] In this sense, motion, knowledge, and moral virtue and vice can be intensified. Only modes of things are intensible by imaginary intensification in the general sense.[31] After some other conclusions, Swineshead states in his last conclusion on alteration that neither sound nor substantial forms can be intended or remitted.[32] Neither the broadening of the terms 'intension' and 'remission' from their proper senses to a more general sense, nor the division of the latter into real and imaginary subcategories is enough, apparently, to allow intension and remission of things like substantial forms.

Roger Swineshead thus seems to move away from Aristotelian realism in creating concepts such as imaginary intensification in the general sense for cases in which what is intensified is not a thing but a mode of a thing. His imaginary intensification does not seem to be a really existing external entity waiting to be abstracted from physical processes, so much as a concept that is fitted to entities by the mind. Whereas Dumbleton might be said to see himself as trying to *discover* the correct mathematical physics, Roger Swineshead gives more the impression of trying to *invent* such a mathematical physics with a greater role for human reason or even creativity implied.

Some further insight into what is involved here can be gained by examining Marshall Clagett's edition of Nicole Oresme's *On the Configurations of Qualities and Motions*, where the concept of modes of things also appears briefly. Time, Oresme says, "in one signification . . . is . . . the enduring succession of mutable things according to before and after."[33] It is not motion, nor is it a "temporal thing," but "still it is not a thing separable from a temporal thing, for it could not exist in the absence of such a temporal thing . . . even by divine absolute power. Whence properly speaking time so assumed is not some *thing* but is rather a *mode of a thing*."[34] In explanation of the passage in which these statements occur, Clagett cites Averroës' Commentary on Chapter 10 of Book IV of Aristotle's *Physics*, where Averroës says,

> Since it is thus with motion, for motion is comprehended by sense, and no part of it is in act, but any given part recedes. Therefore motion is composed of what exists no longer and of what has not yet come into existence. Such things do not have complete being, but their being is put together by the mind from components that are outside the mind . . . as will be said below concerning time, insofar as time is one of the entities whose act is completed by the mind.[35]

On this understanding, then, modes of things are concepts not matched on a one-to-one basis with existing things, but rather concepts building on combinations of things, qualities, or quantities, including past or future things. Although they are not things, they are inseparable from things. Changes in these modes may well be considered imaginary in the sense of involving concepts or reason as well as things. Heytesbury implies something not entirely dissimilar when he rejects an analogy between measures of change in velocity and measures of augmentation, saying that, "Motion and other similar qualities of this sort, which are conceded to intend and remit in a general sense, are not augmented and diminished in the same way as magnitudes that are truly continuous and per se quantities."[36]

Two poles often used for distinguishing types of attitudes toward the relation of mathematics and physics are the "realist" versus the "conceptualist." George Molland uses these poles in characterizing Albertus Magnus's attitude as tending toward the conceptualist.[37] Using this framework, I would judge that Roger Swineshead's position also tends toward the conceptualist pole.[38] Possibly more useful, however, is the dichotomy used by J. M. Thijssen in his recent article on Buridan's mathematics.[39] Thijssen contrasts a mathematical approach, associated with Bradwardine, with an approach using semantic analysis, associated with the Parisian school. In Buridan's view, according to Thijssen, the objects of natural philosophy and mathematics are terms, and the relation of mathematical objects to reality is of a semantic nature and not a matter of abstraction.[40] Given that, according to medieval supposition theory, there are many ways that terms can supposit for things, there might be much more flexibility in the relation between mathematical quantities and real things on the semantical view than on the abstractionist view. Rather than claiming that mathematical entities are not real, as Molland's conceptualists may do, this approach would turn one's attention to the different ways that things can be combined or distinguished when terms are used in propositions to stand for things, sets of things, or parts of things. Latitudes would have a foundation in reality, but there would not be a one-to-one match between, for instance, the latitude of the latitude of motion and some single extended thing in nature. Just as motion may be conceived mentally by combining past, present, and future events, so too a latitude may be conceived by considering qualities at different instants or points of a body. In this way, the fourteenth-century emphasis on propositions and second intentional analysis may support a quantitative or mathematical approach rather than contrasting with it.

To pursue this matter a bit further, I want to point out two additional passages in which Swineshead makes use of the concept of imagination. The first is related to a passage that helped to convince me that the *On Natural Motions* may be *after* Heytesbury's *Rules for Solving Sophismata*. Heytesbury and Roger Swineshead disagree about how augmentation or rarefaction should be measured. Heytesbury wants to measure it using the ratio of the added quantity to the original quantity, whereas Swineshead wants to measure augmentation simply by the maximum quantity gained. A difficult case for Heytesbury's view involves augmentation of a point into a line, of a line into a plane, or of a plane into a solid, for then, at the start of augmentation, the ratio of the new quantity to the old would appear to be infinite. This case is made more complicated by supposing that the increase begins from zero degree of velocity of local motion of the parts and accelerates over time, thus increasing the rate of augmentation, for how can there be a velocity greater than the infinite velocity of augmentation with which the motion is supposed to begin? Heytesbury responds to this complicated case by saying that it is impossible because it contradicts itself, or at least it is impossible given the position that velocity of augmentation is measured by the ratio of the new to the old quantity.[41] A proof of the case is proposed, however, positing that a magnitude may be diminished to nothing, and saying that although this is not possible *de virtute sermonis*, it may be accepted for the sake of disputation, since it does not include a contradiction.[42]

In his commentary on this passage in Heytesbury, Gaetano of Thiene explains that the case is impossible, physically speaking, since prime matter and its quantity are eternal and incorruptible, but that the case does not include a contradiction and is imaginable and so should be faced up to by calculators.[43] Roger Swineshead also thinks that the case is imaginable and, moreover, points out that it has in fact been posited as a basis for disputation by many important thinkers. Therefore, because the case has been posited, he says, Heytesbury's position is ruled out.[44]

Here Swineshead seems to have a rule or convention based argument: If many important people have in the past posited a case, then one ought not to adopt a position which makes the case unimaginable. It is because Swineshead's rejection of this position fits so well with Heytesbury's defense of it that I assume Swineshead likely read Heytesbury's *Rules* before he wrote his own work, even though in these particular passages Heytesbury talks of the case as being impossible, whereas Swineshead refers to his opponent as saying it is unimaginable. A bit earlier, concerning the same opinion, Heytesbury says that a previous similar case is both impossible and not imaginable.[45]

In this passage, then, the term "imagination" is used not with reference to categories that contain things, but modes of things; nor with reference to differences not in things, but in ways of conceiving them. It is used, rather, with reference to hypothetical cases that are not supposed to exist in the real world at all but are posited as part of disputations.

So also in a later passage concerning maxima and minima, Swineshead refers to imaginary cases in relation to the disputational moves he makes, saying that since a case involving a body *A* that doubles in size every other proportional part of an hour and returns to its original size in the intervening time periods has been posited solely as imaginary, he will admit it as imaginable.[46] This is a typical fourteenth-century use of imagination. Yet Swineshead's argument, just discussed, against a position like Heytesbury's goes beyond the familiar fourteenth-century use of imaginary cases in disputation in ruling out a position because it makes a commonly imagined case (augmentation from zero quantity) unimaginable: Here an imagined case becomes a criterion for choosing between conceptions intended to describe or measure reality. This is related to the Ockhamist use of imaginary cases to decide what are and what are not really distinct things, but instead of involving the contents of an imaginary case, it involves the propriety of the moves one might make vis-à-vis imaginary cases in disputation.

Thus, to summarize, imaginary motions and cases and so forth play a not insignificant part in Roger Swineshead's mathematical physics, but not in the way usually taken as typical of fourteenth-century nominalists. Perhaps this occurs in part because Roger Swineshead does not, like Ockham and most of the Oxford Calculators, aim for minimalist ontology and does not therefore use imaginary cases to discover what things are really distinct and what things are not. Indeed, imaginary distinctions and motions have an ontological status different from that of real distinctions and things, but not one which is therefore only conceptual or without basis in the real world. Nevertheless, in Swineshead's positing of latitudes distinguished in imagination into other latitudes and the like, there seems to be more room for theoretical creativity and for positing useful concepts that go beyond the bare substances, qualities, and quantities under consideration, than was apparent in Dumbleton's mathematical physics with its tight ties to physical reality. Finally, in a disputational context there were apparently rules for the positing and acceptance of imaginary cases, as distinguished from what is naturally true. In Gaetano of Thiene's view, a calculator should not refuse to deal with a possible and imaginable case, even if it is physically impossible. Interestingly, Swineshead seems to go farther than this in arguing that an opinion should not be adopted if it follows from the opinion that an otherwise imaginable case becomes unimaginable. In this last role, imagination seems to have a significant impact on what theories of measurement may be adopted.

In Swineshead's use of "imagination," then, we have a further indication of the nature of the broader framework for his mathematical physics. Let me return from this discussion of imagination to the more general consideration of the nature of Swineshead's mathematical physics.

Swineshead on measures of alteration

In Swineshead's conclusions concerning measures of motion of alteration, there is further evidence concerning the nature of his mathematical physics. Swineshead prefaces his conclusions about the causes of the excess of one degree of a latitude over another by a list of thirteen suppositions to clarify terminology, defining such terms as "latitude," "highest degree," "intense degree," "remiss degree," "degree in a latitude," "uniform degree," and so forth. He applies his definitions to heat for clarity, but means them to apply to any intense quality.[47] To a modern eye his concepts seem idiosyncratic. This is in part the case because he does not completely disentangle independent dimensions of quality in intensity and extension, but rather always considers qualities with both intensity and extension. He has, then, a uniform degree of heat that is a heat of which no part is more intense than another.[48] He also has a latitude of heat which is a uniformly difform heat[49] and a degree in a latitude, which is a part of a latitude more or less intense than another part of the latitude.[50] All of Swineshead's concepts thus apply to qualities that are extended in space as well as intense, and he defines a latitude of heat in more or less the same way as a uniformly difform heat. The latter he defines as a difform heat such that of any two parts immediate to each other, the most intense uniform degree that is not in one part is the most remiss uniform degree that is not in the other part.[51] He has to define uniformly difform heat in terms of uniform degrees that are *not* in the adjacent parts, because, given his definition of uniform degrees, there are no uniform degrees in a uniformly difform quality.[52]

The three conclusions that follow these suppositions seem to arise out of a Galenic medical interest in the temperate. So, for instance, the third conclusion is that every degree of the latitude of heat equally distant from the maximum degree and from zero is just as intense as it is remiss.[53]

When Roger Swineshead turns, then, to the exposition of his views on the measure of velocity of alteration, he says that since motion cannot exist without magnitude, magnitude is acquired in motion of alteration, so that both the quality and the latitude of quality may be acquired, where, as I have quoted above, quality and latitude of quality are magnitudes distinguished not really [*re*], but rationally. There are, therefore, according to Swineshead, two rationally distinct motions of alteration corresponding to these two magnitudes, motion of extension, corresponding to quality, and motion of intension, corresponding to latitude of quality. He has two conclusions concerning each of these two types of alteration.[54] Velocity of extension, he says, depends on or can be measured by the quantity of the magnitude through which the quality extends. This is just what one might expect. The second conclusion, however, is the surprising one mentioned above, namely that "Some altered body will be altered for an hour with infinite velocity."[55] This conclusion would hold true, Swineshead says, of a foot-long medium *A* toward which a distant light approaches with continuous motion for an hour, for then at each instant of the hour a new degree of light is suddenly extended through the whole of *A*. This will be infinite velocity of extension, since it is faster than any part by part extension.

A similar pair of conclusions is offered concerning motion of intension. The first conclusion is more or less what might be expected, namely, that velocity of intension or remission depends on the latitude or degree of quality acquired or lost. The second conclusion, however, points out what was implicitly unconventional about the first, saying that of two motions of intension unequal in velocity, one velocity may have no ratio to the other.[56] This would happen, according to Swineshead, if in two motions of intension both bodies gained the same latitude of quality but one gained in addition a single uniform degree. If two bodies beginning at the same degree were altered so that one ends up uniformly of *C* degree and the other ends up uniformly difformly qualified with an extrinsic maximum of the uniform degree *C*, then the conclusion would hold.

Why, one might ask, is Swineshead particularly interested in cases in which the velocity is infinite or beyond all measure or in which a difference of velocity exists but is immeasurably small? One possibility is that he wants to test the validity of his proposed measures of velocity by considering extreme cases.[57] Another possibility is simply that such paradoxical cases interest him. Typically among the Oxford Calculators one would expect a proposed measure of velocity to be rejected, however, if it leads to such immeasurably large velocities or immeasurably small velocity differences, and yet Swineshead seems to want to emphasize his improportional results. Rather than Swineshead simply enjoying paradoxical results, one might suppose that he considered them of potential usefulness in creating and resolving the sophismatical cases that played a prominent role in the students' disputational training, as discussed above with reference to the idiosyncratic nature of his concepts. If this is the case, however, his proposed measures seem to make paradox too easy. Thus his second conclusions concerning extension and intension of qualities might easily appear unchanged as sophismata sentences. Is it satisfactory to admit such paradoxical-sounding statements as true or to define concepts in such a way that such conclusions follow? Perhaps Swineshead thought it was, but if so, then he might just as well have accepted Heytesbury's proposed measure of augmentation, discussed above, along with the inference from it that a motion of augmentation from zero quantity begins infinitely fast.

Similar or related points might be raised concerning Swineshead's further discussion of measures of the velocities of augmentation and of the relations of forces, resistances, and velocities. Swineshead's conclusions often seem more peculiar than is typical for the other Oxford Calculators. Of course, a further possibility is simply that Roger Swineshead is theoretically inept and that his odd conclusions are the most reasonable ones he can come up with or the most reasonable ones he can come up with that are also his own and not simple copies of Bradwardine or Heytesbury, but any possible ineptitude would not likely extend to necessitating his acceptance of infinite or improportional velocities and the like.

More importantly, however, the juxtaposition of the more metaphysical-physical parts of the *On Natural Motions* and the mathematical parts tends to support the view that Swineshead is attempting to create a unified mathematical-physical science. Moreover, Swineshead's seemingly bizarre second and conclusions concerning motions of extension and intension are the natural result, not only of the way he conceptualizes uniform degrees, latitudes, and so forth, but also of his ideas of how intension and remission occur and in what sorts of qualities.

Similarly in Part 5, Swineshead's view on the measure of augmentation fits naturally with his conclusion that in augmentation resulting from nutrition (i.e., in augmentation in the most proper sense) not every part of the augmented body is increased. He therefore does not visualize the whole body expanding uniformly, but rather a part being added to one side, and consequently compares augmentations not by the ratio of the ratios of the total final quantities to the original quantities, but rather by the ratio of the maximum amounts acquired by the two bodies. Similarly, concerning local motion Swineshead expounds physical views that lead to, fit with, and support his mathematical conclusions.

In my opinion it is likely that several of the factors I have mentioned played a role in leading to the sort of mathematical physics found in Roger Swineshead's *On Natural Motions*. Roger Swineshead intended to create a conceptual system for measuring motions in the real world. He was not the most brilliant mathematically of the Oxford Calculators. His mathematical physics was affected by his nonminimalist (non-Ockhamist) attitude toward ontology. He was also affected by the recent successes of Bradwardine and Heytesbury and was copying them in his attempt to write a work combining mathematics and natural philosophy. The logical-disputational or sophismatical nature of Oxford undergraduate education made paradoxical-sounding conclusions not totally unwelcome or unfamiliar.

Although all of these factors probably played a role, I want to conclude by emphasizing the role of Swineshead's apparent attitude toward originality or the status of theory in natural philosophy, which seems related both to his nonminimalist views on ontology and to a sort of semantic approach to science. However modest Roger Swineshead's scientific talents may have been, he nevertheless tried to create a unified system that was his own, incorporating bits and pieces of previous work, but adding to them his own somewhat idiosyncratic ideas.

He reflects this in the preface to his work, which says something like this:

> Here begins the treatise of Master William [!] Swineshead given at Oxford for the use of students.
>
> Having first invoked the Prime Mover, I will attempt to treat motions according to the limit of my capacity, with a way of writing that is brief, however much unpolished. With very many people frequently, and with effect, arousing me to this result, and, most of all, with the zeal for truth inclining me, I am led to the undertaking of the present work with the highest good feelings (? *suppremis affectibus*). And yet my private thinking about such things will be threatened by ancient tradition, which is such that only those things that have acquired approval through it are considered to be authentic by the sects of believers. But what does the small authority of the speaker matter if the reader sees that what is said agrees with the truth? Not even the most careful farmer collects what is to be had without some ear containing grain escaping, nor does the poor peasant woman search through the countryside without finding grain hidden in less obvious places.
>
> As I begin writing the work to follow concerning motion, therefore, I ask first that no one who is about to read the work presume to condemn it because it is not elegantly polished or because it is unusual (*insuetum*), since artful constructions of words and the polished phrases of some authors, just like idols, often stretch out tentacles of falsity behind the pretence of truth. If, indeed, of the things to be said, some may be found to be wrong, the rest are not therefore refuted. The sterility of a field is in no way shown by the emptiness of one stalk any more than the abundance

of another stalk disproves the sterility of the same. Nor is one who reduces to silence those who look at him straight overthrown by those who look askance, sneering, with senile laughter.

The body of the following work is divided into eight parts or distinctions, of which the first plays the role of a preface. The second defines motion and time. The third declares certain unusual, although not new, propositions concerning generation.[58]

Of course, the remarks with which authors open their works are often assumed to be standard *topoi* or platitudes. Nevertheless, I think it is interesting that Roger seems to think he is diverging from or adding to past accomplishments at least in a small way. The image of the old peasant woman gleaning the remaining grains after the harvest of even the most careful farmer is not unlike, in its implications, the image used by Bernard of Chartres and Isaac Newton of the dwarf standing on the shoulders of giants: One's contribution may be small and yet go beyond the achievements of the past. Roger is aware that a number of his positions are unusual or nontraditional, but he offers them as worthy of consideration because they may be true.

Whatever we may think about the truth of Roger Swineshead's conclusions or about the potential usefulness of his concepts, he has an attitude toward the nature of scientific endeavor that needs to be included in our picture of the nature of fourteenth-century Oxford science, and in particular of the mathematical science of the Calculators.

Interestingly, what seems to encourage him in his attempt to create a mathematical system are factors not usually associated with a move toward mathematics: for instance, a nonminimalist approach or a semantical or logical slant. Granted the close connection between logic, physics, and mathematics in Swineshead's work, it is possible that his system-building attitude is a reflection of properties of many of the standard sorts of disputations: Within any given disputation an opponent or respondent is supposed to adopt his own consistent point of view and to maintain it for the course of the disputation. Possibly there is some echo of this situation, then, in the attitude Roger Swineshead takes toward the writing of the *On Natural Motions*. In this way, the disputational context of much of medieval natural philosophical work need not always have led to disconnected conclusions.

Roger Swineshead and Nicole Oresme

It should be noted, moreover, that Roger Swineshead's attitude is similar to that which Nicole Oresme takes in his more successful work *On The Configurations of Qualities and Motions*. Oresme begins by saying, "When I began to set in order my conception of the uniformity and difformity of intensities, certain other things occurred to me to add to the topic so that the treatise might be useful not only as an exercise *but also as a discipline*."[59] When he begins to set forth his concepts in Part I, he says,

> Every measurable thing except numbers is imagined in the
> manner of continuous quantity. Therefore, for the
> mensuration of such a thing, it is necessary that points,
> lines, and surfaces, or their properties, be imagined. . . .
> Although indivisible points, or lines, are non-existent, still it
> is necessary to feign them mathematically for the measures
> of things and for the understanding of their ratios.
> Therefore, every intensity which can be acquired
> successively ought to be imagined by a straight line. . . .
> Therefore, the measure of intensities can be fittingly
> imagined as the measure of lines, since an intensity could be
> imagined as being infinitely decreased or infinitely increased
> in the same way as a line. . . . Of course, the line of
> intensity of which we have just spoken is not actually
> extended outside of the point or subject but is only so
> extended in the imagination, and it could be extended in any
> direction whatever except that it is more fitting to imagine it
> standing up perpendicularly on the subject informed with the
> quality.[60]

Like Swineshead, then, Oresme considers his concepts not to be abstracted from the real quantities of physical bodies, but rather to be "feigned" by him for the purpose of understanding natural things more clearly. At the same time, he claims, as Swineshead seems to assume, that his concepts are more suitable than any others that might be proposed, and he argues in their favor on pragmatic as well as absolute grounds, saying, for instance, that intensities should be called longitudes rather than latitudes, but since tradition has it otherwise, he will conform to the common practice,[61] and that "there is no other way by which the species and diverse modes of difformity could be recognized and otherwise assigned."[62]

Similarly, in his earlier *Questions on the Geometry of Euclid*, question 10, Oresme defends his system, saying, "But one might say: 'Master, it is not necessary for it to be so imagined.' I answer that the imagination is a good one. This is evident by Aristotle who imagines time by means of a line. . . . Further, following this imagination I can more easily understand those things which are said about qualities uniformly difform and so on."[63]

Thus Oresme, like Swineshead but more successfully, aims to create a unified theoretical science on the basis of imagined quantitative concepts tied, but not limited, to existing physical things. Such a science is assumed to describe physical reality, but it leaves some room for originality and creativity on the part of the individual scientist, who develops his system within a tradition but goes beyond it. It is, finally, just possible that Swineshead actually influenced the more successful Oresme in this endeavor. Indeed, we know that Giovanni Casali, who produced another similar system for configuring qualities, was influenced by Swineshead to the extent of adopting his basic definitions almost verbatim from the *On Natural Motions*.[64] And Oresme, after he has introduced his concepts of the latitude and longitude of quality, says that, "it is obvious from the things said that certain moderns do not speak in the best way when they call the whole of the quality its latitude, just as it would be an abuse [of terminology] to understand by the breadth of a surface the whole surface or figure."[65] This other, criticized view, I would argue, may well have been the natural descendant of Roger Swineshead's conceptual scheme in which a latitude of heat and a uniformly difform heat have the same definition on the presupposition that qualities are never found without extension. In this way, by proposing his own modest and idiosyncratic system for measuring qualities and motions, Roger Swineshead may well have helped to incite Nicole Oresme as well as Giovanni Casali to propose their later and superior systems.

Appendix

Roger Swineshead, De motibus naturalibus,
Differentia 4, Capitulum Tertium.

This transcription is based on Erfurt, Amplonian MS F135 (called E), which is a reasonably clear manuscript but with a number of poor readings. Whenever the Erfurt text does not appear to make sense, I have tried to emend the text using Paris, Bibliothèque Nationale MS lat. 16621 (called P), which contains large parts of the work as copied by a fourteenth-century Parisian student. As a copy by someone who understood what he was copying, the Paris manuscript seems to have fewer nonsense readings. On the other hand, it is written in a sloppy script and is now very tightly bound, so that readings near the inner margin are often illegible on microfilm. The Paris manuscript also seems to do some slight condensing. I have not reported the many inversions of words and other differences in the Paris manuscript when the Erfurt manuscript seemed to make good sense. Variant readings are indicated in footnotes. My editorial remarks are in square brackets. Despite comparisons of the two manuscripts, a few passages still seem garbled.

[E, fol. 38ra; P. fol. 60r]

1   Modo super est in isto tertio causam excessus intensionis
vel remissionis unius gradus latitudinis super aliam gradum
eiusdem investigare.

Opinionibus igitur erroniis circa materiam presentem
5   brevitatis causa pretermissis, tres [E, fol. 38rb] conclusiones
sentencie veritatis contentivas restat manifestare. Pro
modum tamen loquendi multipliciter in hac materia varietatis
ne in posterior lateant in processu 13 suppositiones sunt
premittende, et ut planiori stilo procedam conclusiones
10 sequentes una cum suppositionibus specialiter ad caliditatem
applicabo singula tamen de caliditate dicenda de qualitate
intensa intelligantur.

Prima suppositio est hec: latitudo caliditatis est caliditas
uniformiter difformis.[1] Secunda: gradus summus caliditatis
15 est caliditas secundum se totam equaliter maxime distans a
non gradu. Tertia: gradus intensus caliditatis est caliditas
secundum se totam vel secundum extremum[2] eius intensius
distans a non gradu per latitudinem. Quarta: gradus remissus
caliditatis est caliditas[3] secundum partem eius intensissimam
20 indivisibiliter vel per latitudinem vere[4] distans a summo
gradu. Quinta: gradus in latitudine nihil aliud[5] quam pars
latitudinis[6] alia parte eiusdem intensior vel remissior existit.
Sexta: gradus[7] caliditatis[8] est caliditas uniformis vel
uniformiter difformis. Septima: gradus uniformis caliditatis
25 est caliditas cuius nulla pars alia parte eiusdem intensior vel
remissior eiusdem[9] existit. Octava: caliditas uniformiter
difformis est caliditas difformis cuius quarumlibet duarum
parcium sibi invicem immediatarum[10] gradus uniformis
intensissimus qui non est in una est remississimus qui non
30 est in alia. Nona: intensissimus gradus uniformis[11] qui non
est in *A* est gradus uniformis qui non est in *A* quo quolibet

---

[1] P; de formis E.
[2] versus P.
[3] *post* caliditas *add.* P que.
[4] *post* vere *add.* P est.
[5] alius est, P.
[6] *post* latitudinis *add.* P que.
[7] *om.* P.
[8] caliditas P.
[9] *om.* P.
[10] P; mediatarum E.
[11] *om.* P.

gradu uniformi intenciori gradus remissior est[12] in *A* et
eodem gradu nullus gradus remissior vel ita remissus est in
A. Decima: remississimus gradus uniformis qui non est in *A*
35 est ergo[13] gradus uniformis qui non est in *A* quo quolibet
gradu uniformi remissiori[14] gradus intensior est[15] in *A* et
eodem gradu nullus[16] intensior vel ita intensus[17] est in *A*.
Undecima: in tota latitudine caliditatis nullus gradus
uniformis reperitur. Duodecima: inter omnes duos gradus
40 uniformes caliditatis inequales intensive mediat latitudo.
Ultima: quilibet gradus[18] remissus caliditatis est gradus
intensus et non econtra.

Prima conclusio est hec: medio gradu maxime latitudinis[19]
subduplo ad gradum summum, nullus gradus eiusdem
45 latitudinis in duplo remissior existit. Prima siquidem parte
conclusionis de se manifesta, secunda pars sic ostenditur. Si
aliquis gradus latitudinis caliditatis gradu medio foret
remissior in duplo, cum gradus medius per maximam
medietatem tocius latitudinis caliditatis recedat a summo
50 gradu, gradus in duplo remissior per totam latitudinem
caliditatis recederet a summo gradu et ita gradus ille non
foret uniformis neque uniformiter difformis, consequens
falsum et contra sextam suppositionem. Et consequentia
prima[20] probatur sic. Gradus remissus caliditatis secundum
55 partem eius intensissimam indivisibiliter seu per latitudinem
distat a gradu summo iuxta suppositionem quartam. Ergo
graduum remissiorum per latitudinem distancium a summo
gradu gradus in duplo remissior distat per duplam
latitudinem, nec aliquis gradus latitudinis caliditatis distat a
60 summo gradu per latitudinem duplam ad medietatem tocius
latitudinis caliditatis. Ergo medio gradu latitudinis caliditatis
nullus est gradus in duplo remissior. Item quilibet gradus
remissus caliditatis est gradus intensus et non econtra per
ultimam[21] suppositionem. Ergo tota latitudo intencionis
65 caliditatis totam latitudinem remissionis eiusdem excedit.

---

[12] P; *om.* E.
[13] *om.* P.
[14] *om.* P.
[15] P; *om.* E.
[16] *post* nullus *add.* P gradus.
[17] vel ita intensus *om.* P.
[18] *post* gradus *add.* E intensus et.
[19] *post* latitudinis *add.* P caliditatis.
[20] *om.* P.
[21] P; duodecimam E.

Cum ergo ad medium gradum caliditatis latitudinis nullus gradus intensive transcendit duplum, ergo ad medium gradum in remissione[22] nullus gradus duplum attinget.

Secunda conclusio[23] est hec: tocius latitudinis caliditatis
70   quelibet pars ad extremum eius intencius terminata tam intensione quam remissione toti latitudine coequatur. Ista sic ostenditur: totius latitudinis caliditatis quelibet pars ad extremum eius terminata intencius secundum partem sui intensissimam equaliter recedit a summo gradu sicut tota
75   latitudo. Ergo per suppositionem quartam tocius latitudinis caliditatis quelibet pars ad extremum intencius terminata toti latitudini equalis est remissio, ac eiusdem latitudinis quilibet pars ad extremum intencius terminata secundum extremum sui intensius equaliter receditur a non gradu sicut total
80   latitudo, ergo per secundam suppositionem quelibet talis pars intensive toti latitudini coequatur se. Sequitur conclusio. Iterum remissio et intencio [E, fol. 38va] latitudinis penes gradum medium vel penes extremum intencius oportet attendi. Sed penes gradum medium attendi
85   non poterunt, ergo conclusio vera. Si enim penes gradum medium remissio et intensio attenditur, omnes partes latitudinis quarum cuiuslibet medius gradus foret idem cum medio gradu tocius latitudinis foret equales remissive, et ita due latitudines quarum una a gradu summo recedat per
90   latitudinem duplam et alia per subduplam forent equeremisse. Et ita penes recessum a gradu summo non attendetur remissio quod est contra quartam, nec intencio penes recessu a non gradu quod est contra secundam.

[P, fol. 60v] Tertia conclusio[24] est hec: omnis gradus
95   latitudinis caliditatis equaliter[25] a[26] summo et a non gradu ita[27] intensus sicut remissus et econtra[28] existit. Ponatur enim per adversarium quod *A* sit gradus latitudinis caliditatis per *B* latitudinem precise recedens a gradu summo et per *C* latitudinem equalem *B* recedens a non gradu, et quod *A* sit
100 magis intensus quam remissus. Quo posito arguitur sic: tota latitudo caliditatis et tota latitudo intencionis latitudinis caliditatis cum tota latitudine remissionis eiusdem sibi

---

[22] in remissione, *om.* E.
[23] P; *om.* E.
[24] P; *om.* E.
[25] *post* equaliter *add.* P distans.
[26] *post* a *add.* E gradu medio.
[27] eque P.
[28] et econtra *om.* P.

invicem sunt equales. Ergo demptis equalibus relicta erunt
equalia. Duabus igitur partibus *BC* correspondentibus a tota
105 latitudine intencionis ablatis, demptisque aliis duabus
partibus *BC* correspondentibus a tota latitudine remissionis,
relicte partes erunt equales. Omnibus vero partibus predictis
ablatis, *A* cum intencione latitudinis intencionis una cum
remissione latitudinis remissionis sibi correspondentibus
110 remanebit. Ergo intencio *A* et remissio *A* sibi invicem
coequantur et ita *A* non est magis intensus quam remissus
quod est oppositum positi.

Sed falsitas septime suppositionis et undecime apparenter
sic poterit ostendi: medius gradus latitudinis est gradus
115 uniformis, ergo in tota latitudine caliditatis gradus uniformis
reperitur. Sed istud sic dissoluitur: medius gradus latitudinis
caliditatis in tota latitudine caliditatis non reperitur et ita
nullus gradus uniformis in tota latitudine caliditatis
consitit.[29]

120 Secunde vero conclusioni sic obicitur: si tocius latitudinis
caliditatis et cetera, caliditas summa tota latitudine
caliditatis indivisibiliter foret intensior, sed caliditas summa
medio gradu latitudinis caliditatis divisibiliter est intensior,
ergo excessus summe[30] caliditatis supra gradum medium
125 excessum eiusdem supra latitudinem totam caliditatis in
infinitum excedet quod est falsum.

Iterum data conclusione datoque passo per totum
uniformitei difformiter calido mediante intensione totius
latitudinis caliditatis, totum istud passum foret ita intense
130 calidum sicut aliqua eius pars ad extremum passi intencius
calidum terminata, et ita in calido uniformiter difformi per
totum omnes partes ad extremum intencius terminate forent
eque calide, quod est falsum.

Item[31] data conclusione caliditas remissa et caliditas
135 summa intensive forent equales. Capto enim passo cuius
prima quarta sit summe calida alieque tres quarte mediante
intencioni totius latitudinis caliditatis sint calide, tota
caliditas passi erit ita intensa sicud aliqua eius pars ad eius
extremum intencius terminata. Sed tota ista caliditas est
140 remissa et aliqua pars ad extremum intensius terminata est
summa, ergo et cetera.

Pro solutione prime rationis ultima consequentia non
debet admitti. Si enim excessus caliditatis summe supra

[29] E; reperitur P.
[30] P; *om.* E.
[31] P; *om.* E.

medium excessum eiusdem supra totam latitudinem in[32]
145 infinitum excederet et non extensive ergo intensive, et ita
aliqua pars excessus caliditatis supra medium ad excessum
eiusdem supra totam latitudinem dupla intensive et alia
tripla et sic in infinitum, quod est falsum. Ideo ex assumptis
prime rationis, non ultimum consequens eiusdem, sed ista
150 conclusio sequitur manifeste: scilicet quod excessus
caliditatis summe supra medium excessum eiusdem supra
totam latitudinem non proportionaliter excedit, quod est
concedendum, sed corpora equaliter calida secundum se tota
consistunt quorum omnes due partes sibi invicem equales
155 caliditatum equaliter sunt active.

Pro secunda ratione sciendum quod uniformiter difformiter
calidi quelibet pars ad extremum intensius terminata est ita
intense calida sicud totum vel sicut alia pars ibidem
terminata. Nec sequitur ergo omnes partes ibidem terminate
160 sunt equaliter calide nisi corpora dicantur equaliter calida
quia caliditatem ita intensam sufficit unum agere sicut
sufficit aliud et econtra, quod non est ponendum, quoniam
tunc passum cuius una quarta foret summe calida et alie tres
remisse calide cum quarta summe calida [E, fol. 38vb] fore
165 equecalidum, quod est falsum. Que enim corpora sunt
equaliter calida in fine precedentis solutionis dicitur.

Pro tertia ratione neganda est prima conclusio adducta.
Nec dato passo cuius una quarta sit summe calida et alie
tres mediante intensioni totius latitudinis caliditatis sint
170 calide, tota caliditas passi caliditati quarte summe equalis est
intensive, sed remissior, quia ista tota est remissa et
caliditas extreme quarte est summa. Neque cuiuslibet
caliditatis quelibet pars ad extremum intencius terminata toti
illi caliditati equalis est intensive, sed solum in latitudine
175 caliditatis ubi tota caliditas est uniformiter difformis dicta
conclusio declaratur.

Circa velocitates in alteratione distinguendas antiquam
radicem erroris primitus extirpabo. Cum enim motus absque
magnitudine fieri non poterit, magnitudo per motum
180 alterationis adquirenda consistit. In alteratione vero tam
qualitas quam latitudo qualitatis poterit adquiri et qualitas a
latitudine secundum rationem distinguitur. Duo igitur
magnitudines quamvis non re, ratione tamen, differentes:
qualitas, scilicet, et latitudo qualitatis motibus alterationis
185 manent adquirende. Duo ergo motus alterationis quorum

---

[32] P; *om.* E.

primus extensio, secundusque intencio[33] consistit, secundum
rationem inter se distincte, illis duabus magnitudinibus
correspondent. Et ecce primam radicem cuiuslibet erroris in
hac materia penitus elklsam[?]. Inter hos siquidem motus
190 nulla differentia per prius concepta, quidam penes
latitudinem et qualitatem et tamen quidam penes latitudinem
tantum, quidam vero penes gradum, et quidam penes nihil
velocitatem in alteratione ponebant. De quorum numero
quondam quamvis exstiteram, predictas tamen opiniones
195 licite potero declinare. Ad veritatem centro sperici fixam per
punctum preter centrum pervenietur.

Primo igitur penes quid attenditur velocitas in motu
extensionis, secundo penes quid in motu intencionis facilius
declarare. Duas autem conclusiones velocitatem extencionis
200 terminantes primitus sunt ostendende.

[P, fol. 61r] Quarum hec est prima. Penes quantitatem
magnitudinis secundum quam fit extencio velocitas
extencionis attenditur. Ista sic probatur: per extensionem
sola qualitas extensa adquiritur, ergo extensio mediante qua
205 continue per horam qualitas per duplum extensa adquiritur,
ad extensionem mediante qua continue per horam qualitas
extensa subduplo, ad istam erit dupla. Et cum duplicitas et
subduplicitas extensionis qualitatis per solas magnitudines in
quibus extensio fit cognoscantur, penes quantitatem
210 magnitudinis secundum quam fit extensio velocitas
extensionis attende.

Secunda est hec: aliquod passum per horam alterabitur
infinita velocitate. Sit enim *A* pedale medium receptivum
luminis a *B* luminoso mediante *C* medio, quod in hoc
215 instanti secundum se totam illuminabitur ab *B*. Et quiescente
A medio, ymaginetur *C* medium corrumpi per accessum *B*
luminosi. Et sit hora mensurans accessum. Quo posito in
quolibet instanti accessus, ad punctum omnem *A* novum
lumen agetur a *B*. Istud ex capitulo proximo una cum casu
220 satis apparet. Ergo in quolibet instanti hore per totum *A*
medium lumen novum extenditur. Et cum subita extensio
luminis per pedalem quantitatem sit velocitas infinita, *A*
continue per horam alterabitur infinita velocitate. Quod
autem subita extensio luminis per pedalem sit infinita
225 velocitas arguitur sic: omni extensione partibili luminis
possibili subita extensio luminis per pedalem velocior est.
Extensionum vero luminis partibilium, aliqua poterit esse

[33] *post* intencio *add.* E que.

alicuius velocitatis et aliqua duple ad istam et aliqua triple
ad istam et sic deinceps. Ergo subita extensio luminis per
230 pedale velocitas infinita consistit.

Ad velocitatem quippe intencionis terminandam,
crementum latitudinis inter gradus proportionales per
crementum latitudinis inter quantitates proportionales per
prius manifestabo. Cuiuslibet igitur latitudinis quibuscumque
235 tribus gradibus proportione maioris inequalitatis continue
proportionalibus assumptis, latitudo inter gradum
intensissimum et medium ad latitudinem inter medium et
remississimum constat esse duplam. Ista conclusio tam in
quantitatibus continuis quam discretis sufficienter ostenditur.
240 Ad quantitatem enim inter [E, fol. 39ra] pedalem et
semipedalem quantitas inter bipedalem et pedalem dupla
reperitur. Ad numerum inter 4 et duo, numerus inter 8 et 4
dupla proportione se habet. Nec alicubi in quantitatibus
continuis et discretis contra[?] instancia poterunt assignari.
245 Cuiuslibet igitur latitudinis quibuscumque tribus gradibus et
cetera.

Tunc[34] due conclusiones intensionem et remissionem in
velocitate terminantes quarum hec est prima. Omnis
velocitas intensionis penes adquisitionem latitudinis aut
250 gradus, omnis motus remissionis penes deperditionem
consimilis latitudinis et gradus per respectum ad tempus
debet attendi. Incipiat[35] *A* ignis purus agere in *B* aquam
puram et usque ad ultimum instans hore continue agat sic
quod in hora usque ad gradum summum exclusive fiat
255 intensio caliditatis. Et simul incipiat[36] *C* ignis purus maioris
potentie quam *A* in duplo transmutare aliquam aquam puram
equalem *B* sic quod in medio instanti hore fiat intensio
caliditatis in ista aqua ad gradum summum exclusive.
Ymaginetur[?] per idem tempus precise mediante lumine per
260 totum uniformiter[37] per totam materiam[38] *D* aque a non[39]
gradu usque ad summum gradum[40] inclusive caliditatis
uniformiter intendi. Quo posito per motum intensionis *C*[41]
tanta latitudo qualitatis adquiriri mediante *C* quam[42]

---

[34] P; *om.* E.
[35] P; incipiant E.
[36] P; maneat E.
[37] ?; E uni^mi yma^em.
[38] *post* materiam *add.* E B.
[39] P; *om.* E.
[40] P; *om.* E.
[41] *sic* E, P.
[42] E; quantum P.

mediante *A*. Omnis autem velocitas per quam latitudo
265 adquiritur est motus intensionis. Ergo velocitas motus
intensionis C ad velocitatem motus intensionis *A* dupla
consistit. Motu etiam[43] intensionis quo in prima medietate
hore per materiam *D* uniformiter attenditur[44] caliditas ad
gradum summum inclusive tantum unus gradus inducitur in
270 *D* cui nullus equalis sibi inducitur in aquam alteram. Ad *C*
ergo velocitas istam velocitatem alteram tantum per unicum
gradum excedit.
Consimiliter in motu remissionis terminus[45] velocitatis
poterunt ostendi.
275 Secunda conclusio est hec: duorum motuum intensionis
inequalium in velocitate, velocitas unius velocitate alterius
non proportionatur. Sint enim *AB* due uniformiter calida per
totum quorum neutrum alio sit calidius et ab hoc instanti
ymaginetur *A* intendi per totum *C* gradu uniformi quousque
280 fuerit in duplo magis calidum quam iam existit. Sit *D* instans
terminans intensionis *A*. Per idem tempus precise
ymaginetur *B* intendi tota latitudine intensionis sub *C* gradu.
Quo posito, arguitur sic: cuilibet latitudine caliditatis
adquisite *C* gradu in aliqua parte dati temporis, equalis
285 latitudo caliditatis mediante tota latitudine intensionis sub *C*
adquiri in eadem parte et econtra. Ergo toti latitudini
caliditatis adquisite *C* gradu in toto tempore terminato ad *D*
instans, latitudo equalis mediante tota latitudine intencionis
sub *C* adquiritur in eodem tempore precise et econtra. [P,
290 fol. 61v] Cum igitur ultra latitudinem equalem latitudini
adquisite tota latitudine sub *C* tantum unus gradus uniformis
adquirat mediante *C*, nec aliquis gradus ultra datam
latitudinem mediante latitudinem sub *C* adquiritur, ergo per
conclusionem proximam motus intensionis cuius intenditur *B*
295 excedit in velocitate. Iste vero excessus solum per unum
gradum uniformem existit. Ergo inter velocitatem unius
motus et alterius nulla proportio vel minima maioris
inequalitatis reperitur. Minimam autem proportionem
inequalitatis dare non contingit. Ergo conclusio vera.
300 Istis autem intellectis pro motibus alterationis plenius
cognoscendum tres latitudines inter se distinctas ponam,[46]
quarum quamlibet in duas latitudines non re sed ratione
diversas distinguam. Prima quippe trium latitudinum latitudo

[43] P; *om.* E.
[44] P; attenditur E.
[45] E; *om.* P.
[46] E; ponantur P. *post* ponam *add.* E quem.

qualitatis existit que spacium et magnitudo in motu
305 alterationis verissime poterit appellari. Quolibet enim motu
magnitudo poterit adquiri vel deperdi sicut inductive satis
apparet. Nec magnitudo preter latitudinem qualitatis motus
alterationis adquisita erit vel deperdita. Ergo et cetera. Hec
autem latitudo in duas latitudines tantum ratione distinctas
310 divisibilis existit, quarum prima intensio qualitatis, secunda
remissio eiusdem convenienter nominatur. Illarum siquid
utraque finitur quarum prima uno gradu tantum secundam
excedit sicut duodecima[47] suppositione satis liquit.

Secunda trium latitudinum predictarum latitudo motus
315 alterationis existit. Talem siquid latitudinem penes motum
alterationis coget ponere vel nullum alterabile altero
alterabili velocius aut tardius alterari concedere. Hec autem
latitudo sicut latitudo qualitatis secundum rationem in duas
latitudines dividitur,[48] quarum prima latitudo motus
320 intensionis, secundaque latitudo motus remissionis existit,
illarum itaque[49] latitudinum utraque existente infinita
quelibet pars latitudinis motus remissionis ad quietem
terminata est infinita. Etiam[50] latitudo que est[51] pars
latitudinis motus intensionis ab alico gradu finito incipiens
325 omnem gradum finitum velociorem vel sibi equalem
continens infinita [E, fol. 39rb] latitudo consistit. Quelibet
autem finita latitudo remissionis est finita remissio, quamvis
quelibet latitudo intensionis infinita intensio consistat.[52]
Intensione namque non per extremum debilius sed per
330 extremum intensius fortius terminata, remissio per
extremum remissius debilius quod similiter est extremum
forcius, finietur.

Tertiam quippe latitudinem trium predictarum latitudinum
latitudinem latitudinis motus appellabo,[53] cuius essentiam sic
335 ostenditur. Latitudo motus intensionis a quiete incipiens et
ad A gradum uniformem terminata in hora potest adquiri, et
in medietate hore similiter, et ita deinceps. Ergo preter
istam latitudinem motus latitudinem aliam distinctam[54]
mediante qua latitudo motus velocius et tardius poterit talis

---

[47] *sic*; should be 13. Some of the later numbering has been changed in E.
[48] E; distinguitur P.
[49] E; siquidem P.
[50] P; *om.* E.
[51] que est P; quelibet E.
[52] E; existat P.
[53] P; appelle E.
[54] P; repeated in E.

340   latitudo acquiri[55] oportet[56] assignari. Hec autem latitudo in
duas latitudines ymaginarie distinguitur quarum prima
latitudo adquisitionis latitudinis motus, secunda latitudo
deperditionis latitudinis motus convenienter poterit appellari.
Istarum vero latitudinum utraque sicut latitudo motus tendit
345   in infinitum.

     Specialiter tamen contra istam latitudinem latitudinis
arguitur sic. Si enim respectu latitudinis motus secunda
latitudo penitus distincta poneretur consimili ratione
respectu secunde latitudinis latitudo tertia[57] foret ponenda et
350   sic deinceps. Sed infinite latitudines distincte a latitudine
motus superfluunt. Ergo latitudo latitudinis non est ponenda.

     Iterum tunc foret aliquis gradus velocitatis qui in nullo
motu poterit reperiri. Quilibet enim gradus secunde
latitudinis est gradus velocitatis, nec secunda latitudo poterit
355   esse motus quia[58] tunc non foret distinctus a latitudine
motus. Ergo aliquis est gradus velocitatis qui nullius motus
existit. Et ita aliquis foret gradus velocitatis quo nihil posset
moveri.

     Iterum tunc aliquo gradu velocitatis poterit mobile aliquod
360   incipere motum a quiete uniformiter difformiter intendendo
motum continue. Sit enim A gradus latitudinis mediante quo
latitudo motus intensionis a quiete incipiens et ad B[59]
gradum terminata, qui sit C, uniformiter adquiratur in hora.
Adquirat, scilicet, C latitudinem. Quo posito, totam C
365   latitudinem sibi[60] adquiret uniformiter in hora A gradu
uniformi. Ergo quamlibet partem, cum iam A gradu[61] C
latitudinis sibi adquiret C gradu. Cum igitur in hoc instanti
quod nunc pono[62] sibi incipiat adquirere aliquam partem C,
iam A gradu incipiat adquirere aliquam partem C, et quelibet
370   pars C est motus. Ergo A gradu velocitatis sibi incipit
motum. Cum ergo solum a quiete motum incipiatur
uniformiter difformiter motum intendendo, sequitur
conclusio.

[55] P; adequari E.
[56] P; quem E.
[57] P; ita E.
[58] E; cum P.
[59] E; *om*. P.
[60] ? This and subsequent readings of 'sibi' are based on what seems to be a single
    S perhaps with a dot in both manuscripts. Possibly the name "Socrates" is
    intended.
[61] cum iam A gradu *om*. P.
[62] quod nunc pono P.

Ad primam consequentia prima est neganda. Latitudine
375 namque latitudinis motus nihil poterit moveri, neque latitudo
latitudinis motus[63] aliqua res adquirenda per motum existit.
Igitur respectu latitudinis[64] motus latitudo[65] cum illarum
duarum latitudinum una per aliam et econtra adquiratur.

Pro secunda ratione conclusio prima debet concedi cuius
380 est hec [P, fol. 62r] ratio: adquisitionem motus distinguitur[66]
motu quemadmodum adquisitio magnitudinis distinguit a
magnitudine. Nec adquisitio motus distincta foret a motu
nisi velocitas adquisitionis motus a velocitate distinguaretur.
Ergo aliquis est gradus velocitatis qui nullius motus sed
385 tantum adquisitionis motus existit, nec illo gradu poterit
moveri aliquid. Poterit tamen illo gradu motus adquiri quo
gradu mobile movere poterit.

Pro tertia ratio conclusio adducta est concedenda. Gradu
enim latitudinis latitudinis motus poterit aliquid mobile
390 incipere motum a quiete uniformiter difformiter intendendo
continue, nec propter hoc gradus latitudinis que est motus
poterit fieri. Graviora vero sophismata de intencione et
remissione motus per istam positionem poterit facillime
dissolui, quibus latitudinis motus latitudinem ignotes velud
395 bruta funiculus captive laqueantur.

Notes

1 Edith Sylla, "The Oxford Calculators," in Norman Kretzmann, Anthony
 Kenny, and Jan Pinborg, eds., *The Cambridge History of Later Medieval
 Philosophy* (Cambridge: Cambridge University Press, 1982), 540–63.
2 The work is available in Erfurt Amplonian MS F135, fols. 25va–47rb. Parts
 are also found in Paris, Bibliothèque Nationale, MS lat. 16621, fols. 39r,
 40r, 40v–51v, 54v–62r, 66r–84v, and in Venice, San Marco, MS Lat. VI 62,
 fol. 111r. The Paris manuscript sometimes has a text that makes more
 sense, but because it appears in a student notebook that is sloppily written
 and presently very tightly bound, it is often difficult to read on microfilm.
 (For the writer of the notebook, cf. Zénon Kaluza, *Thomas de Cracovie,
 contribution à l'histoire du Collège de la Sorbonne* (Wroclaw, 1978), 60–64,
 84–94, where the student is identified as Etienne Gaudet.) Because of the
 difficulty of reading the Paris manuscript, I have transcribed the Erfurt text
 (cited as E) and used the Paris text (cited as P) for corrections when they
 seemed called for. The definitions concerning measurement of qualities and
 alteration found in the Venice manuscript are adopted by Giovanni di Casali
 with minor modifications. See Marshall Clagett, *The Science of Mechanics*

[63] nihil . . . motus *om.* P.
[64] *post* latitudinis *add.* P latitudinis.
[65] *post* latitudo *add.* P tertia non est
[66] P; word deleted E.

in the Middle Ages (Madison: University of Wisconsin Press, 1959), 382–3, 386–7.

3 Cf. James Weisheipl, "Roger Swyneshed, O. S. B., Logician, Natural Philosopher, and Theologian," in *Oxford Studies Presented to Daniel Callus*, Oxford Historical Society, new series, 16 (Oxford: Clarendon Press, 1964), 231–52.

4 Paul Spade, "Roger Swyneshed's *Obligationes*: Edition and Comments," *Archives d'histoire doctrinale et littéraire du moyen âge* 44 (1977):243–85; Spade, "Roger Swyneshed's *Insolubilia*: Edition and Comments," *Archives d'histoire doctrinale et littéraire du moyen âge* 46 (1979):177–220.

5 E, fol. 27ra–b, "Ultimo: numquid omnis motus actio vel passio motoris aut motorum [insert in margin: vel mobilis existat] dubitatur. . . . He answers in the affirmative. E, fol. 27rb, "Quia vero presens materia non realis sed pocius logicalis existit, nec in isto negocii more logicorum notionibus et non rebus operam adhibebo, forma vivente non autem dilatante materiam predictam intendo terminare."

6 Edith Sylla, "The Oxford Calculators and Mathematical Physics: John Dumbleton's *Summa logicae et philosophiae naturalis*, Books I and II," in Proceedings of International Workshop on *The Interrelations Between Physics, Cosmology, and Astronomy. Their Tension and Its Resolution 1300–1700*, Tel-Aviv and Jerusalem, April 29–May 2, 1984 (forthcoming).

7 See note 58 for Roger Swineshead's list of the parts of the work.

8 Cf. Edith Sylla, *The Oxford Calculators and the Mathematics of Motion, 1320–1350: Physics and Measurement by Latitudes*, unpublished Ph.D. dissertation, Harvard University, August 1970, 120.

9 H. Lamar Crosby, Jr., ed., *Thomas of Bradwardine. His Tractatus de Proportionibus. Its Significance for the Development of Mathematical Physics* (Madison: University of Wisconsin Press, 1955).

10 E, fol. 44rb; P, f. 77v. "Et quoniam predicte differentie proportionum in quoddam tractari [P: tractatu] de proportionibus intitulato sufficienter ab arsmetricis extracte sunt et unite, non propter falsitatem capituli tertii prefati tractatus, veritas capituli primi reliquanda consistit, causa maioris expeditionis capitulam [P, gratia brevitatis] primum illius tractatus supponam in presente."

11 See Edith Sylla, "Medieval Quantifications of Qualities: The 'Merton School,'" *Archive for History of Exact Sciences* 8 (1971):9–39, and Sylla, "Medieval Concepts of the Latitude of Forms: The Oxford Calculators," *Archives d'histoire doctrinale et littéraire du moyen âge* 40 (1973):223–83, at 233–38.

12 Curtis Wilson, *William Heytesbury. Medieval Logic and the Rise of Mathematical Physics* (Madison: University of Wisconsin Press, 1960), chap. 4.

13 See Edith Sylla, "The Oxford Calculators and the Mathematics of Motion" [note 8], and James A. Weisheipl, "The Place of John Dumbleton in the Merton School," *Isis* 50 (1959):439–54.

14 E, fol. 38vb, "Aliquod passum per horam alterabitur infinita velocitate."

15 See note 58.

16 E, fol. 26rb. "Secundum differencias rerum generalissimas tantum tria sunt predicamenta . . . substantia, qualitas et quantitas." 26va, "secundum differencias modorum rerum positivorum generalissimas tantum septem sunt predicamenta . . . relatio, actio, passio, ubi, quando, habitus, tempus, et positio."

17 See note 5.

18 E, fol. 28ra, "Temporalitas siquidem est entis mutabilis temporalis entitas."

19 These *secundum imaginationem* arguments are often connected with the Condemnation of 1277 and the distinction between God's absolute and ordained powers. See John Murdoch, "From Social into Intellectual Factors: An Aspect of the Unitary Character of Late Medieval Learning," in John Murdoch and Edith Sylla, eds., *The Cultural Context of Medieval Learning* (Dordrecht, Holland: D. Reidel, 1975), 280–81, 291–92, 297. See also in the same volume, Heiko Oberman, "Reformation and Revolution: Copernicus's Discovery in an Era of Change," 409 and related notes.

20 E, fol. 38vb, "Duo igitur magnitudines quamvis non re, ratione tamen, differentes, qualitas, scilicet, et latitudo qualitatis." For the context of this remark, see Appendix, lines 166ff.

21 E, fol. 39ra; P, fol. 61v, transcribed in Appendix, lines 300–03.

22 E, fol. 39rb, in Appendix, lines 303–45.

23 E, fol. 39rb, in Appendix, lines 346–51.

24 Ibid., lines 374–78.

25 E, fol. 36vb, "Tantum quatuor qualitates tangibiles et prime proprie intensibiles et proprie remissibiles existint."

26 E, fol. 36rb, "Proprie siquid qualitas intenditur quando sola actione naturali mediante qualitate simili in specie vel effectu absque aliquo adveniente fortificatur."

27 E, fol. 36vb, "Secunda est hec: gravitas et levitas sapor et odor communiter intendi poterunt et remitti."

28 E, fol. 36rb, "Communiter vero non tantum res sed modi rerum intendi poterunt et remitti, quando namque res aut modus rei secundum debilius et forcius aut peius et melius vere aut imaginarie mutatur intenditur communiter vel remittitur."

29 E, fol. 37ra, "qualitas siquidem vera intencione communiter dicta intenditur quando per qualitatem in effectum similem, in specie tamen sibi dissimilem, absque alico adveniente fortificatur."

30 E, fol. 37ra, "modus vero rei ymaginaria intencione communiter dicta dicitur intendi quando res cuius est, per illum modum forcius immutatur."

31 E, fol. 37ra, 37va, "sic enim motus, scientia, virtus moralis, et vicium poterit intendi. Solus autem modus rei ymaginaria intensione communiter dicta intensibilis existit." "quoniam intencio ymaginaria in solis modis rerum reperitur."

32 E, fol. 37rb, "Quinta conclusio est hec: neque [deletion] neque sonus neque forma substantialis intendi poterit neque remitti."

33 Marshall Clagett, ed., *Nicole Oresme and the Medieval Geometry of Qualities and Motions* (Madison: University of Wisconsin Press, 1968), 273.

34 Ibid., 273, 275.

35 Ibid, 463, "quoniam ita est de motu, motus enim comprehenditur sensu et nulla pars eius est in actu, sed quaelibet pars demonstrata recessit: ergo est compositus ex hoc, quod iam deficit et ex hoc quod nondum est. Sed talia non habent esse completum, sed esse eorum componitur ex actione animae in eo quod est in eis extra animam . . . ut post declarabitur de tempore, secundum quod est de numero entium, quorum actus completur per animam." It is worth noting that in a rare comment on the ontology of intension and remission, Oresme says that in the intension and remission of qualities there is continually a different quality, while in the whole time of alteration there is one successive quality, whose nature I take it is like that

of motion. Cf. ibid., 36–37, 299, 301. His theory, therefore, is more like that of Burley than that of the core Oxford Calculators. Cf. Sylla, "Medieval Concepts of the Latitude of Forms" [note 11].

36 William Heytesbury, *Regule solvendi sophismata* (Venice: Bonetus Locatellus, 1494), fol. 49v, "motus enim et alie huiusmodi qualitates que communiter concedunter intendi vel remitti non consimiliter augentur vel diminuuntur sicut magnitudines vere continue et per se quante."

37 A. G. Molland, "Mathematics in the Thought of Albertus Magnus," in James Weisheipl, ed., *Albertus Magnus and the Sciences. Commemorative Essays 1980* (Toronto: Pontifical Institute of Mediaeval Studies, 1980), 467: "The realist places his focus on the actual existence of mathematical objects in the outside world, and expects mathematics to tell him quite a lot about the world; he often hints at mathematical design in nature. The conceptualist on the other hand pays particular attention to the fact that the mathematician operates on objects pictured in the imagination, and he often seems to lose sight of their anchorage in external bodies. In this matter Albert veers very much towards the conceptualist pole."

38 He does not fit Molland's characterization [see note 37], however, in that he does not "seem to lose sight of their anchorage in external bodies."

39 J. M. Thijssen, "Buridan on Mathematics," *Vivarium* 23 (1985):55–78.

40 Ibid., 75–76.

41 Heytesbury, *Regule* [note 36], fol. 48rb, "Ad tertium et ultimum dicitur quod conclusio ibidem proposita est omnino impossibilis. Et ad casum ibidem suppositum, dicitur quod est impossibilis, quoniam non est possibile quod si [*sic*] incipiat aliqua magnitudo crescere vel maiorari a [non] quanto quin in infinitum velociter inciperet illa crescere, sicut probat argumentum sufficienter saltem iuxta responsionem positam. Et ideo tenendo illam responsionem et illum modum loquendi negatur casus tanquam impossibilis."

42 Ibid., fol. 48va, "Sed ad probandum casum arguitur et evidenter sic: posito gratia argumenti quod eadem magnitudo numero poterit diminui per partem ante partem usque ad non quantum. Quamvis enim hoc non sit possibile de virtute sermonis tamen ex quo casus non claudit contradictionem satis poterit admitti gratia disputationis."

43 Ibid.,
   et dicit quod hoc est tamen impossibile quia materia prima que est in illa magnitudine est incorruptibilis et [in?] anihilabilis physice loquendo. Et cum quantitas sit eterna materie est impossibile quod illa magnitudo condensetur usque ad non quantum. Sed dicit ille magister bene scis hoc, sed quia non implicat contradictionem et est satis imaginabile, ideo calculatores non debent fugere casum, quia est fuga baranorum.

44 [E, fol. 41ra; P, fol. 69r] Item casu ymaginabili posito et admissio, positio predicta est inymaginabilis. Posito enim iuxta imaginationem quod superficies crecat in corpus, etiam linea in superficiem, etiam punctus in lineam, et quodlibet illorum a non gradu velocitatis ymaginatur incipere crementum per horam intendendo per motum. Iuxta positionem predictam secuntur contradictoria. Secuntur enim quod post quodlibet instans crementi aliud ab hoc immediate, illud quod iam est linea vel punctus et superficies velocius crescet quam in hoc instanti incipit crescere, et tamen illud idem in hoc instanti incipit crescere infinita velocitate. Ex quibus immediate duo contradictoria consequuntur. Quod autem prima duo secuntur satis patet positionem intuenti. Ideo positio predicta negat casum tamquam inymaginabilem. Cum tamen casus predictus sit et fuerat positio [multorum *add.* P]

magnorum, nec est verisimile philosophante quemcumque positionem aliam ponere quam impositionem non poterit ymaginare casu vero ymaginato, pars predicta locum non habet.

Although Swineshead's exact argument here is hard to construe – is he saying that *after* a case has been imagined, then it is not plausible to adopt a position which would make this case lead to contradictory consequences? – this may be a passage worth comparing with Roger Swineshead's ascribed position on obligations to see whether the two views are consistent. See Kretzmann, Kenny, and Pinborg, eds., *The Cambridge History of Later Medieval Philosophy* [note 1], 335–41.

45 Heytesbury, *Regule.*, fol. 46rb. Should a distinction be made between impossible and unimaginable? In these contexts the two seem to be more or less conflated.

46 E, fol. 47ra; P, fol. 84r, "Et quia casus ibidem suppositus solum est ymaginarius, ideo ipsum tamquam ymaginabilem admitto."

47 These definitions are edited in the Appendix, lines 6–42.

48 E, fol. 38rb, Supposition 7, in the Appendix, lines 24–26.

49 Appendix, Supposition 1, lines 13–14.

50 Ibid., Supposition 5, lines 21–22.

51 Ibid., Supposition 8, lines 26–30.

52 Ibid., Supposition 11, lines 38–39.

53 E, fol. 38 va, P, fol., 60v, in the Appendix, lines 94–96.

54 Interestingly, Swineshead prefaces his conclusions by saying that previous errors about the measure of alteration have arisen because of a failure to distinguish between intension and extension, some proposing to measure alteration by the latitude alone, some by both the latitude and the quality, some by the degree gained, and some by nothing, and that although he formerly was among those holding these other opinions, he is now justified in rejecting them. See Appendix, lines 185–96.

55 E, fol. 38vb, P, fol. 61r. See Appendix, lines 212–213.

56 E, fol. 39ra, P, fol. 61r. See Appendix, lines 275–76.

57 Cf. John Murdoch, "Philosophy and the Enterprise of Science in the Later Middle Ages," in Yehuda Elkana, ed., *The Interaction Between Science and Philosophy* (Atlantic Highlands, N.J.: Humanities, 1974), 64–70.

58 The preface to the *On Natural Motions* is found only in the Erfurt manuscript and contains a number of words that I have been unable to read, although I think I have been able to read enough to get the general sense of the passage. I would like to thank Sten Ebbeson for suggesting several possible readings of dubious passages. The errors that remain are, of course, my own (and those of the manuscript's scribe!):

[E, fol. 25va] Incipit tractatus Magistri Wilhelmi Swineshep datus oxonie ad utilitatem studencium.

Motore primo primitus invocato, iuxta capacitatis mee, modulum et decursum stilo brevi quamvis rudi tractare de motibus temptabo. Cum effectu plerisque siquidem pluries ad hoc excitantibus zeloque veritatis praeceteris inclinante presentis negocii suppremis affectibus inducor ad agressum.

Verumtamen int⟨er⟩ius mihi talia cogitanti acerime minabitur antiqua consuetudo, cuius solius subiecta famositati iuxta sectas credencium autentica reputantur. Sed lectori quid oberit minima dicentis auctoritas si dicta veritati videat consonare? Nec incola cautissimus habenda [?] sic colligit quin nequid effugere spica cum granis, nec quavis [?] anicula per rura rimabitur quin grana reperiet occulta latibulis.

De motibus igitur scripturus ad presens intentum opus inspectures petam in primis ne quicquam in eo quia non palliatum aut insuetum condempnare presumant. Cum curiose verborum compagines [in? *add.* MS, *sed non bene*] culteque quorumdam famositates velud quid ydola quedam falcitis [?falsitatis] tentamina pretende⟨re⟩ soleant [*correxi ex* soliant] solam effigiem veritatis. Sed [?Si] quid dicendorum ratione superdemu [!? illegible] quod allatum [!?] inveniatur non propter hoc cetera confutentur. Sterilitatem enim agri vacuitas unius stipule non ostendit, quemadmodum sterilitatem eiusdem alterius copia non sufficit approbare [?]. Nec menniles [!?] aut sub riso senile pervertantur cum labia contrahent obliquius aspi⟨ci⟩entes, qui per incessum rectum in [E f. 25vb]spiciendo stupefacti conticessent.

Tocius autem operis substantiam octo distinctionibus sive differentiis assignabo. Quarum prima prohemii vicem habet. Secunda motum et tempus diffinet. Tertia quedam insueta quamvis non nova de generacione declarat. Quarta de alteratione determinat. Quinta circa augmentum et diminutionem versatur. Sexta de motu locali perscrutatur. Septima proportionales possibiles in motibus manifestat. Octava maximum a minimo dividit et limitat. In iniciis autem subsequencium differenciarum determinanda plenius numerando processus planiores expressius recitabo.

59 Clagett, *Nicole Oresme* [note 33], 159, my italics. Clagett points out the tentative tone of many of Oresme's conclusions (Ibid., 35–36), but it should be noted that part of Oresme's purpose is merely to show that there is likely *some* physical explanation of many phenomena that might otherwise be considered miraculous or magical. For this end he need not demonstrate the particular causes of specific phenomena. This does not, however, indicate that he was not confident of the general value of his system of configurations. See Bert Hansen, *Nicole Oresme and the Marvels of Nature. The De causis mirabilium* (Toronto: Pontifical Institute of Mediaeval Studies, 1985), 49–85, 96–101. I do not see in Oresme the desire to humble philosophy in the face of theology as described by Edward Grant, "Scientific Thought in Fourteenth-Century Paris: Jean Buridan and Nicole Oresme," in Madeleine Pelner Cosman and Bruce Chandler, eds., *Machaut's World: Science and Art in the Fourteenth Century* (Annals of the New York Academy of Sciences, vol. 314, New York, 1978), 111–116.

60 Clagett, *Nicole Oresme*, 165, 167, 169.

61 Ibid., 169–73.

62 Ibid., 175, cf. 207.

63 Ibid., 535.

64 Cf. Edith Sylla, "Medieval Quantifications of Qualities: The 'Merton School'" [note 11]: 34–38.

65 Clagett, *Nicole Oresme* [note 33], 173.

4

Thomas Bradwardine: mathematics and continuity in the fourteenth century

JOHN E. MURDOCH

Our knowledge of the history of mathematics and the utilization of mathematics in natural philosophy in the Latin Middle Ages has grown immensely over the past forty years, and no one would deny that a major share of the credit for that growth must be given to Marshall Clagett. The subject of the present chapter is a fourteenth-century work that provides an instance of the application of mathematics within late medieval natural philosophy, and as such it is quite fitting that it should appear here. However, this fit with Marshall Clagett's scholarly concerns is not the only reason why I thought it appropriate on this occasion to write on Thomas Bradwardine's *Tractatus de continuo*. It is also because more than thirty years ago Marshall Clagett placed in my hands microfilms of the two basic manuscripts of this *Tractatus*, counseling me that, if I was seriously interested in the medieval treatment of infinity and continuity, then I should investigate the contents of these two films. One might say, then, that the connection of the following with Marshall Clagett is not only intellectual or spiritual, but material.

Manuscripts, authorship, and date of the *Tractatus de continuo*

There are three extant manuscripts of the *De continuo:*[1]

1. Toruń, Poland: Gymnasial-Bibliothek (Książnica Miejska im Mikołaja Kopernika): MS R 4° 2, pp. 153–192.
2. Erfurt, German Democratic Republic: Wissenschaftliche Bibliothek der Stadt: MS Amploniana 4° 385, fols. 17r–48r *and* 55v.
3. Paris: Bibliothèque Nationale: MS nouv. acq. lat. 625, fol. 71v.

The only complete copy of the *De continuo* is preserved in the Toruń manuscript. Folios 17r–48r of the Erfurt codex omit more than half of the discussion of the two definitions (23 and 24) dealing with infinity and terminate in the middle of Conclusion 134, of which there are 151 in all.[2] The fragment in the Paris manuscript, which is also found on fol. 55v of the Erfurt codex, presents what appears to be a second redaction[3] of only the 24 opening definitions and a few sentences beyond. All three manuscripts are written in continental hands of the latter half of the fourteenth century.

In the Erfurt codex, our work (including the fragment) appears anonymously. The name "Bradwardin" is written in the margin at the beginning of the Paris fragment, while the Toruń copy carries the following explicit "Sic igitur primus liber, qui est de compositione continui quantum ad sua essentialia, finem capit. Explicit tractatus Bratwardini de continuo." The words *primus liber* are something of a puzzle, since the *De continuo* clearly seems to be complete as it stands in the Torun manuscript.[4] In any event, the two manuscript ascriptions to Bradwardine provide quite reasonable evidence for his authorship. This is somewhat strengthened by the fact that Bradwardine's unquestionably genuine *Tractatus de proportionibus velocitatum in motibus* is cited no less than seven times in the *De continuo,* it being the only medieval Latin work mentioned.[5]

However, it is somewhat surprising that Bradwardine finds no reference anywhere in the rather substantial fourteenth-century literature on infinity and continuity or that the *De continuo* is not cited, even without mention of Bradwardine as author, in this material. Indeed, I have been able to find only one very minor and anonymous quotation from the *De continuo* in an anonymous question on the infinite.[6]

It seems clear, nevertheless, that the *De continuo* is one of Bradwardine's works. Because his *Tractatus de proportionibus* was written in 1328, its citation by the *De continuo* provides us with a *terminus post quem* for the latter work. Further, considering Bradwardine's career in general, and especially his move to theological preoccupations later in his life, it would seem reasonable to date the *De continuo* sometime between 1328 and ca. 1335.[7]

As we shall see, the *De continuo* is axiomatic in form, but one other feature needs mention at this point: It is the frequent occurrence (some seventeen times) of the third person singular, occurrences that cannot be reasonably construed as impersonal or as third-person references of an author to himself. Furthermore, with but one exception these occurrences are found directly after the annunciation of a Conclusion, serving to remind us of what the author has just done and at the same time to explain what he is about to do next.[8] This suggests, not that the *De continuo* as we have it is some kind of *reportatio,* but rather that one of its copiers saw fit to introduce "guidepost" sentences into the text in order to draw attention to various areas and divisions of the author's concern that would otherwise be somewhat obscured by the treatise's axiomatic format.[9]

Finally, one should realize that Bradwardine's *De continuo* is but one work among many belonging to an extensive controversy in the earlier years of the fourteenth century over the problem of the composition of continua, the heart of the controversy consisting in the rise of a number of indivisibilists opposing the usual Aristotelian, *semper divisibilia* position and the subsequent extensive criticism of these indivisibilists.[10] The apparent failure of medieval writers to cite the *De continuo* may make it seem as if Bradwardine was not a major protagonist in this debate. Again, the axiomatic presentation of his point of view somewhat conceals the part he did play in the debate, since the usual "give and take" flavor that one derives from a more ordinary scholastic form is missing. Yet Bradwardine does specifically mention two Oxonian indivisibilists as targets of his criticism: Henry of Harclay and Walter Chatton (whom he refers to simply as Henricus modernus and Waltherus modernus).[11] At least in England, then, Bradwardine should be seen as very much part of the continuum controversy and should, accordingly, be given due attention.[12]

The structure of the *De continuo*

Although any outline fits uncomfortably into the axiomatic format of the *De continuo,* it may be helpful to attempt to provide one if for no other reason than to point out more clearly the various areas covered by the work, while at the same time rendering more evident Bradwardine's particular geometric approach to the problems of continuity. The relevant numbers of the Definitions, Suppositions, and Conclusions are indicated in parentheses, to provide easy reference to their texts, which are given in the Appendix below.

I. Definitions
    A. Kinds of continua (Def. 1–6)
    B. Indivisibles (Def. 7–8)
    C. Time and motion (Def. 9–14)
    D. Superposition and imposition (Def. 15–16)
    E. Temporal limits (Def. 17–22)
    F. Two types of infinity (Def. 23–24)
        1. Digression: Discussion of the two infinites of Def. 23–24
II. Suppositions (Supp. 1–10)
III. Preliminary positive conclusions
    A. Geometrical preliminaries
        1. The nature and relations of indivisibles (Concl. 1–6)
        2. The permissible relations between separation, super-

VIII. Refutatory conclusions V: On the existence of indivisibles and
their relation to continua (Concl. 142–151)

It should be noted that many of the divisions of this outline (especially
of the Conclusions) derive in one way or another from remarks made
in the text itself of the *De continuo*. In most instances, the relevant
textual passages occur in the third person singular remarks referred to
above and therefore may not be Bradwardine's own doing. Thus, although the first thirty-four Conclusions are not specifically called preliminary or positive, the transition to the business of refutation is duly
noted directly after the annunciation of Conclusion 35: "Istis omnibus
prelibatis, incipit reprobare falsas opiniones de continuo." Similarly,
the division of preliminary Conclusions into geometrical (III, A: Concl.
1–20)[13] versus physical (III, B: Concl. 21–34) is also reflected in the
text: "Post preambula geometricalia ponit alia preambula naturalia."
And the division of IV, A: Concl. 35–42 from IV, B: Concl. 43–54 is
noted in a similar fashion.[14]

The division according to twelve different disciplines of the Conclusions directed against the finitist indivisibilist view of continua (V, A–
L: Concl. 57–114) receives even more support from the text. It is, to
begin with, marginally annotated by the appearance of the name of
each discipline. But attention is also explicitly drawn to it through a
tolerably long introductory passage that points out the distinction and
order of the disciplines or "sciences" intended.[15] Finally, a concluding
passage reminds the reader that these twelve *scientie* – termed *legittime* by Bradwardine – have led to the complete overturning of the
finitist position.[16]

It may appear that some of the divisions set forth in the above outline
do not seem properly to characterize the Conclusions involved. Thus,
the annunciations of Conclusions 18–20 (III, A, 1) do not appear to
speak of a concern with the infinite divisibility of a finite straight line.
However, since these Conclusions are all directed toward the Corollary
that follows Conclusion 20 – namely, "that every straight line can be
divided into many straight lines" – it is evident that the division of
lines is indeed in question. Further analysis of the text also reveals that
Bradwardine had infinite divisibility in mind when he claimed in the
Corollary that one could divide any finite straight line *in multas rectas*.[17] Just which kind of infinite divisibility he had in mind is not quite
clear. For although the longest digression (I, F, 1) in the *De continuo*
concerns the distinction of the actual/categorematic/absolute infinite
from the potential/syncategorematic/relative infinite, Bradwardine
does not specify in any later passage which of these two infinites he
intends. However, relative to the case in question of the divisibility of
a finite straight line, it seems all but certain that, Aristotelian that he

is in such matters, Bradwardine would opt for a potential infinite divisibility.

The axiomatic format of the *De continuo*

The extreme importance of Euclid's *Elements* in the Middle Ages would naturally suggest that any geometrical work should be composed in the axiomatic, deductive fashion of this central Greek mathematical work. Such is exactly what we find Bradwardine doing in his elementary compendium entitled *Geometria speculativa*.[18] The selection of an axiomatic form for his other works is not, however, such an expected thing. Yet that is precisely what we find. Not only are the *Tractatus de continuo,* his *Tractatus de proportionibus velocitatum in motibus,*[19] and his *De incipit et desinit*[20] written in this form, but so also is his major theological work, the *De causa Dei contra Pelagium*.[21]

The axiomatic structure of the *De continuo* requires – or at least strongly suggests – that all of its Conclusions should follow directly, that is, deductively, from the Definitions and Suppositions set down at the outset. It seems clear that this is what Bradwardine intended. He makes no remark about the Definitions,[22] yet he does claim that his ten Suppositions would be evident to everyone (*10 Suppositiones pono que per se patent omnibus*),[23] which is to say he believed they would also be admitted by those who disagreed with him relative to the problem of the composition of *continua*. One should notice, however, that one of his Suppositions effectively extends the "axiomatic base" for Bradwardine. It claims that (Supp. 4) "All sciences are true in which a continuum is not assumed to be composed of indivisibles." A third-person gloss then explains the function of this Supposition: In certain instances, in addition to his own first principles (i.e., Definitions and Suppositions), Bradwardine will utilize the conclusions of other sciences as additional "principles" in his refutation of indivisibilism, since it would be far too time-consuming to prove each of the appropriated conclusions "from scratch," as it were.[24] The Supposition is therefore a kind of labor-saving device. For example, Bradwardine can, and will, use various theorems of Euclid without having to prove such theorems himself. The gloss on the Supposition adds, however, that when it is a question of such a theorem's involving the indivisibilist composition of continua, then the theorem will not be taken over as proved or true. This caution is taken in order to avoid, we are told, a *petitio principii*. But more of such matters below.[25]

The primary advantage of the axiomatic format in which Bradwardine cast the *De continuo* was presumably that by so doing, he would refute the position of his opponents and at the same time establish his own point of view with the rigor and definitiveness characteristic of mathematical science. The Conclusions of the *De continuo* would thus be as undeniable as the theorems of Euclid's *Elements,* provided, of course, that the logic was sound and Bradwardine had not explicitly or tacitly introduced some dubious or false assumption. However, even assuming that Bradwardine had introduced nothing dubious or false and that his logic was impeccable, there is still a decided disadvantage arising from his use of an axiomatic presentation. It is simply that such a format makes it quite difficult to see exactly how Bradwardine fit into the controversy over the continuum that was veritably raging in the earlier years of the fourteenth century. Almost all other treatments of continuity were then in the more usual scholastic *quaestio* form. This allowed, indeed required, that one set forth such and such an argument – for example, of Aristotle for the infinite divisibility of continua – and then present one's indivisibilist reply to that argument. For an indivisibilist, things might have ended there, but if one supported Aristotle's position, there then followed a response to the indivisibilist reply, and so on.[26]

Nothing of this sort occurs in the *De continuo.* In fact, Bradwardine's work, taken alone, would render it exceedingly difficult, if not impossible, to discover what almost any of the arguments were (on either side) in the controversy over continua. For example, a central argument of Aristotle against indivisibilism(*Physics* VI. 1) was that points or indivisibles, because they have no parts, cannot constitute a continuum either by having common limits or by being in contact. Bradwardine cites neither this crucial Aristotelian argument as such nor even the arguments of his named adversaries (Henry of Harclay and Walter Chatton) in reply to it.[27] It is true, however, that Bradwardine's invocation of the superposition of indivisibles and its systematic dissociation from imposition (= continuity) (I, D: Def. 15–16; III, A, 2: Concl. 7–15) relates directly to Aristotle's "contact argument" and the replies of such indivisibilists as Harclay and Chatton to that argument; but only after one is quite familiar with the medieval history of Aristotle's argument and the various responses to it can one begin to see just how Bradwardine fits into the whole picture. It is in part the axiomatic form of the *De continuo* that makes it exceedingly difficult to realize his role in this regard.

It is equally problematic to see how Bradwardine fits into the debate over how, on the basis of an indivisibilist hypothesis, one can, or cannot, account for the existence of different speeds among mobiles.[28] There are elements in the *De continuo* that can be related to the arguments within this debate, but Bradwardine nowhere makes it clear either that or how these elements are relevant.

On the other hand, it is easier to see from within the axiomatic structure of the *De continuo* how Bradwardine fit with some of the specific *mathematical* arguments against indivisibilism and their presumed resolutions. It was standard procedure, for example, to argue against indivisibilism by revealing the contradictions between such a hypothesis and accepted geometrical truths, where this revelation was based on the absurdities that arose when, under an indivisibilist hypothesis, one applied various techniques of radial and parallel projection to the simplest of geometrical figures.[29]

Thus, if one drew all of the radii from the center of two concentric circles to the circumference of the outer circle, since these radii would intersect both circles in the same number of constituent indivisibles, the two circles would be equal. Such an absurdity was held effectively to refute the indivisibilist position. Now Bradwardine has, in effect, the bare bones of such an argument in his Conclusion 74. Further, one of the indivisibilist replies – of Chatton, for example – to that argument was that the radii proceeding from the outer concentric circle meet *before* they reach the smaller inner circle, thus intersecting the latter in fewer points and hence accounting for its smaller size. This response is effectively cut off by Bradwardine's Corollary to his Conclusion 15 (which claims that the radii of a circle do not meet before reaching the circle's center). Similarly, Bradwardine presents (Concl. 87) the parallel projection argument against indivisibilism that derives the equality of a diagonal of a square with its side through constructing all of the parallels between all of the indivisibles in opposite sides of the square.[30] But he also destroys the indivisibilist reply to this argument based on the contention that lines intersecting obliquely "take more" of each other than lines intersecting at right angles. This he accomplishes by his Conclusion 14: "Any straight line intersecting another straight line cuts it in some point and not in more than one."

In sum, it seems that when the arguments within the current controversy over continua were geometrical arguments, the axiomatic form of the *De continuo* was not as detrimental to seeing something of the details of this controversy as were other kinds of arguments belonging to the debate. Yet this is not at all surprising given the geometrical basis of most of what Bradwardine was doing in the *De continuo*. It is to that basis that we must turn next.

The central position of mathematics in the *De continuo*

In one of the more literary moments in the *De continuo,* Bradwardine makes it quite clear that one must not neglect mathematics in doing natural philosophy.[31] Because the issue of the composition of continua is a central problem in natural philosophy, it therefore follows that mathematics, and geometry in particular, should provide the basis for unraveling the truth relative to that problem. It is natural, then, that Bradwardine should have set down a geometrical "preamble" and have been aware of so doing.[32] This does not mean that Bradwardine did not view various fundamental doctrines within natural philosophy as similarly basic to his enterprise (Cf. III, B: Concl. 21–34), nor should we ignore the fact that he is comfortable in appealing to natural philosophy in the course of proving a thoroughly geometric Conclusion.[33] This "mix" fits well with his contention that the proper view of the composition of continua is important for both mathematics and natural philosophy.[34]

If one turns, then, to the preliminary positive Conclusions of the *De continuo,* it comes as no surprise that the opening Conclusions dealing with the nature and relations of indivisibles (III, A, 1: Concl. 1–6) are treated from a geometrical point of view. That is to say, although these introductory Conclusions begin by speaking of indivisibles and their relation to continua in general (Concl. 1–3), all is soon translated into terms of points within, and related to, straight lines (Concl. 4–6, with Corollaries). What Bradwardine has here pointed out is that the most important relations that can obtain between indivisibles – namely, the likes of "being in the same indivisible place," "being the same distance from," and "being immediate to" – can and should be interpreted as essentially geometrical relations.[35]

The very next battery of Conclusions (III, A, 2: Concl. 7–15) make the beginning "geometrization" more explicit by connecting the last-named relation of "immediacy" to that of the well-known and respectable geometrical notion of superposition. Let us see how and why. Begin by recalling Aristotle's argument that indivisibles cannot constitute continua because they can neither have common limits nor be in contact since either possibility would require that they possess parts which, being indivisible, they obviously do not.[36] It would appear that in the eyes of many of the scholastics who were protagonists in the fourteenth-century continuum debate, what Aristotle had done in this opening chapter of Book VI of the *Physics* was to show that the answer to the question of how *any* parts might constitute a continuum lay in the manner in which these parts were *connected to* one another. Indivisibles fail the test because they cannot be connected to one another. Let us grant that two indivisibles cannot have common limits. But what if indivisibles could, in some way, at least be in contact with one another, even though they have no parts? This is exactly what an indivisibilist such as Henry of Harclay did maintain: Indivisibles could indeed touch or be in contact in a special way.[37] What Bradwardine did was to interpret this contact between indivisibles in a strictly geometrical way: namely, as equivalent to the perfectly respectable geometrical notion of superposition (perfectly respectable, that is, because it is exactly that notion that Euclid utilizes in proving his congruence theorems).[38] Thus, to say that two indivisibles are in contact or are

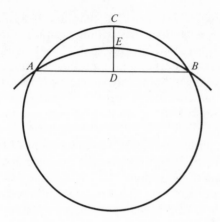

Figure 1

immediately joined to one another is the same as saying that they are superposed to one another.[39] However, having thus bestowed upon the indivisibilists a new level of geometrical respectability for the connection of his presumably continuum-composing indivisibles, Bradwardine shows in another sequence of Conclusions that this is indeed mere presumption. For continuity is equivalent to (the geometric notion of) imposition (Def. 16), and these Conclusions (7–13) are held effectively to demonstrate that there can be no association of superposition with imposition, which is to say (translating back from the geometrical level), no association of contact or immediacy of indivisibles with continuity. Therefore, indivisibles in contact cannot compose continua; Aristotle's conclusion has been established in a new way.[40]

The invocation of superposition may well be the most important entry of a geometrical notion into the *De continuo*,[41] but it is only a fragment of Bradwardine's geometrization of his subject, some of which occurs in a most unexpected fashion. The demonstration of the infinite divisibility of a finite straight line is a case in point. One would expect that the establishment of such a divisibility might be held to follow from the fact that any finite straight line can be bisected (Euclid *Elements* I.10). Bradwardine does not, however, satisfy this expectation.[42] On the contrary, what he does is quite *un*expected. He takes (Concl. 20) the finite straight line *CD* whose divisibility is in question as the perpendicular bisector of the chord *AB* of a segment of a circle less than a semicircle (Figure 1). He then proves that the finite straight line *CD* can be divided (at *E*) by constructing an arc *AEB* of a larger circle over the same chord. Further, it is a Corollary that follows from this Conclusion with no further ado that "every straight line can be divided into many straight lines," which, we have seen, is to be taken as a division into infinitely many straight lines.[43]

Why Bradwardine adopted this rather unusual way of establishing the infinite divisibility of a straight line, he does not say. (Indeed, he is often silent relative to questions most in need of explanation.) One can only speculate. It is true that Euclid does not specify as an assumption that any finite straight line can be divided. Yet he does have as one of his postulates that a circle can be constructed with any point as center and any distance as radius.[44] One might also parenthetically note that Bradwardine also includes this particular postulate in his *Geometria speculativa*.[45] Of course, this so-called circle postulate is what allowed Bradwardine in his Conclusion 20 to construct an arc of a larger circle over the same chord and to do so no matter how small the segment determined by the chord might be. That is to say, the indeterminacy guaranteed by the circle postulate (that a circle can be drawn with *any* distance as radius from any point as center) gives rise to an indeterminate number of shorter and shorter arcs over the same chord and, at the same time, further and further divisions of the finite straight line that was the perpendicular bisector with which the whole construction began. If this is a possible interpretation of why Bradwardine established the infinite divisibility of a finite straight line in the way that he did, then one should presently credit Bradwardine with seeing just what "indeterminacy" assumption Euclid did make that could be of use to Bradwardine in his refutation of the indivisibilist composition of continua.

Yet, whatever motives may have caused Bradwardine to prove the infinite divisibility of a finite straight line in such an unexpected manner, directly after establishing this divisibility he moves to the topic of the continuity of motion (III, B, 1: Concl. 21–26). The central contention relative to this topic is Conclusion 24: "Given any local motion whatever, a uniform and continuous local motion can be found that is faster or slower than it by any ratio [that can obtain] between finite straight lines," together with its Corollary: "Any finite space whatever can be traversed in any finite time whatever." All of this is demonstrated, geometrically, to be sure, by the three immediately preceding Conclusions: (1) A finite straight line can be moved continuously and uniformly with a circular motion about one of its ends remaining at rest (Concl. 22); (2) Such a motion is motion without deformation (Concl. 21); (3) The velocities of points within such a rotating finite straight line can be related to the metrical properties of the segments determined by those points within that line (Concl. 23).

Several observations about Bradwardine's procedure are in order. First, the required motion without deformation is established on grounds of the dissociation of superposition from imposition that was demonstrated in Conclusions 9 and 10. Secondly, it is clear that the proportionality existing between the velocities and segments in Conclusion 23 (which in turn leads directly to the continuity of motion specified by Conclusion 24 and its Corollary) is obtained through Bradwardine's choosing the motions involved as those of lines revolving about their end points. For in selecting just that type of motion, he has automatically established a one-to-one correspondence between the *motion* of any point and the *magnitude* of the line segment determined by that point. But if we know that these magnitudes can be found to stand to one another as any finite straight line to any finite straight line and, further, that any finite straight line is infinitely divisible, it follows that we have established the infinite divisibility and hence continuity of the range of possible motions. Put in another way, one can say that it is then shown that all possible degrees of motion can be made to stand in a one-to-one correspondence to all possible finite straight lines, both the degrees and the lines similarly ordered according to magnitude.

Bradwardine may generalize his results to cover motions of alteration and augmentation and diminution (Concl. 25) and may also, expressing himself in the language of the intension and remission of forms, claim (Concl. 32) that there is no slowest or most remiss degree of motion (which some might have taken as an indivisible of motion), but the core of his investigation of the continuity of motion is to be located in the proportionality between motions and geometrical magnitudes set down in Conclusions 23 and 24. The interpretation is, therefore, once again quite thoroughly geometrical.[46]

The refutatory conclusions

The primary burden of the *De continuo* is the definitive refutation of any and all hypotheses claiming that continua can be composed of indivisibles. Just which hypotheses these were is succinctly indicated by a classification Bradwardine himself gives:[47]

I. The nonindivisibilist position (Aristotle, Averroës, al-Ghazali, and "many moderns"). Continua composed of parts divisible without end.

II. The indivisibilist position:
   A. The corporeal indivisibilist (Democritus). Extended indivisibles.[48]
   B. The indivisibilist maintaining the composition of continua out of *points*. Nonextended indivisibles.
      1. Composition out of a *finite* number of points. (Pythagoras – the father of this sect – Plato, and Walter the modern).

    2. Composition out of an *infinite* number of points, which
       points are:
       a. *Immediately conjoined* to one another. (Henry the
          modern).
       b. *Mediate* to one another (Robert Grosseteste).[49]

Bradwardine believed that he had refuted all of these indivisibilist positions, but it is apparent that his major effort was directed toward only three possibilities:

    1. consecutive or immediately joined indivisibles (IV: Concl. 35–56);
    2. a finite number of indivisibles (V: Concl. 57–114);
    3. an infinite number of indivisibles (VI: Concl. 115–137).

The hypothesis of an infinite number of mediate indivisibles, which Bradwardine ascribes to Grosseteste,[50] is summarily dismissed (Concl. 140 and its Corollary) on the basis of a previous Conclusion (120) which had rather questionably established that the composition of a continuum out of an infinite number of indivisibles would mean that these indivisibles would have to be immediate to one another. Of course, this Conclusion in itself automatically makes the hypothesis of an infinity of mediate indivisibles untenable.

In any event, it is directly on the heels of his dismissal of infinite mediate indivisibilism that, likening his success to that of an alchemist creating gold,[51] Bradwardine announces his true view of the composition of continua (Concl. 141 and Corollary): "No continuum is made up of atoms, since every continuum is composed of an infinite number of continua of the same species." It is after demonstrating this positive Conclusion to his satisfaction that Bradwardine draws attention to his having thus far ignored in his refutation the corporeal indivisibilism of Democritus. Nevertheless, he believes that his own divisibilist view of continua adequately disposes of Democritus' atoms, although he admits that this Greek atomist most likely meant something else by an indivisible body than what had been the object of criticism in the *De continuo*.[52] In any case, it was not Democritus, but rather fourteenth-century atomists who were Bradwardine's primary concern. We have already seen that, in refuting these medieval indivisibilists, Bradwardine felt that one could reveal that their positions caused difficulties in all manner of branches of knowledge.[53] These "difficulties" amounted to establishing that this or that indivisibilist hypothesis led to a series of absurdities within, or relative to, a given *scientia*. When a new hypothesis is being entertained, it is appropriately spelled out in the antecedent clause of the Conclusion;[54] when further difficulties were held to follow from the same hypothesis, it is not repeated but simply referred to in the antecedent by a brief "If this is so, . . ." (*Si sic, . . .*). In any event, these absurdities are of various sorts in the *De continuo:* (1) Some would be, or would amount to, self-contradictory statements: that some mobile is simultaneously moving and at rest (Concl. 104), that an indivisible is divided (Concl. 49), that the laws of noncontradiction and excluded middle will not hold (Concl. 111), and so forth. (2) Some would contradict established truths in a given science (this is the most frequently occurring type of *reductio*): in arithmetic, that there would be incommensurable numbers (Concl. 57); in geometry, that parallels would meet (Concl. 42), that a horn angle could be divided by a straight line (Concl. 78), that the circumference of a circle would be twice its diameter (Concl. 73); in music, that the octave would not be composed of the intervals of a fourth plus a fifth (Concl. 62); in astronomy, that all celestial spheres would be of equal size and would move with equal speed (Concl. 94); and so forth. (3) Some would run counter to the occurrence of known phenomena: that there is no motion at all (Concl. 106), that there is no condensation or rarefaction (Concl. 99), and so forth. (4) Some contradict other Conclusions that have been demonstrated in the *De continuo:* thus, that only a finite number of points can be immediately joined to a given point (Concl. 38) is inconsistent with the claim that an infinite number can be so related (Concl. 36), and so forth. (5) Some simply contradict the hypothesis in question: that immediate indivisibles imply mediate indivisibles (Concl. 37), that composition out of an infinite number of indivisibles implies the composition out of a finite number of indivisibles (Concl. 139), and so forth.

If one canvasses all of the refutatory Conclusions of the *De continuo,* it is clear that the hypothesis of finitism is that species of indivisibilism to which Bradwardine devotes most attention. This is not unexpected, since it is that hypothesis from which inconsistencies more easily flow, both for Bradwardine and for other critics of the indivisibilist stance. The most notable reflection in the *De continuo* of this fact is that it is finitism that Bradwardine takes through the paces relative to the twelve different sciences (V, A–L: Concl. 57–114). Of course, it is with respect to geometry that the greater number of the inconsistencies in question obtained.[55] And the same is true of the hypothesis of immediacy for constituent indivisibles, since either the existence of merely a finite number of points in a line or of consecutive points in a line are assumptions that lead to any number of geometrical absurdities. In this part of his refutations Bradwardine was on reasonably firm ground. This is not the case, however, in his attempt to refute the composition of continua out of an infinite number of indivisibles. For almost all of the Conclusions (115–137) in which Bradwardine reputedly derives absurdities from the hypothesis of an infinity of indivisibles make use of the notion that the extensions of continua are proportional to the numbers of constituent indivisibles within them. In fact, this proportionality between the number of constituents and the magnitudes of the things they constitute is Conclusion 2 of the *De continuo*. Of course, the proportionality does not hold when the number of constituents is infinite. But precisely the contrary of this is asserted by the very first Conclusion (115) that Bradwardine directs against the infinitist hypothesis. Indeed, in proving this (false) Conclusion, Bradwardine expressly claims that the only way to account for the ratio of the magnitudes of continua to one another is by appealing to the ratio of the atoms composing them.[56] In a few words, unlike some of his contemporaries,[57] Bradwardine did not adequately understand the properties and relations of infinite sets [58] and it is this which, for the most part, renders his criticism of the infinitist hypothesis ineffective.[59]

One final note should be taken of Bradwardine's refutation of his indivisibilist opponents: It relates to the concluding section of the *De continuo* that asks the question of whether indivisibles can exist as entities separately from the continua to which they are connected (VIII: Concl. 142–151). Bradwardine's reply to this question is the Ockhamist one[60] that indivisibles never have such separate existence. In particular, the existence of indivisibles that is presumably required to account for the termination of continua and for the continuation of their parts is totally unnecessary because these functions are adequately accounted for by continua *qua* continua alone (Concl. 151 and Corollary). Further, indivisibles are only accidental to the continuum in which they occur and are hence quite unnecessary for the existence of continua as such (Concl. 143). Finally, Bradwardine shows that the assumption of the existence of indivisibles is not only unnecessary, but that assuming such existence leads to the contradicting of Conclusions that had already been established in the *De continuo*.

The continuity commitments of geometry

There is no doubt that geometry functions as a kind of fulcrum in the *De continuo* enabling Bradwardine to lever his indivisibilist opponents into this or that inconsistency. But if geometry is so central to his overturning of his adversaries, what is one to say of the *assumptions* geometry might make about continuity? That Bardwardine was aware of this kind of question is adequately established by the fact that, at the conclusion of his labors, he returns to the problem of a *petitio principii* he had first mentioned relative to his fourth Supposition.[61] This time he is only concerned with a *petitio* with respect to his use of geometry. His brief discussion is indeed one of the most interesting in the *De continuo*.[62] He introduces the discussion with a reference to Averroës.[63] The point is that someone like Averroës might criticize Bradwardine's procedure in the *De continuo* by claiming that geometry everywhere *assumes* continua not to be composed of indivisibles but cannot demonstrate that this is so. This assumption, the criticism would continue, makes any attempt to disprove indivisibilism by means of geometry guilty of a *petitio*. Bradwardine answers this criticism by saying, to begin with, that to suppose that geometry *assumes* that a continuum is not composed of indivisibles is false. Nor, he adds, can we find this denial among the geometrical demonstrations of Euclid's *Elements*. Moreover, "it is not logically needed everywhere, since it is not in Book V of the *Elements*"[64] and, Bradwardine continues, "nor is it geometrically supposed in some demonstration that a continuum is not composed of an *infinite* number of *mediate* indivisibles, since, if the opposite be given, any demonstration whatsoever proceeds, as is inductively evident by knowing how to demonstrate geometrical conclusions." In other words, we can show that Euclid's geometry is consistent with the affirmation of the composition of continua out of an *infinite* number of *mediate* indivisibles by proving, one by one, all of the propositions of the *Elements*. We shall find that even with this assumption all are capable of proof.

But this does not mean that there are absolutely no claims concerning the composition of continua in the *Elements*. For Bradwardine hastens to add that "none the less Euclid supposes in his geometry that the continuum is not composed of a *finite* number of *immediate* atoms, although he does not expressly posit this among his Suppositions."

Should one ask why the finitist and immediatist brand of indivisibilism is excluded by Euclid, Bradwardine's only reply is that certain theorems of the *Elements* could not be demonstrated without this assumption.[65] Thus, the first theorem of Book I of the *Elements* could not be demonstrated if the continuum is said to be composed of a finite number of indivisibles. This theorem sets as a problem the construction of an equilateral triangle over any given finite straight line. Now, if straight lines can be composed of a finite number of indivisibles, then at least one finite straight line which we may select could be composed of only two indivisibles. Or, at least, so Bradwardine maintains. But over such a line no equilateral, or any other, triangle could be constructed. For there would be no space between the lines drawn from the ends of the given, two-point, line to some point lying outside the line. Consequently, these lines could not be said to include an angle, and thus no triangle can be constructed.

Secondly, Bradwardine maintains that the *immediacy* of indivisibles in continua is inconsistent with the proof of the fourth and eighth theorems of Book I of Euclid, and with the proof of theorem 23 of Book III. This is so, he relates, because these theorems are proved by means of *superposition*. These three theorems are the so-called congruence theorems in Euclid. As such, they all employ Euclid's fourth common axiom, that is, his axiom of congruence, which contains the idea of superposition. Accordingly, Bradwardine contends that it is this use of superposition in these theorems which is the impediment to the possible admission of immediate indivisibles.[66]

Thus far Bradwardine has claimed that geometry *assumes* the denial of certain kinds of indivisibilism. One could, however, *demonstrate* this denial on the basis of some of the Conclusions established in the *De continuo*. If one objects that these Conclusions in turn depend on the very geometry that *assumes* the denial in question, Bradwardine counters by pointing out how one could prove the needed Conclusions of the *De continuo* without appealing to Euclid at all. One need not analyze Bradwardine's suggestions beyond a reference to the relevant text[67] to realize that he was sensitive to issues we would now describe as the consistency and independence of first principles or axioms.

Even though one can take exception to the conclusiveness of some of Bradwardine's refutatory claims and can remain somewhat unconvinced by some of the elements of his account of the relations between Euclidean geometry and the continuity of geometrical magnitudes, the *Tractatus de continuo* is an important document within the history of fourteenth-century natural philosophy. Not only does it provide us with an outstanding example of the application of mathematics that is increasingly apparent as one uncovers the history of later medieval natural philosophy, but it addresses the issue of the justification of applying the mathematics in question. That is a much rarer fourteenth-century occurrence, and Bradwardine deserves considerable credit for broaching the issue, whatever his success may have been in resolving it.

Appendix: The enunciations of the *De continuo*

Although V. P. Zoubov has already published the text of these enunciations in his Russian article on the *De continuo* [note 12], it has been deemed appropriate to provide a corrected and more complete text here. At the same time, it was thought that easy reference to these enunciations would be helpful in filling out and substantiating some of the points made above. No attempt, save one, has been made to record variant readings (some of which have, however, been given by Zoubov). In this regard, one might note that two added Conclusions are given in the Erfurt MS [note 2].

Thomas Bradwardine, Tractatus de continuo

[PARS I] [DEFINITIONES]
1. Continuum est quantum cuius partes ad invicem copulantur.
2. Continuum permanens est continuum cuius partes singule manent simul.

3. Continuum successivum est continuum cuius partes succedunt secundum prius et posterius.
4. Corpus est continuum permanens longum, latum et profundum.
5. Superficies est continuum permanens longum, latum sed non profundum.
6. Linea est continuum permanens longum, non latum nec profundum.
7. Indivisibile est quod nunquam dividi potest.
8. Punctus est indivisibile situatum.
9. Tempus est continuum successivum successionem mensurans.
10. Instans est terminus athomus temporis.
11. Motus est continuum successivum tempore mensuratum.
12. Motum esse est indivisibile motus.
13. Materia motus est quod per motum acquiritur.
14. Gradus motus est illud materie motus suscipientis magis et minus quod acquiritur per aliquod motum esse.
15. Lineam linee superponere partialiter vel totaliter est ipsam lineam secundum longitudinem totius vel partis simpliciter sine medio adherere alteri.
16. Lineam linee secundum partem vel secundum totam imponi est ipsam secundum longitudinem ipsius totius vel partis in aliam continuari.
17. Aliquod post aliud mediate esse, fuisse vel fore est ipsum cum medio inter illa esse, fuisse vel fore.
18. Aliquod immediate post aliud esse, fuisse vel fore est ipsum sine medio esse, fuisse vel fore.
19. Incipere esse per affirmationem de presenti et negationem de preterito est nunc esse et immediate ante hoc non fuisse.
20. Incipere esse per negationem de presenti et affirmationem de futuro est nunc non esse et immediate post hoc fore.
21. Desinire esse per negationem de presenti et affirmationem de preterito est nunc non esse et immediate ante hoc fuisse.
22. Desinire esse per affirmationem de presenti et negationem de futuro est nunc esse et immediate post hoc non fore.
23. Infinitum cathegorematice et simpliciter est quantum sine fine.
24. Infinitum sinkathegorematice et secundum quid est quantum finitum et finitum maius isto et finitum maius isto maiori et sic sine fine ultimo terminante; et hoc est quantum et non tantum quin maius.

> [Note: The Paris MS and the fragment in the Erfurt MS have the following variants for Def. 13–14, 23–24: (13) Finis motus ⟨est⟩ quod per motum acquiritur. (14) Gradus est secundum quod motus suscipit magis et minus quod acquiritur per aliquem motum. (23) Infinitum privative [et] simpliciter est

quantum sine fine. (24) Infinitum privative secundum quid est quantum finitum, finitum maius illo et maius, et sic sine fine; et hoc est quantum, et non tantum quin maius. These two MSS also omit Def. 18.]

[PARS II] [SUPPOSITIONES]

1. Omne maius posse dividi in equale et in differentiam qua excedit.
2. Si finitum addatur finito, totum erit finitum.
3. Ubi diversitatis vel dissimilitudinis nulla est causa, simile iudicatur.
4. Omnes scientias veras esse ubi non supponitur continuum ex indivisibilibus componi.
5. Omnia mediata distare, omnia divisa mediari.
6. Omne corpus, superficiem, lineam atque punctum uniformiter et continue posse moveri.
7. Omnium duorum motuum localium eodem tempore vel equalibus temporibus continuatorum velocitates et spatia illis pertransita proportionales existere.
8. Omnium duorum motuum localium super idem spatium vel equalia deductorum velocitates et tempora proportionales econtrario semper esse.
9. Quacumque velocitate vel tarditate potest unum mobile moveri vel spatium pertransiri potest quodcumque.
10. Esse vel non esse finitum certo tempore mensuratur.

[CONCLUSIONES]
[PARS III] [CONCLUSIONES PREPONENDE]

1. Nullum indivisibile maius alio esse.
2. Si duo continua eiusdem speciei ex indivisibilibus equalibus numero componantur, ad invicem equalia esse.
3. Nullius continui multa indivisibilia in eodem situ indivisibili situari.
4. Nullius recte multa puncta ab aliquo eius termino equaliter posse distare.
   Coroll. Cuiuslibet continui quelibet duo athoma a quocumque eius fine habere distantias inequales.
5. In continuo multa eius indivisibilia uni eius indivisibili ex eadem parte immediata esse non posse.
   Coroll. Nullius recte plura puncta quam duo uni eius puncto ex diversis partibus immediate iunguntur.
6. A nullo puncto ad unam rectam plures quam duas rectas equales ad invicem duci posse.
   Coroll. Nullam rectam in pluribus punctis duobus tangere lineam circularem.

7. Si inter duas lineas una mediet vel plures et tamen finite, illas equedistare.
8. Inter nullas rectas sibi superpositas puncta alica mediare.
9. Lineam rectam secundum totum vel partem magnam recte alteri superponi et habere aliquod punctum intrinsecum commune cum ista non contingit.
10. Linee recte unam partem magnam alie recte imponi et aliam partem magnam superponi eidem vel ad latus distare ab illa impossibile comprobatur.
11. Unius recte duo puncta in alia continuari et per partem eius magnam superponi eidem vel ad latus distare ab illa non posse.
12. Linee recte unam partem magnam recte alteri superponi et aliam ad latus distare ab ista est impossibile manifestum.
13. Unius recte duo puncta alteri superponi vel unum imponi, aliud vero superponi et magnam eius partem ad latus distare ab ista non posse contingere.
14. Quelibet recta secans rectam secat eam in aliquo sui puncto et non in pluribus quam in uno.
15. Nulle recte in aliquo puncto concurrentes aliud punctum intrinsecum illis habent.
    Coroll. Semidyametros circuli non concurrere ante centrum nec rectas ductas a basi trianguli ad angulum illi oppositum se tangere citra illum.
16. Angulum rectilineum assignatum in duos angulos rectilineos et datum latus trianguli rectilinei in duas rectas et triangulum rectilineum totum datum in triangulos rectilineos per rectam partiri.
17. Angulum contingentie quemlibet in angulum contingentie et angulum periferie super rectam, quoque basim trianguli contingentie oppositam angulo contingentie in duas rectas, et totum triangulum contingentie in triangulum contingentie minorem et triangulum a portionibus circumferentie et recta contentum per circulum maiorem secare.
18. Super datam rectam finitam quantumcumque volueris circuli ⟨seu⟩ circumferentie quantumcumque describere portionem circuli seu circumferentie medietate minorem.
19. Si super eandem cordam vel cordas equales portiones inequales circulorum vel circumferentiarum medietate minores consistant, minor portio est maioris circuli circumferentieque maioris, maior vero minoris. Si vero circulorum vel circumferentiarum inequalium super eandem cordam constituuntur portiones medietate minores, iste erunt necessario inequales, et maior circulus et circumferentia minorem portionem habebit, minor vero maiorem.

Coroll. Vas concavum resupinum positum equedistantis orizonti supra locum elementi fluxibilis plus istius elementi continere in loco humo quam alto. Tali vero vase pleno elementi huius ascendente affluere quasdam partes; descendente vero contentum fluidum congregari, et maxima vasis latera vacua derelinqui, atque liquidi summitatem ultra vasis dyametrum continue elevari. Rursum tale vas talis elementi semiplenum ascendens fieri aliquotiens magis plenum, aliquotiens vero plenum et superius cumulatum; et aliquando in tantum quod affluent quedam partes descendens effici minus plenum. Si vero tale vas ponatur simpliciter infra locum huius elementi, per totum contraria prioribus evenire.

20. Rectam perpendiculariter exeuntem a puncto medio corde ad punctum medium archus portionis circuli medietate minoris per circulum in duas rectas dividere et utrumque angulum portionis minoris et angulum circumferentie partiri, ipsam insuper portionem minorem linealemque secare.

    Coroll. Omnem rectam im multas rectas posse dividi.

21. Si linee recte punctum aliquod vel pars aliqua moveatur localiter, quamlibet partem magnam et quodlibet medium punctum quod est cum eius uno extremo necessario commoveri.

22. Cuiuslibet recte linee finite uno termino quiescente potest reliquus eius terminus circulariter uniformiter et continue transferri, tota recta et qualibet parte eius magna ad terminum eius immobilem terminata circulum describente et quolibet eius puncto moto circumferentiam circuli faciente.

23. Si recta finita super unum eius terminum quiescentem circulariter moveatur, omnes duas rectas terminatas ad punctum immotum et alia puncta mota et velocitates istorum punctorum proportionales certissime scias esse.

24. Quocumque motu locali signato potest motus localis uniformis et continuus omni proportione recte finite ad rectam finitam velocior et tardior inveniri.

    Coroll. Quodcumque spatium finitum quocumque tempore finito posse uniformiter et continue pertransiri.

25. Quocumque motu successivo signato potest motus successivus eiusdem speciei in omni proportione recte finite ad rectam finitam velocior et tardior reperiri.

26. Si quid continue localiter moveatur, in eodem instanti non acquirere multos situs nec in eodem situ in diversis instantibus potest esse.

    Coroll. In aliis motibus similiter esse constat.

27. Omnis inceptio vel desinitio non mensuratur tempore, sed instanti.

28. Omne quod non est aliquale et erit tale nunc incipit vel aliquando incipiet esse tale.
29. Omne quod est aliquale et non semper erit tale nunc desinit vel aliquando desinet esse.
30. Si unum continuum habeat athoma immediata et infinita sive finita, quodlibet sic habere.
31. Si unum continuum ex indivisibilibus componitur secundum aliquem modum, et quodlibet sic componi, et si unum non componitur ex athomis, nec ullum.
32. Nullius forme suscipientis magis et minus remississimum graduum esse.
33. Si forme intensibilis et remissibilis sit remississimus gradus possibilis, indivisibilia in omni continuo immediate coniunguntur.
34. Si sic, tota latitudo talis forme tantum finitos gradus habere et omne continuum solum finita athoma continere.

[PARS IV] [DE INDIVISIBILIBUS IMMEDIATA CONIUNCTIS]

35. Si athoma in continuo immediata ponantur, immediata puncta centro circuli et quadrati sive cuiuslibet corporis punctis circumferentie circuli et quadrati lateris atque superficiei corporis exteris equaliter correspondent.
36. Si sic, cuilibet centro dato infinita puncta immediate coniungi.
    Coroll. Nulla duo puncta in superficie plana nec indivisibilia in ullo continuo sibi sine medio coniungi.
37. Si sic, inter quelibet duo indivisibilia continui cuiuscumque infinita eius indivisibilia mediare.
    Coroll. Omne continuum habere athoma infinita.
38. Si sic, puncto in medio superficiei plane sito octo puncta immediata esse et non plura.
    Coroll. Puncto in medio corporis situato 26 puncta et non plura immediata ease.
39. Si sic, nullam lineam circularem plura octo puncta habere nec finitam rectam plura tribus, nec extremam superficiem corporis alicuius ⟨plura 26⟩.
40. Si sic, angulum rectum esse minimum angulorum nec angulum esse acutum, et omnes obtusos equales esse ad invicem nec aliquem angulum obtusum penitus reperiri.
41. Si sic, nullum triangulum nec circulum nec omnino angulum esse posse, et quinque famosa corpora et omnia geometricalia, non absque grandi geometrie et mathematice totius iniuria, deperire.
42. Si sic, linee equedistantes concurrerent.

43. Si sic, motus uniformis uno gradu velocior alio in equali tempore acquirit plus illo, et per nullum divisibile, sed per indivisibile tantum.

44. Si sic, quodlibet agens naturale indivisibiliter fortius ⟨alio⟩ in equali tempore ageret in equale passum plus illo, sed plus indivisibiliter tantum et in minori tempore ageret equefortiter.

45. Si sic, quodlibet agens naturale in aliquo tempore aliquod passum transmutans in tempore indivisibiliter maiori indivisibiliter plus faciet et in tempore indivisibiliter minori tantum indivisibiliter minus ageret.

46. Si sic, quodlibet agens naturale equale alteri transmutanti aliquod passum in aliquo tempore transmutatione signata, passum indivisibiliter minus transmutabit in equali tempore transmutatione indivisibiliter maiori, passum indivisibiliter maius transmutatione indivisibiliter minori, si illud valeat transmutare.

47. Si sic, omnes motus et omnia agentia sive passa indivisibiliter se tantum superantia adequari.

48. Si sic, omnes motus et omnia agentia atque passa equari ad invicem, excedere et excedi.

49. Si sic, indivisibile dividetur.

50. Si sic, omne quod incipiet esse aliquale vel desinet esse tale secundum utramque significationem incipiet vel desinet esse tale.
     Coroll. Cuiuslibet et qualiscumque rei talis esse primum intrinsecum et postremum.

51. Si sic, cuiuslibet forme suscipientis magis et minus remississimum gradum dare.

52. Si sic, aliquem tardissimum motum esse.

53. Si sic, continuum ex athomis integrari.

54. Si sic, substantia et qualitas naturalis ex substantiis et qualitatibus indivisibilibus componuntur.

55. Si sic, omne continuum componitur ex indivisibilibus infinitis et tamen ex finitis, et neque ex infinitis neque finitis, et componitur ex athomis et non componitur ex illis.

56. In nullo continuo athoma immediate coniungi.

[PARS V] [DE INDIVISIBILIBUS FINITIS]

57. Si continuum ex finitis athomis componatur, sicud numerus athomorum unius continui ad numerum athomorum alterius, ita illud continuum ad aliud se habere.
     Coroll. Si aliquod continuum sit incommensurabile alteri, et numerum incommensurabilem numero reperiri.

58. Si sic, athoma in continuo immediate iunguntur.

Coroll. Omnia continua habere athoma infinita et ex athomis non componi.

59. Si sic, debilissimus gradus soni se habet sicud unitas et ceteri sese sine medio consequentes ut sequens series numerorum.

60. Si sic, omnis gradus soni ad omnem gradum soni se habet ad proportionem in habitudine numerali.

61. Si sic, tonum et omnem proportionem superparticularem mediat proportio numeralis.

62. Si sic, omnis dyapason ex dyatessaron cum dyapente ⟨non⟩ componitur.

63. Si sic, consonantie musicales ex certo numero vocum, intervallorum, tonorum et semitonorum non constant, nec debitis proportionibus modulantur.

64. Si sic, nullis vel paucissimis sonis pro basi suppositis possunt concorditer adaptari alie musicales consonantie.

65. Si sic, tonus partiri non potest.

Coroll. Totius musice sonoritatem iocundam non sine sonorum suorum desolatione inimica perpetuo silentio condempnari.

66. Omnis recta linea habet particulares lineas infinitas.

67. Omnem angulum rectilineum sive contingentie in tales angulos dividere infinitos.

68. Omnem triangulum rectilineum sive contingentie in infinitos tales triangulos posse dividi vel partiri.

69. Omnis superficies habet superficies et lineas infinitas et puncta similiter infinita.

70. Omne continuum componitur ex infinitis continuis eiusdem speciei et habet athoma propria infinita.

71. Si sic, dare recte finite equalem portionem circumferentie circuli et portionem circumferentie circuli rectam esse.

72. Si sic, tantum circulum assignare quo maior esse non potest et idem de quolibet continuo permanente.

73. Si sic, periferiam circuli esse duplam dyametri.

74. Si sic, omnes circulorum periferie et omnes circuli sunt equales.

75. Si sic, alique partes circumferentie circularis sunt recte et angulum rectilineum continentes.

76. Si sic, multi sunt circuli centra non habentes.

77. Si sic, aliquem circulum dyametrum et aream non habere.

78. Si sic, angulus contingentie dividetur per rectam.

79. Si sic, omnes bases triangulorum subtense eodem angulo sunt equales.

Coroll. Omnes rectas lineas equales esse.

80. Si sic, numerus punctorum, ordo et proportio triangulorum ysopleurorum est secundum numerum unitatum, ordinem et proportionem numerorum productorum ex additione primi numeri ad individuam unitatem, et secundi numeri ad primum numerum sic productum, et tertii numeri ad secundum sic formatum, et ita semper deinceps.

Coroll. Primum triangulum ysopleurum ex tribus punctis componi, secundum vero ex sex, tertium vero ex decem et ita de aliis.

81. Si sic, aliquis est triangulus nullum angulum habens et aliquis tantum unum.

82. Si sic, aliquis triangulus tres angulos rectos habet et linee equedistantes concurrunt.

83. Si sic, aliquis triangulus est subsesquitertius ad quadratum qui est subduplus ad idem.

84. Si sic, aliquis trigonus est circularis.

85. Si sic, numerus punctorum, ordo et proportio quadratorum sequitur numerum, ordinem et proportionem talium numerorum.

86. Si sic, non ⟨potest⟩ cuilibet trigono dato quadratum equum describere.

87. Si sic, omnis quadrati dyameter suo lateri est equalis.

88. Si sic, aliquod quadratum est circulus.

89. Si sic, dyameter quadrati est commensurabilis eius coste et quelibet linea cuilibet alteri.

90. Si sic, aliquod quadratum angulo et dyametro et area simul caret et aliquod quadratum unum angulum tantum habet.

91. Si sic, triangulus et quadrangulus super basim quadrati inter lineas equedistantes contenti in qualibet magna proportione super illud quadratum excedere.

92. Si sic, pyramidem et cubum speras esse, cum aliis heresibus geometricis infinitis.

93. Si sic, nullam esse multiplicationem vel visionem rectam, fractam sive reflexam lucis vel coloris.

94. Si sic, omnes speras celestes et stellas et elevationes earum a terra esse quantitatis equalis et equevelociter circumferri.

95. Omne continuum habere infinita minima et partes magnitudinis.

96. Si substantia composita ex finitis substantiis athomis componatur, condensationem materie prime non fieri per athoma prioribus pauciora.

97. Si sic de substantia, rarefactionem materie prime non fieri per athoma materie plura primis.

98. Si sic de substantia, condensationem non fieri per pauciora puncta prioribus nec rarefactionem per plura.
99. Si sic de substantia, condensationem et rarefactionem non esse possibilem.
100. Si sic de substantia, velocitatem in motibus proportionem motorum ad sua mota non sequi.
    Coroll. Substantiam naturalem compositam ex finitis non componi.
101. Si sic de continuo, substantiam naturalem continuam ex indivisibilibus substantiis finitis componi.
102. Si sic, impartibile in medietates patietur.
103. Si sic, aliquod mobile in locis variis et a se distantibus simul esse.
104. Si sic, aliquod motum simul quiescere et moveri.
105. Si sic, omnes motus similis speciei in velocitatibus adequari.
106. Si sic, motum non esse omnino.
107. Si sic, sanitatem humanam non servare nec perditam restaurare.
108. Si sic, omnes substantias a materia separatas esse equales ad invicem in virtute.
109. Si sic, nullam substantiam a materia separatam esse infinite virtutis.
110. Si sic, non contingit recte scribere nec recte loqui.
111. Si sic, contradictoria simul esse vera et idem verum et falsum.
112. Si sic, idem est iustum et iniustum.
113. Si sic, non est recte diligere nec odire, delectari congrue nec tristari.
114. Si sic, nullum posse virtuosum fieri nec felicem.

[PARS VI] [DE INDIVISIBILIBUS INFINITIS]
115. Si omne continuum ex indivisibilibus infinitis componitur, omne continuum eiusdem generis et athoma propria eodem genere proportionalia reperiri.
116. Si substantia naturalis continua ex infinitis indivisibilibus substantiis componitur, condensationem materie prime non fieri per puncta substantialia materie pauciora prioribus nec rarefactionem per plura.
117. Si sic, condensationem et rarefactionem non fieri per puncta accidentalia quantitatis continue pauciora prioribus nec plura.
118. Si sic, in continuatione seu in discontinuatione liquidorum nullam materiam priam corrumpi nec generari.
119. Si sic, condensationem et rarefactionem non esse.
120. Si sic, athoma cuiuscumque continui immediate coniungi.

121. Si sic, cuiuslibet forme suscipientis magis et minus remississimum gradum dare.

122. Si sic, aliqua superficies erit summe alba et simul summe nigra.

123. Si sic, athomum naturale posse movere et moveri motu successivo et continuo omni velocitate possibili et similiter omni tarditate.

124. Si sic, nullum agens naturale corporeum posse movere subito superficiem naturalem, lineam, punctum.

125. Si sic, aliquod agens naturale corporeum est equalis activitatis cum athomo naturali, et aliquod minoris et quodlibet infinite.
    Coroll. Motum successivum nullum esse continuum nec subitum.

126. Nullam substantiam sive qualitatem ex substantiis sive qualitatibus integrari.
    Coroll. Omnem substantiam materialem et qualitatem similiter esse corpus, et nullam esse superficiem compositam nec aliquam lineam radiosam nec punctum aliquod luminosum.

127. Si sic de continuo, substantiam et qualitatem ex infinitis substantiis et qualitatibus athomis integrari.

128. Si sic, athoma infinita in omni proportione finita et infinita ad alia infinita procul dubio se habere.
    Coroll. Omnia athoma infinita et quelibet alia athoma infinita proportionari contingit.

129. Si sic, omnia athoma infinita quibuscumque infinitis athomis adequari, excedere et excedi, omnia continua consimilis generis equalia esse, excedentia et excessa.

130. Si sic, omnes velocitates et tarditates motuum equales esse.

131. Si sic, multa puncta in eodem situ indivisibili situari.

132. Si sic, aliquod continuum in eodem situ indivisibili situari.

133. Si sic, superficies composite ex lineis equalibus numero et longitudine sunt equales; si vero componantur ex lineis equalibus numero et inequalibus in longitudine, que ex longioribus componitur excedet; et idem de corporibus ex superficiebus compositis consequens scias esse.

134. Si sic, omnis quadrati medietas est maior toto quadrato.

135. Si sic, una superficies terminata in omni proportione finita excedit aliam sibi equalem.

136. Si sic, quarta pars circuli sive trianguli et medietas eius sunt equales.

137. Si sic, omnis linea circularis est equalis cuilibet linee circulari et costa quadrati dyametro, et omnis recta omni recte necessario erit equalis.

[PARS VII] [SUMMA OPINIONUM PREDICTARUM
REPROBATIONIS DE COMPOSITIONE CONTINUI]

138. Nullum continuum ex indivisibilibus infinitis componi.

139. Si continuum componitur ex infinitis indivisibilibus immediatis ad invicem, componitur ex finitis.
  Coroll. Nullum continuum ex infinitis indivisibilibus immediatis componi.

140. Si continuum componitur ex infinitis indivisibilibus mediatis, componitur ex immediatis.
  Coroll. Nullum continuum ex indivisibilibus mediatis componitur.

141. Nullum continuum ex athomis integrari.
  Coroll. Omne continuum ex infinitis continuis similis speciei cum illo componi.

[PARS VIII] [UTRUM INDIVISIBILIA SINT
DISTINCTA A CONTINUO]

142. In continuatione sive discontinuatione corporum liquidorum nullam materiam primam nec aliquam substantiam primam nec qualitatem primam vel secundam corrumpi; et de quantitate et indivisibilibus quantitatis similiter esse constat.

143. Omnem substantiam esse per se possibile carere omni accidente.

144. Omne quod non est pars nec causa alterius potest corrumpi altero toto salvo.

145. Potest esse continuum et finitum sine aliquo indivisibili continuante et finitante.

146. Si indivisibilia continuorum sint realiter ut ponuntur, substantia materialis indivisibiles substantias habet.

147. Si sic, indivisibilia omnis continui immediate coniungi.

148. Si sic, continuum ex athomis integrari.

149. Si sic, aliquod accidens subiectum primum non habere.

150. Si sic, potest non improbabiliter apparere omne corpus esse tenacitatis et resistentie infinite et dare seu furari esse meriti et demeriti infiniti.

151. Superficiem, lineam sive punctum omnino non esse.
  Coroll. Continuum non continuari nec finitari per talia, sed seipso.

Notes

1 For an analysis of the contents of the Toruń and Erfurt codices, see (1) Maximilian Curtze, *Analyse der Handschrift R. 4° 2, Problematum Euclidis explicatio, der Königl. Gymnasialbibliothek zu Thorn*, Leipzig, 1868, Separatabdruck aus der *Zeitschrift für Mathematik und Physik*,

Supplementheft zum 13. Jahrgang, 60 pp. (2) W. Schum, *Beschreibendes Verzeichniss der Amplonianischen Handschriften-Sammlung zu Erfurt*, Berlin, 1887.

2 There are two added Conclusions in the Erfurt MS: (107A) "Si sic, acutarios criticare laudabiliter 6 die sicud diebus aliis aptatis ad hoc." (121A) "Si sic, cuiuslibet forme intensibilis et remissibilis gradus quilibet propter primum intensionis est fortificationis infinite et similis ⟨de remissione⟩."

3 See the variant readings listed after the Definitions in the Appendix.

4 The puzzle noted by Curtze [note 1], which arises by the words *primus liber*, still remains unsolved. Was the *De continuo* part of another, larger work by Bradwardine? Or, what seems more likely, was it part of a collection of works dealing with mathematics and science, the words being added by the copier of such a collection?

5 One might also consider the penchant (on which, see below "The axiomatic format of the *De continuo*") Bradwardine had for writing works that, like the *De continuo*, were axiomatic in form as indication of his authorship. Furthermore, in Conclusion 19 of the *De continuo* it is asserted that, if arcs of two unequal circles are constructed over the same chord, which arcs are less than half the circumference of their respective circles, we must conclude that the arc enclosing the smaller portion (in area) belongs to the greater circle, while that enclosing the greater portion belongs to the smaller circle. Immediately following this enunciation Bradwardine elicits a "natural porism" to the effect that if we take a concave dish lying on its back parallel to the horizon and filled with a liquid element, then the surface of the liquid will be spherical and will progressively mount to a higher peak in the dish the closer it is taken to the center of the earth, and, conversely, will decrease in its peak – in fact, even flow out over the sides of the dish – the higher we lift the dish into the heavens. Now after the proof of the third Conclusion in Chapter 4 of Bradwardine's undoubtedly genuine *Geometria speculativa* (see note 18), which Conclusion asserts the same geometrical proposition as Conclusion 19 of the *De continuo*, we find written: "Ista propositio sumitur in naturalibus ad probandum quod idem vas in numero plus capit in celario quam in solario, et generaliter plus inferius quam superius." Naturally, we cannot prove anything of Bradwardine's authorship from this curious coincidence, but it certainly suggests that the two works may have been written by the same man.

6 Kraków, Biblioteka Jagiellońska: MS. 1578, in a *questio* (56v–60r) beginning "Queritur una questio, supposito actu infinito, utrum unum infinitum sit maius alio vel minus," we find the following exposition (60rb) of the syncategorematic infinite. "Infinitum sinkategorematice et secundum quid est finitum et finitum maius isto et sic sine fine ultimo terminante." This may be an (unascribed) quotation of the *De continuo;* cf. Definition 24 in the Appendix.

7 For the most recent bio-bibliographical material on Bradwardine, see the relevant entry in vol. 2 of the *Dictionary of Scientific Biography*.

8 The following examples will serve to illustrate the kind of references in question: "Postquam reprobavit opiniones Pytagori et Henrici per rationes geometricas, hic incipit facere idem per rationes naturales; Post scientias speculativas ad aliquas regulas practicas stilum vertit." These two examples occur after, respectively, the enunciations of Conclusions 43 and 113. For the most part, references to the text of the *De continuo* will specify, not MS

foliation, but rather where the quotations occur in the series of Definitions, Suppositions, and Conclusions indicated in the Appendix.

9 Such "divisions of the author's concern" are also indicated in the outline of the *De continuo* (see below "The Structure of the *De continuo*").

10 On this rise of indivisibilism and its criticism in the fourteenth century, see J. Murdoch, "Infinity and Continuity," in N. Kretzmann, A. Kenny, and J. Pinborg, eds., *The Cambridge History of Later Medieval Philosophy* (Cambridge, 1982), pp. 564–581; and "Naissance et développement de l'atomisme au bas moyen âge latin," in *Cahiers d'études médiévales II: La science de la nature: théories et pratiques* (Montreal/Paris, 1974), pp. 11–32.

11 See J. Murdoch, "Henry of Harclay and the Infinite," in *Studi sul XIV secolo in memoria di Anneliese Maier,* ed. A. Maierù and A. Paravicini-Bagliani (Rome, 1982), pp. 219–261; J. Murdoch and E. Synan, "Two Questions on the Continuum: Walter Chatton (?), O.F.M. and Adam Wodeham, O.F.M.," *Franciscan Studies* 26 (1966):212–288; V. P. Zoubov, "Walter Catton, Gérard d'Odon et Nicholas Bonet," *Physis* 1 (1959):261–278.

12 The following represent the history of work on Bradwardine's *De continuo*: (1) M. Curtze, as in note 1 above; (2) E. Stamm, "Tractatus de continuo von Thomas Bradwardina," *Isis* 26 (1936):13–32; Anneliese Maier, *Die Vorläufer Galileis im 14. Jahrhundert* (Rome, 1949), pp. 160–161, for the first correct identification of Bradwardine's opponents Henricus and Waltherus modernus; V. P. Zoubov, "Traktat Bradwardina O Kontinuume," *Istoriko-matematicheskiie Issledovaniia* 13 (1960):385–440, in Russian with appendix of Latin text; J. Murdoch, *Geometry and the Continuum in the Fourteenth Century: A Philosophical Analysis of Thomas Bradwardine's Tractatus de continuo,* unpublished Ph.D. dissertation, University of Wisconsin, 1957 (contains complete text of the *Tractatus*).

13 References to the *De continuo* will refer simply either to the numbers of the Definition, Supposition, or Conclusion as given in the Appendix or, as in the present case, both to the relevant numbers and to the corresponding entry in the foregoing outline.

14 See the division of geometrical from physical arguments quoted in note 8.

15 Directly following the enunciation of Conclusion 57 and Corollary:

> Hic pro sequentibus est advertendum, quod duplex est scientia speculativa, scilicet realis et sermocinalis. Realis est triplex, ut patet 6° *Methaphysice,* scilicet mathematica, naturalis et divina. Mathematica autem multas habet partes, quarum prima est arismetica, ut primo *Arismetice* Boetii diffuse probatur; et hoc est quod dicitur in prologo huius conclusionis, quod ipsa est mater totius mathematice. Secunda pars est musica, que subalternatur arismetice, tertia geometria, quarta perspectiva subalternata geometrie, quinta astronomia, que sine prima, secunda, tertia, quarta haberi non potest. Ille precedunt scientiam naturalem et hec subalternat sibi astrologiam. Et has sequitur methaphysica sive divina. Ille reales scientie sermocinalibus naturaliter sunt priores, ideo ab eis incipit tenens ordinem naturalem. Scientie sermocinales sunt 3, grammatica, rethorica, dyalectica. Igitur ordine has omnes scientias speculativas sequitur moralis scientia, que non pure speculativa, sed forte aliqualiter practica est. Moralis, ut patet secundo *Ethicorum,* "non contemplationis gratia," sed magis operationis: "Non enim scrutatur quid est virtus ut sciat, sed ut bonus efficiatur." Et per

omnes scientias istas et secundum ordinem recitatum arguit contra opinionem Pytagore.

16 At the end of Conclusion 114, which is the last of the Conclusions (57–114) treating of finitism: "Huius autem 12^e legittime concordem scientiam reverenter ausculta: Nullum omnino continuum ex indivisibilibus finitis componatur. (Vocat duodecimam legittimam 12 scientias per quas istam conclusionem probavit.)"

17 In particular, the utilization of this Corollary in establishing Conclusion 66, which deals with *lineas infinitas*, clearly reveals Bradwardine's intent.

18 The *Geometria speculativa* was published in Paris in 1495 and later dates, and has been edited by A. G. Molland in his unpublished doctoral dissertation *Geometria speculativa of Thomas Bradwardine: Text with Critical Discussion* (Cambridge University, 1967).

19 Edited and translated by H. Lamar Crosby as *Thomas of Bradwardine: His Tractatus de Proportionibus. Its Significance for the Development of Mathematical Physics* (Madison, Wis., 1955).

20 Lauge Olaf Nielsen, "Thomas Bradwardine's Treatise on 'incipit' and 'desinit.' Edition and Introduction," *Cahiers de l'Institut du moyen-âge grec et latin* (Université de Copenhague), 42 (1982), pp. 1–83.

21 Edited, with a long introduction by Henry Savile, London, 1618; reprinted Frankfurt, 1964.

22 The long digression following Definitions 23–24 does not speak of the function of these Definitions but only gives an exposition of their content.

23 This remark occurs directly after the first Supposition itself.

24 "Hoc dicit quia aliquando utitur declaratis in aliis scientiis quasi manifestis, quia nimis longum esset hec omnia declarare. Ubi autem tractant de compositione continui ex indivisibilibus non supponit eas veras esse propter petitionem principii evitandam."

25 See the final section of this chapter, on the continuity commitments of geometry.

26 Of course, the pattern could be much more complicated. Indivisibilists often replied to their own contentions, responding to these replies in turn, and in general the series of objections and replies could continue, back and forth, for lengths that often seem excessive.

27 On the replies of Harclay and Chatton, see J. Murdoch [note 11], pp. 242–248; Murdoch–Synan [note 11], pp. 254, 259.

28 The later medieval history of this debate is recounted in J. Murdoch, "Atomism and Motion in the Fourteenth Century," in E. Mendelsohn, ed., *Transformation and Tradition in the Sciences* (Festschrift for I. B. Cohen), Cambridge University Press, 1984, pp. 45–66.

29 See Murdoch–Synan [note 11], pp. 216–217 and references therein. These arguments received considerable impetus and respectability in the Middle Ages because they occurred in Duns Scotus and the *Metaphysica* of Al-Ghazali.

30 Note that this Conclusion as well as Conclusion 74 dealing with concentric circles occur in the section devoted to the refutation of finitism. But the same difficulties are held to follow if an infinite number of indivisibles are involved (Concl. 137), which is clearly in error (see discussion below on Bradwardine and the infinite).

31 The passage serves to introduce Bradwardine's refutation of the finitist hypothesis (V, A–L: Concl. 57–114):

Assertio ponens continuum ex indivisibilibus finitis componi est omnibus scientiis inimica, omnes impugnans, et ideo concorditer ab omnibus inpugnatur, cum qua mathematica primo congrediatur et vincat. Ipsa enim suis ceteris sororibus acutius contemplatur; inflexibilius telum iacit et se protegit clyppeo tutoris. Nullus enim physico certamine se speret gavisurum triumpho nisi mathematice utatur consilio et auxilio confortetur. Ipsa est enim revelatrix omnis veritatis sincere et novit omne secretum absconditum at omnium litterarum subtilium clavem gerit. Quicunque igitur ipsa neglecta physicari presumpserit, sapientie ianuam se nunquam ingressurum agnoscat. Arismetica igitur prima totius mathematice mater et ianua sic ordinatur certamine.

32 See text cited above in note 8, and, in particular, that cited directly after note 13.

33 Thus he appeals to *scientia naturalis* and to *perspectiva* in providing an additional proof for Conclusion 10. Cf. also the "natural" Corollary to Conclusion 19.

34 Following directly after Conclusion 141 and its Corollary: "Sic namque nature componitur fundamentum, mathematice columpna firmatur et totius physice fabrica solidatur."

35 One can even reasonably argue that Bradwardine's "Aristotelian" definition of a continuum (Def. 1), which allows for overlapping of parts in place of common limits for parts, was geometrically "planned" because it fits with the notion of two lines having a common segment, at the same time allowing Bradwardine's opponents greater latitude in explaining how their indivisibles might be connected to one another when they compose a continuum.

36 *Physics* VI.1.231a21–231b6.

37 Thus Henry of Harclay maintains that indivisibles can touch one another *secundum distinctos situs* (Cf. J. Murdoch [note 11], pp. 243–244), while the indivisibilist Gerard of Odo allows such touching because indivisibles do have "parts" *secundum differentias respectivas loci* (Cf. J. Murdoch, "Superposition, Congruence and Continuity in the Middle Ages," in *Mélanges Alexandre Koyré*, vol. I, *L'aventure de la science* [Paris, 1964], pp. 431–434).

38 The so-called congruence theorems of Euclid's *Elements* are I.4; I.8; and III. 24 (which Bradwardine considers in his treatment of the continuity assumptions of geometry; see note 65). The notion of superposition appears explicitly in the Adelardian translation of Euclid's axiom of congruence (a translation that Bradwardine used in his own *Geometria speculativa*): "Si aliqua res alicui *superponatur,* appliceturque ei, nec excedat altera alteram, ille sibi invicem erunt equales." On the whole history of the importance of the idea of superposition for continuity, see the article cited in note 37.

39 See Definition 15; although the superposition of lines is here in question, the same consideration applies to superposed indivisibles.

40 For fuller details, see the Koyré Festschrift article cited in note 37. The central Conclusion in Bradwardine's treatment of superposition and continuity is Conclusion 10.

41 Superposition also plays a role in assuring the rigidity of motion involved in Conclusion 21 as well as the divisibility of lines at hand in Conclusion 20.

42 This does not mean, of course, that he does not appeal to the operation of bisection in other regards in the *De continuo*. Cf. Conclusions 16–17, for example.

43 See above, note 17.

44 Euclid, *Elements,* Postulate 3 (in its most frequently used medieval Latin

version): "Super centrum quodlibet, quantumlibet occupando spacium, circulum designare."

45 *Geometria speculativa*, Tract. I, cap. 2, petitio 2.

46 Conclusions 27–29 set forth Bradwardine's contentions about temporal limits. Conclusion 50 maintains that the standard distinctions relative to these limits (see Def. 17–22) are destroyed if one adopts indivisibilism.

47 Directly after the enunciation of Conclusion 31:

> Pro intellectu huius conclusionis est sciendum, quod circa compositionem continui sunt 5 opiniones famose inter veteres philosophos et modernos. Ponunt enim quidam, ut Aristoteles et Averroys et plurimi modernorum, continuum non componi ex athomis, sed ex partibus divisibilibus sine fine. Alii autem dicunt ipsum componi ex indivisibilibus dupliciter variantes, quoniam Democritus ponit continuum componi ex corporibus indivisibilibus. Alii autem ex punctis, et hii dupliciter, quia Pythagoras, pater huius secte, et Plato at Waltherus modernus, ponunt ipsum componi ex finitis indivisibilibus. Alii autem ex infinitis, et sunt bipartiti, quia quidam eorum, ut Henricus modernus, dicit ipsum componi ex infinitis indivisibilibus immediate coniunctis; alii autem, ut Lyncul ⟨niensis⟩, ex infinitis ad invicem mediatis. Et ideo dicit conclusio: "Si unum continuum componatur ex indivisibilibus secundum aliquem modum," intendendo per 'modum' aliquem predictorum modorum; tunc sequitur: "quodlibet continuum sic componi ex indivisibilibus secundum similem modum componendi."

48 Note that Bradwardine's Definition 7 of an indivisible as that which can never be divided allows extended corporeal indivisibles such as those of Democritus. The definition of an indivisible as that which has no parts would not.

49 Indivisibles *mediate* to one another are those such that, between any two of which, there is always another.

50 Grosseteste did hold with different infinite numbers of indivisibles being *contained* within continua, but not, it seems, *composing* continua. The latter, however, was held to be Grosseteste's view by Henry of Harclay (see article cited in note 11), and Bradwardine most likely was following Harclay's contention.

51 Preceding, and introductory to, Conclusion 141: "Ut igitur alchimista post multos ignes aurea gaudet massa, victor quoque, terminatis laboribus pluribus, gaudet de triumpho, sic et tu, post tot scrutinia studiorum, carissimis amplectere affectibus sinceram que sequitur veritatem."

52 Following Conclusion 141:

> Omnes igitur opiniones erronee specialiter reprobantur, preter opinionem Democriti ponentem continuum componi ex corporibus indivisibilibus, que tamen per illam conclusionem et eius corollarium sufficienter reprobatur. Non tamen est verisimile quod tantus philosophus posuit aliquod corpus indivisibile, sicud corpus in principio est diffinitum, sed forte per corpora indivisibilia intellexit partes substantie indivisibiles et voluit dicere substantiam componi ex substantiis indivisibilibus.

53 See above, note 15.

54 See the antecedents, for example, in Conclusions 35, 57, 115, 116, 139, 140, 146.

55 This is one of the most frequent kind of indications of the central role played by geometry in the *De continuo*. To it one can add the geometrical manner in which motion is treated (see remarks above relative to Concl.

21–24) as well as the fact that geometry often provides the "moving force" in proving Conclusions concerning other sciences. Thus, Conclusion 93 about optics is based on the thoroughly geometrical Conclusion 41, the "astronomical" Conclusion 94 relies on the geometrical Conclusion 39, and so on.

56 Following the proof of Conclusion 115: "Et non est alia via salvandi proportionem continuorum ad invicem quam secundum proportionem athomorum componentium ipsa primo."

57 See the first article cited in note 10, pp. 569–573, on other fourteenth-century opinions about the relations of infinite sets.

58 Compare Bradwardine's application of infinite sets in his *De causa Dei* to refute the possible eternity of the world (see J. Murdoch, *"Rationes mathematice": Un aspect du rapport des mathématiques et de la philosophie au moyen âge* [Paris, Palais de la Découverte, 1961], pp. 18–20) with the views treated in the reference in the previous note. One may also note Bradwardine's *Geometria speculativa,* Tract. III, cap. 3, quinta regula:

Quantitates sunt equales que ad unam quantitatem comparate proporciones habent equales. Quoniam si habent equalem proporcionem ad terciam equalis est excessus earum super illam terciam ex premissis: et si est equalis excessus earum super idem commune, ipse quantitates erunt equales inter se per quintam communem scientiam. Ex ista potest sumi argumentum ad probandum quod unum infinitum non sit maius alio infinito, quoniam omnium infinitorum ad unam magnitudinem vel multitudinem finitam est equalis excessus, quoniam infinitus, et per consequens equalis proporcio.

59 However, when Bradwardine treats of an infinity of indivisibles that are *immediate* to one another, then his refutation is much more successful.

60 See J. Murdoch, "Infinity and Continuity" [note 10], pp. 573–575.

61 Above, note 24.

62 Following the proof of Concl. 141 (and all but directly after the text cited in note 52):

Posset autem circa predicta fieri una falsigraphia: Avroys in commento suo super *Physicorum,* ubi dicit, quod naturalis demonstrat continuum esse divisible in infinitum et geometer hoc non probat, sed supponit tanquam demonstratum in scientia naturali, potest igitur inpugnare demonstrationes geometricas prius factas dicendo: Geometriam ubique supponere continuum ex indivisibilibus non componi et illud demonstrari non posse. Sed illud non valet, quia suppositum falsum. Non enim ponitur inter demonstrationes geometricas continuum non componi ex indivisibilibus nec dyalecticer indiget⟨ur⟩ ubique, quoniam ⟨non⟩ in 5to *Elementorum* Euclidis. Et similiter, nec geometer in aliqua demonstratione supponit continuum non componi ex infinitis indivisibilibus mediatis, quia, dato eius opposito, quelibet demonstratio non minus procedit, ut patet inductive scienti conclusiones geometricas demonstrare. Verumtamen Euclides in geometria sua supponit, quod continuum ex [in] finitis et immediatis athomis non componitur, licet hoc inter suas suppositiones expresse non ponat.

63 Averroës, *In phys.,* III, comm. 31

64 The emendation giving the negative here (see text in note 62) is necessary since (1) it fits with the foregoing *nec dyalecticer indiget⟨ur⟩ ubique,* and (2) there is good evidence to the effect that indivisibilism is *not* allowed in Book V, not that it is (since Book V excludes the infinitely large and the

infinitely small, but includes incommensurables, within its theory of proportion).

65 Directly following text in note 62:

Si falsigraphus dicat contrarium et ponat aliquam lineam ex duabus punctis componi, Euclides non potest suam conclusionem primam demonstrare, quia super huius lineam non posset triangulus equilaterus collocari, quia nullum angulum haberet, ut patet per 16^{am} et eius commentum. Similiter, si dicat falsigraphus, continuum ex athomis immediatis componi, 4^{am} suam conclusionem et 8^{am} non probat, ambe enim per su⟨per⟩positionem probantur. Similiter in probatione ⟨23⟩ 3^i. Iste autem conclusiones non demonstrantur per aliquas conclusiones priores, sed ex immediatis principiis ostenduntur. Per has autem conclusiones relique demonstrantur, et ex hiis 3^{bus} quasi tota geometria Euclidis dependet et in ipsa omnis alia geometria fundatur, quare geometria supponit ⟨continuum⟩ ex [in]finitis et immediatis athomis non componi.

66 Unfortunately, Bradwardine does not state why this use constitutes an obstacle to the admission of finite and immediate indivisibles. Yet I believe he may well have meant something of the following nature: If we maintain that geometrical continua – in this particular instance, solids – are composed of immediate indivisibles, then, if (as in I.4) we superpose one triangle upon another in order to determine their congruence, we should no longer have *two* triangles, but rather *one* triangular right prism composed of two indivisible surfaces. In more general terms, therefore, the inconsistency Bradwardine cites between Euclid's congruence theorems and the assumption of immediate indivisibles, amounts to the fact that the immediacy of the indivisibles, interpreted as superposition, renders Euclid's axiom of congruence ineffective.

67 Directly following text in note 65:

Sed illud quod est per omissionem axiomate introductum, insufficientie arguentis inputari non debet. Est enim geometria sufficiens ostensive et per impossibile ex propriis principiis demonstrare nullum suum continuum ex indivisibilibus finitis immediatis componi: ostensive, ut patet in 66^a precedente et per 58 et eius corollarii primam partem; per inpossibile potest probari, ut patet per 71^{am} et multas sequentes. Si autem dicat falsigraphus adversariis, quod in propositione quarundam conclusionium illarum inititur 15^e et eius corollario, que probatur per 11^{am}, et hec probatur per 10^{am}, que probatur per geometriam Euclidis, quare adhuc indiget geometria Euclidis; dicendum, quod hoc sit verum de quibusdam conclusionibus supradictis, tamen de omnibus non est verum. Preterea, 10 probatur non supponendo aliquam conclusionem Euclidis, ut patet in prima et 3^a probatione. Si autem obicitur quod in 3^a probatione eius statim supponitur perpendicularem educi, quod fit per 11^{am} primi Euclidis, quare et cetera; dicendum, quod per istam 11^{am} docetur artificialiter extrahere perpendicularem, sed sine illa est illud principium, omni intellectui manifestum, quod a quolibet puncto recta contingit perpendicularem extrahere (licet eius talis extractio pluribus sit ignota). Et illud principium est sufficiens sine illa 11^a supradicta. Est ulterius advertendum, quod per istud principium potest probari 11^a primi *Elementorum* Euclidis, et per illam 15$^{a[m]}$ et ista[m] 10$^{a[m]}$. Hoc probari ut in commento eius apparet; quare illa 10, et omnes que sequuntur ex ea, sine commento goemetrie Euclidis, maxime sicud eas probat Euclides possunt manifeste demonstrari.

PART II. ASTRONOMY AND COSMOLOGY

5

Plinian astronomical diagrams in the early Middle Ages

BRUCE EASTWOOD

Introduction: The nature of astronomy as a discipline

In the early thirteenth century the beginning student of astronomy had Sacrobosco's *Tractatus de spera* as a text to lead him through the elements of celestial motion. Sacrobosco sets out a simple and complete conceptual geometry of the heavenly bodies, wherein successive definitions of the elements of astronomy introduce all the parts of the picture in an ordered and integrative manner. Two centuries earlier – at the death of Gerbert, in the age of Fulbert of Chartres and Notker Labeo of St. Gallen – there was no such text and no such clear and conceptually complete pattern of understanding. Instead the students of astronomy found the theoretical tradition represented essentially by Book VIII of Martianus Capella's *De nuptiis philologiae et Mercurii* and its commentaries[1] plus a set of illustrated astronomical excerpts from Pliny's *Naturalis Historia*.[2] These materials embedded an unintegrated and incomplete set of formal understandings in terminological preciosity and unnecessary grammatical obscurity. The wave of the future in the early eleventh century lay more with the new texts from the Arabic on the construction and use of the astrolabe than with the flawed theoretical materials of the previous two centuries.[3] Indeed, the learning involved in the mastery of these practical astrolabe texts, which required the actual drawing and manipulation of the circles and arcs of stereographic projection for stellar and planetary positions of all sorts, brought the practitioner much closer to a formal, or abstract, comprehension of heavenly motions than the astronomical texts of the preceding era. It is only in the light of the character of astronomical knowledge at the beginning of the ninth century that we can appreciate the teachings offered by the Capellan and Plinian materials and more especially by a set of astronomical diagrams added to the Plinian excerpts for instructional purposes.

141

From the time of Cassiodorus to that of Alcuin the content of astronomy was characterized by series of discrete definitions of terms
which emphasized observation, identification, and occasional tracking
of celestial bodies. Cassiodorus' definition of the West as "the place
where some stars set according to our vantage point"[4] or his definition
of planetary station as the appearance of motionlessness of a star in
some positions though it is constantly moving[5] are statements which
make sense as the results of careful though unmeasured and unstan-
dardized observation. Such definitions are only confusing when approached as foundations for scholarly or classroom study. There is no
coherent conceptual apparatus in Cassiodorus' description of astronomy within which to frame and correct by refinement the definitions
he gives.[6] Instead, astronomy is presented as a series of observational
activities and no more.[7]

Isidore of Seville offers more – much more – of the same. He begins
his chapter on astronomy in the *Origines* with such encouraging (for
moderns) statements as, "Astronomy is the law of the stars,"[8] and
"Astronomy includes the turning of the heavens, the rising, setting and
movement of the stars, and from what cause these things are brought
forth."[9] However, Isidore's reader then begins to learn that the cause
and the *ratio* of astronomy are to be found in definitions, disconnected
descriptions lacking in all mathematical or even qualitative conceptual
framework.[10] The Isidorean view of astronomy as an art finds further
clarification in the *Differentiae*, where the author proceeds from a tri-
partition of philosophy into physics, ethics, and logic to a sevenfold
division of physics into arithmetic, geometry, music, astronomy, astrology, mechanics, and medicine.[11] The definitions given by Isidore
to each of these disciplines reinforce the conclusion we reached after
investigating the meaning of astronomy in the *Origines*. In the *Differentiae*, astronomy is again the law of the stars (*lex astrorum*), with *lex*
meaning authoritative and invariant practice, not a rational concept,
just as in Roman law.[12] As one of seven parts of *physica* astronomy
is categorized with *mechanica*, the methods of making things,[13] and
medicina, the production of cures.[14] Bernhard Bischoff has shown that
the Isidorean sevenfold division of physics was favored in the Irish
scholarly tradition of the seventh and eighth centuries and can be found
in Aldhelm of Malmesbury's *De virginitate*, the Pseudo-Isidorean *De
numeris*, and the grammatical tract of one Clement, an Irish teacher
at the palace school under Louis the Pious, writing during the reign of
Lothar.[15] Clement seems to have used the *anonymus ad Cuimnanum*,
the earliest manuscript of which is the Anglo-Saxon codex at Sankt
Paul in Kärnten MS. 2/1 (25.2.16), fols. 21–42, of the early eighth century. This text was composed ca. 700 or earlier and offers us insight
into a widespread conception of the *artes* at the time. It tells us that

the arts existed in germ but not developed form as early as Adam, the first man. After him the arts unfolded in the world. Thus, Iubal discovered music and Tubalcain, the use of iron and brass, with all the arts appearing in the Wisdom of Solomon. The *anonymus* includes the same sevenfold partition of physics found in Isidore's *Differentiae* and gives examples of the contents of each, with music comprehending all tones and noises down to the underground thunder of earthquakes. Physics even includes plowing, tanning, and masonry according to the *anonymus*.[16] While Bischoff shows that this division of the arts in Clement can be found in many manuscripts of the ninth to twelfth centuries,[17] its significance is much greater for the preceding two centuries, when it had much less competition and was widespread among both insular and continental Irish scholars and their students. With the Carolingian revival of learning the seven arts as represented by Martianus Capella and Boethius came to be the obviously dominant classification, even though the earlier Irish tradition continued with much diminished strength.

The Carolingian Revival was an important turning point for astronomical study. The enlargement of knowledge was accompanied by a change in the nature of the discipline, which had formerly been one of imprecise experiences, based on observing and identifying.[18] Approximately at the time of Alcuin (d. 804) there appeared in the Carolingian world a set of astronomical excerpts from Pliny,[19] which found their place frequently in conjunction with calendrical computus texts and often with other materials of astronomical or cosmological nature. During the first half of the ninth century Martianus Capella's *De nuptiis* became the authoritative textbook for many of the liberal arts, especially for astronomy (Book VIII), and a Plinio-Capellan background combining these two sources seems to have been the most widespread theoretical account of planetary motion from the ninth century until the middle of the eleventh century.[20] To speak of a theoretical account of planetary motions in the ninth and tenth centuries is not to imply that anything comparable to, let us say, Aristotle's theory in the *De caelo* could be found, but rather that a number of doctrines about the stars and planets came to form enough coherence and continuity to give astronomy a conceptual element quite absent in the preceding era, dominated by the work of Isidore. The diagrams for the astronomical excerpts from Pliny witnessed and helped to spread this new level of understanding.

Origin and contents of the Plinian excerpts

The excerpts from Pliny are six in number and were studied in detail by Karl Rück, who identified the locations in the *Naturalis Historia* from which the elements were drawn and then reordered in these six combinations.[21] From Book II excerpts were taken on four subjects: the order and periods of the seven planets, the harmonic intervals between the circular orbits of the planets, the apsides of the planets, and the ascents and descents of the planets primarily with respect to latitude in the band of the zodiac. These four excerpts have accompanying diagrams in many medieval manuscripts, while the remaining two excerpts, on the changes of the seasons and on weather prognostications, have no diagrams in the manuscripts. Not only do the first four have illustrations, they also come closer to providing basic teachings for a developed theory of planetary motions, and it is these illustrated excerpts with which we are concerned.

On the basis of ten manuscripts from the ninth to eleventh centuries[22] Rück argued for the compilation of a large astronomical computistical text, including the Plinian excerpts, in an Anglo-Saxon monastery somewhere in the area under Bede's intellectual influence not too long after his death, about the mid-eighth century.[23] Hence the name York Excerpts for the Plinian texts. The path followed by this astronomical computistical material to the continent, according to Rück's conjecture, was via Irishmen traveling late in the eighth century from Britain to Salzburg, where the compilation became the basis for a new composition of the same sort, including a new computus, in the year 810.[24] These are the roots claimed for the astronomical computistical text surviving in the Munich MS. CLM 210 and Vienna MS. 387, both copied from the same exemplar within a decade after this new compilation.[25] V. H. King, on the other hand, has argued cogently and tellingly against Rück's thesis, concluding that the two Plinian excerpts from Book XVIII could not have been made in England, that the set of excerpts was made on the Continent, and that other components as well bear witness to the compilation of the astronomical computistical materials in CLM 210 and Vienna 387, known as the Three-Book Computus, for the first time on the Continent in or very near the year 810, a date found in the computus itself.[26]

In the light of present knowledge there is no sound basis for maintaining Rück's thesis, and we are left with the conclusion that Bede's England was not the origin of the Plinian excerpts. Beyond this knowledge, however, the localization of their origin remains open to speculation, for King was not able to find convincing reasons for positive assignment of place or time to the excerpts themselves, although he argued with success for the compilation as a whole.

The first excerpt, *De positione et cursu septem planetarum*, simply and briefly lays out the planetary order, pointing out initially that the distances are definite (*certis spatiis*) and that the seven planets are called wanderers even though nothing wanders less than they do. Saturn takes thirty years to travel a path which is a circle, returning to its exact starting point. Jupiter also makes a circle in its more rapid, twelve-year orbit. Mars, close to the sun and fiery, takes two years to return to its initial point. The time of the sun's orbit is $365\frac{1}{4}$, which is 360 *partes* plus $5\frac{1}{4}$ days computed by observation of the sun's shadow. Venus and Mercury have a peculiar relationship (*peculiaris ratio*) with the sun, since they are never found at large angles from or in opposition to it as are the other planets. Venus is large, varies its course, completes a circuit in 348 days, and remains within forty degrees of the sun. Mercury acts likewise but to a lesser extent (*simili ratione sed nequaquam magnitudine aut vi*), moves circularly, takes nine days less than Venus for a full circuit, can be seen shining at times before sunrise and at times after sunset, and remains within twenty degrees of the sun. Finally, the moon, which is the most familiar and most observed of the planets, exhibits a variety of traits. Always waxing or waning, it passes from horned to half to full orb, can be seen spotted and also flawlessly bright, and shines day and night, high in the sky to the north and low in the sky to the south. It requires at least $27\frac{1}{3}$ days in the sky plus two more in conjunction with the sun before emerging on its course anew. The information in the excerpt is only marginally more detailed than that in Isidore of Seville's *De natura rerum* 23,[27] but the Plinian text emphasizes precision and the regularity of the planetary orbits.

The second excerpt, *De intervallis earum*, is the shortest of the four. It records a diversity of attempts to calculate the interplanetary distances and establishes a set of musical, or harmonic, intervals from the earth to the celestial sphere, carrying the zodiac. While some persons have argued, records the excerpt, that the moon–sun distance is nineteen times the earth–moon distance, Pythagoras computed the distance from earth to moon as 125,000 stadia, with the moon–sun distance double that and the sun–zodiac distance as triple the earth–moon. On the other hand, Pythagoras also used musical intervals at times to designate the distances.[28] These intervals are: earth–moon 1, moon–Mercury $\frac{1}{2}$, Mercury–Venus $\frac{1}{2}$, Venus–sun $1\frac{1}{2}$, sun–Mars 1, Mars–Jupiter $\frac{1}{2}$, Jupiter–Saturn $\frac{1}{2}$, Saturn–fixed stars $1\frac{1}{2}$. The total is seven tones.

The third excerpt, *De absidibus earum*, provides more information than Bede's chapter with the same title.[29] The excerptor has chosen to describe variations both in planetary visibility and in longitudinal motion, in each case explaining the apparent motions by the less evident, underlying causes. Planetary apsides are shown to be fundamental to the explanation of each kind of appearance. Considering first the outer planets, Pliny reports that they are controlled over their whole circuits by solar rays. These planets are hidden when traveling in conjunction with the sun. Their morning rising occurs with the sun beyond 11°. At 120° they make their first, or morning, station. At 180° they have an evening rising. Continuing on to 120° from the other side they make a second, or evening, station, and finally these planets return to invisibility within 12° of the sun, which is called their evening setting. Among these outer planets Mars is very close to the sun and so is affected also at 90° from either of its risings, remaining for six months in the zodiacal sign of its station, whereas the other superior planets never have a station of more than four months in one sign. The two inferior planets are obscured in an evening conjunction, as are the other planets. When apart from the sun, the inferiors have morning risings in the same way as the outer planets. From their greatest elongation the inner planets return to the sun, where they make a morning setting and disappear. Again, they rise in the evening and reach their appointed limit from the sun, at which point they retrograde toward the sun and become invisible in an evening setting. Venus has two stations at its greatest elongations after its two risings, morning and evening. Mercury's stations are too short to be seen. The reason for all this is the force of solar rays on the planets at trine, preventing the forward motion of the planets and pushing them farther away by the fiery radial force. We cannot perceive this directly, so that the planets seem to us to stand still, whence the name, station. The violence of the solar ray becomes extreme here and compels the planets by its heat. The effect is much stronger at evening rising when the sun faces the planets fully and expels them out to their apogees (*summae absides*). Here they are least observable because of their maximal distance and minimal speed, so much the slower when this occurs in the farthest signs. The planetary circles are called apsides from the Greek, with each planet having its own circle, unlike those of the celestial sphere. Whereas the zodiac is a band with the earth at its center, each *absis* has its own center, and the *absides* vary in path, length, and motion. The apogees, or highest apsides (*absides altissimae*), are as follows: Saturn in Scorpius, Jupiter in Virgo, Mars in Leo, sun in Gemini, Venus in Sagittarius, Mercury in Capricorn, with each apogee at midpoint in its sign and each perigee at the midpoint of the opposite sign. Thus the planets appear smaller and slower moving at apogee, while they look larger and swifter near perigee. The natural motions do not change but are constant, the appearances resulting necessarily because of changing distances, which can be shown in a diagram with radial lines from the outer circle becoming closer together just as in the spokes of a wheel.

The fourth excerpt, *De cursu earum per zodiacum circulum*, is the longest and describes the most complex phenomena covered by the planetary excerpts. There is nothing at all of this information in Isidore's *De natura rerum* or his *Origines*, and Bede provides only the initial, succinct list of planetary latitudes from Pliny with none of the other material found in this excerpt.[30] As excerpted by the medieval compiler, Pliny describes a variety of appearances during the zodiacal circuits of the planets and explains them as effects of solar radial forces, just as he explained the stations and retrogradations of planetary orbits in the previous excerpt.

Why do the sizes and colors of planets change, and why do they move toward the north and then away to the south? There is latitude and obliquity to the zodiac, whereby these appearances occur. And only those parts of the earth under the zodiac are inhabited, while the remainder perish under the influence of the poles. Venus runs beyond the zodiacal band by two degrees. The moon traverses the full width of the band but never exceeds it. Aside from these, Mercury is the most variable, but of the twelve degrees in the band, this star does not traverse more than eight, of which two are in the middle, four are above the middle, and two are below. The sun travels in the middle, between two *partes*, unequal, on the twisted path of dragons. Mars travels on the four middle degrees; Jupiter, on the middle and the two above; Saturn, on two like the sun. The three outer planets climb in latitude from morning rising and approach the north, then descend to the south from their evening rising. With the solar ray coming from the other side, the planet is depressed toward the earth by the same force by which it was earlier supported in the heavens. Such is the difference between rays coming from below and those from above. Venus, like the outer planets, begins to ascend in latitude from morning rising and to descend in the evening. In its morning setting it drops in altitude, and at evening station it simultaneously retrogrades and loses in latitude. Mercury climbs from its morning rising in both latitude and altitude. However, after its evening rising it loses in latitude, following the sun at a fifteen-degree interval, and remains almost motionless for four days. Then it loses altitude and retrogrades from its evening setting until its next morning rising. This planet descends for the same number of days that it climbs.[31] Venus takes fifteen times as long to ascend as descend. Saturn and Jupiter descend in twice the time it takes them to ascend. And such is the variety of nature that Mars descends in an interval four times as long. Mars never makes a station when Jupiter is 120° away and rarely at a 60° interval. They make a simultaneous rising when in the signs Cancer and Leo. Mercury rarely has evening risings in Pisces, frequently in Virgo; morning risings frequently in Aquarius, very rarely in Leo. It does not retrograde in Taurus or Gemini and not less than 25° within Cancer. The moon has a double conjunction with the sun in no sign other than Gemini; there is no lunisolar conjunction in Sagittarius. The very first day or night of the new moon is visible only in Aries. The moon responds to variation in the approaching rays from the sun. At quadrature the moon is half full. At trine a third appears obscured, and it is full in opposition. As it wanes the moon exhibits the same shapes at the same intervals. The explanation of its motions is the same as that for the three outer planets. Saturn and Mars are invisible for 170 days or more, Jupiter for 36 or at least 26, Venus for 69[32] or at least 52, Mercury for 12[33] or at the most 17. The invisibilities occur because of conjunction with the sun and the crossings of their inclined circles (*absides*) and the extremities of their orbits, since it is at these places they are obscured. The planets are not seen for the greatest number of days when stations coincide with the crossings and extremities of their paths. Likewise they struggle in the heat (*vapor*) of the sun, and though with difficulty, they still descend. But of all these planets the path of Mars is the most unobservable.

Any comparison of the content of these excerpts regarding planetary motion with the passages in Isidore or Bede on planetary motion will show the superiority of the Plinian excerpts as theoretical accounts. The complexity of the phenomena explained is much greater in the last two excerpts, and it is here that the attempt at theoretical explanation is most impressive, when seen in the light of the noncausal descriptions in Isidore and Bede. Longitudinal motion, especially the stations and retrogradations, are considered to be regularly periodic and to be caused by the force of solar rays, also described as the vaporous heat of solar rays. In the fourth excerpt variations in latitude and all the diversity of planetary risings and settings are grouped together along with more data on stations, retrogradations, and invisibilities. The limitation of planetary paths to the zodiacal band is combined with the prior explanation of stations and retrogradations to account for the very great variety in visibility and invisibility of planets, including their motions in latitude. The explanation is hardly successful, but it offers an attempt at an integrated reasoning, however flawed. What the Plinian excerpts on the planets clearly mean to do is to show that apparent diversity can be seen from the simplest cases of the circularity of planetary paths with fixed, though different, periods, distances, and centers, to the complex cases of varying speeds, varying latitudes, and varying visibilities. When compared to the alternative astronomical texts of the beginning of the ninth century, the Plinian excerpts are alone in their claim to provide much precise data and a framework of ordered causality within which to understand the data.

Origin and development of the diagrams

When we investigate more precisely the connections between the four Plinian excerpts on planetary astronomy and the two distinct computistical collections, we are reminded that both the excerpts themselves and the accompanying astronomical diagrams in the Three Book Computus and the Seven Book Computus appear in different groupings. Wien 387 and CLM 210, the sole representatives of the Three

Book Computus, contain the first three of the four planetary excerpts: (a) *De positione et cursu septem planetarum*, (b) *De intervallis earum*, and (c) *De absidibus earum*. This computus also contains only one completed Plinian diagram, that for the planetary harmonic intervals.[34] The computus has as well an unfinished diagram, which follows the preceding illustration and would seem at first to be an incomplete version of a diagram for planetary apsides.[35] However, this is not the case, as I shall argue later.

Looking for the Plinian planetary excerpts in the Seven Book Computus, we can find all four, that is, the same three as in the previous computus plus (d) *De cursu earum per zodiacum circulum*. In the majority of the surviving manuscripts of the Seven Book Computus there appear diagrams for some but not always all of the planetary excerpts. The earliest, Madrid 3307, provides a diagram for each of the four.[36] Only one other manuscript of this compilation contains the full set of diagrams, Monza F.9.176 of the middle of the ninth century.[37] A third ninth-century example, Vat. lat. 645, has the first two diagrams, preceding the Plinian texts,[38] while two other manuscripts, Vat. Regin. lat. 309 and Paris BN n.a.l. 456, have none of the planetary diagrams.[39]

In addition to the two astronomical-computistical compilations, many codices of cosmological-astronomical texts from about the middle of the ninth century onward include one or more of the Plinian planetary excerpts, frequently with diagrams. For his study of the Plinian excerpts Rück used not only CLM 210 and Vienna 387 but eight more manuscripts, which are examples of this third, loosely defined group.[40] There are as well very many more manuscripts having these excerpts, with and without the astronomical diagrams.[41] Finally, there are manuscripts which have one or more of the Plinian diagrams without the appropriate Plinian texts, and these manuscripts grow rapidly in number after the eleventh century, as the diagrams lose scientific significance and may even appear as decorative elements on occasion.

The earliest diagram extant for any of the Plinian excerpts is that for the planetary intervals, found in exactly the same form in the two manuscripts of the Three Book Computus (Figure 1).[42] The first three excerpts follow consecutively in the manuscript, with the excerpt on planetary apsides concluding on the final line of a folio with "sicut in rotis radios ut subiecta figura demonstrat." This is immediately followed on the next folio by the diagram for intervals (excerpt b) on the recto and an incomplete diagram on the obverse. There is some confusion in that the diagram for planetary intervals comes immediately after the excerpt for apsides, so that a reader might initially think that the *subiecta figura* for apsides is the diagram depicting intervals. It appears that the diagrams entered the Three Book Computus as a separate element, taken from a different manuscript and added to the orig-

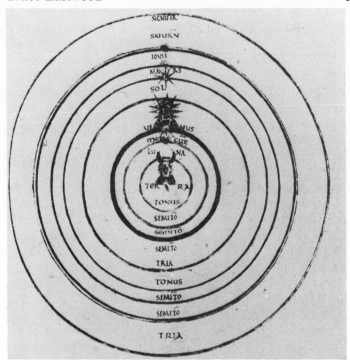

Figure 1. Planetary intervals. (Vienna Nationalbibliothek MS lat. 387, fol. 123r.)

inal computus manuscript, from which CLM 210 and Wien 387 derived. Because we have the closing sentence of the apsidal excerpt referring to a diagram devised with the composition of the excerpt, while there is no reference in Pliny's original text to such a diagram, we recall a conclusion reached above, that the Plinian excerpts cannot be shown to be original with the Three Book Computus. Therefore, the appearance or nonappearance of diagrams with the excerpts in this computistical compilation of ca. 810 is wholly inconclusive. We have the diagram for intervals in a less than proper location in the manuscript, and the succeeding diagram may or may not be that devised for the apsidal excerpt.

The diagram which follows that for planetary intervals presents a number of difficulties (Figure 2). The first is that it is unlike any Plinian apsidal diagram in any other manuscript of the Middle Ages.[43] This difficulty alone does not disqualify Figure 2 as an early attempt to represent a part of the text, subsequently abandoned for another diagram. However, the next difficulty is that the diagram we have in the Three Book Computus does not correspond well to one or another part

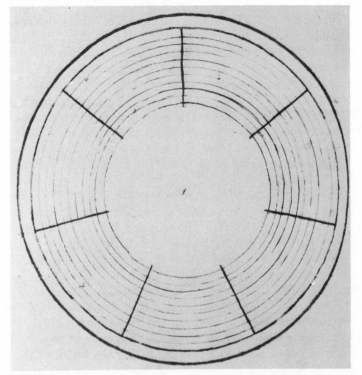

Figure 2. Incomplete diagram. (Vienna Nationalbibliothek MS lat. 387, fol. 123v.)

of the apsidal excerpt. The illustration has no labels whatsoever, which makes it very unlikely that we are looking at a finished diagram. Yet the final clause of the excerpt makes it clear that a diagram had been worked out to accompany the text. If we persist and assume that Figure 2 shows an incomplete apsidal diagram, we must find reasons for the elements of that diagram.

The figure on folio 123v consists of twelve concentric circles, a large empty central circle, and seven equally spaced radial lines, which begin just inside the central circle and end at their intersections with the eleventh circle, leaving the outermost band uncut. The two outermost circles, forming the outer band, are heavier lines than the ten circles within. The outer band may well be meant to be the zodiac, but we may wonder why there are not divisions of that band into twelve. If the inner circles are for planetary orbits, we should ask why there are ten bands for only seven planets. An answer which proposes the seven planets plus the three sublunar elemental bands of fire, air, and water would be quite unsatisfactory, since the elements have nothing to do with Plinian apsides. The number of bands is obviously incorrect. Next we should consider the concentricity of the circles. The Plinian text makes it quite clear that the planetary circles are not concentric and not centered on the earth. Last we consider the seven radial lines. These, at least, seem candidates for properly descriptive elements, the spokes of a wheel referred to at the end of the excerpt. However, it is not apparent why there should be seven radial lines, since the excerpt lists apogees for only six planets. Indeed, if six planets were not the basis for the number of radial lines, twelve signs of the zodiac would be the next most likely basis, since the outer band is evidently the zodiac.[44]

In sum, there is insufficient reason to label this diagram an incomplete or early version of an apsidal diagram. Instead we can say it is the wrong diagram, which is the most probable reason for its being incomplete. However it strayed into the two surviving manuscripts of the Three Book Computus, it did not represent a Plinian apsidal diagram but was somehow mistaken for one by the copyists. Being the same in CLM 210 and Wien 387, this erroneous and incomplete diagram must have been in the original manuscript of the Three Book Computus, from which these were copied but which does not survive. We must look to other manuscripts for the apsidal diagram mentioned at the end of the excerpt, and we must conclude that the origin of the apsidal diagram predates the compilation of the Three Book Computus, although the extant diagrams are later.

Only with the compilation of the Seven Book Computus, whose manuscripts date from the third decade of the ninth century onward, have we good evidence for the appearance of the Plinian diagrams. Comparing the intervals diagram in Wien 387, the Three Book Computus, with that in Madrid 3307, the earliest of the Seven Book Computus manuscripts,[45] we find essential identity. The Madrid manuscript shows four minor differences (see Figure 1): (a) an absence of the pictorial busts for the sun and moon, (b) placement of seven circular spots for the planetary bodies on a horizontal radius to the left rather than on a vertical radius, (c) placement of the planetary names below rather than above their respective circles, and (d) a somewhat more careful measuring of the intervals between planetary circles in order to show better the varying distances. Each change is no more than an individual improvement of a stable image and may well represent the original form of this diagram better (in Madrid 3307) than do the early extant versions in CLM 210 and Wien 387. Lacking only the more careful measurement of intervals between circles, Vat. lat. 645 shows exactly the same image as Madrid 3307, including the other individual improvements noted. Monza F.9.176 is identical with Madrid 3307 in its image for intervals, including the four improvements listed. Par-

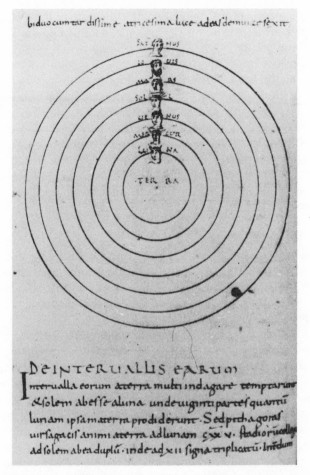

Figure 3.   Planetary order. (Monza Biblioteca capitolare MS lat. F.9.176, fol. 70v.)

enthetically we can add that Harleian 647, one of the very early non-computus texts with illustrated excerpts, also gives a figure for planetary intervals the same as Madrid 3307.[46] In conclusion, there is a commonality to the early diagrams for intervals that suggests a similar commonality for the other Plinian diagrams.

There is no diagram in the Three Book Computus accompanying the first excerpt, *De positione et cursu septem planetarum*. In manuscripts of the Seven Book Computus this illustration is essentially invariant and remains so in much later manuscripts (Figure 3).[47] When we return to the subject of the apsidal diagram, we find considerably more diversity, yet enough stability in the general form to increase our confidence in the judgment that the incomplete diagram of the Three Book Computus (Figure 2) is in no way a version of a Plinian apsidal diagram. Because of the greater variety and complexity of information in the excerpt on apsides, more diversity is evident in the details of the apsidal diagrams. Yet the basic form is so constant that we can define with reasonable certitude what must have been present in the now lost original diagram to which the excerpt refers. The central point emphasized is the zodiacal location of apogees, and this alone necessitates representation of the twelve divisions of the zodiac and the eccentric orbits of the six planets arranged so as to show the correspondence of each apogee with the appropriate sign. Other information may or may not be included in one or another apsidal diagram in the manuscripts.

Two examples (Figures 4 and 5) offer both the sort of diversity and the faithfulness to fundamental concern found in the diagrams. Among the secondary differences between these two we can list: greater or less intersection of the apsidal circles, representation or not of the motions in station and retrogradation, and inclusion or not of planetary bodies with radial lines from those bodies to the middle. Examination of each of these variations reveals the limits of the diagrams as conveyors of precise information.

First, with regard to intersections of the circles, it seems obvious that the diagrams make no pretense of representing planetary order and intervals, given in the first two excerpts. In fact, the variably eccentric orbits mentioned in the excerpt pose a real problem for one who has just finished learning apparently simple regular intervals. Rather than deal with that problem, however, the apsidal diagrams step away from a concern with complete orbits and emphasize only apogee (and perigee), and this emphasis involves accentuating the eccentricities to the point of multiple intersections of the planetary circles. Nothing other than location of apogee and perigee for each planet is clearly intended in the diagrams.

Second, with respect to the representation of stations and retrogradations, Paris 5239 (Figure 5) does not include explicit reference in the diagram, whereas Madrid 3307 (Figure 4) does so. The stations would seem to be the redirection more or less radially from the center, while retrograde appears to be intended by a sharper reversal of direction on the circle. While the stations and retrogrades are correctly placed at apogee and perigee in Figure 4, there is a curious distinction in detail that is nowhere stated or explained in the Plinian excerpts. A careful tracing of the circles shows that Mars and Venus have stations at apogee and retrograde at perigee, while Saturn, Jupiter, Mercury, and the sun have stations at perigee and retrograde at apogee. Most curious of all in this arrangement is the attribution of station and retrograde to the sun, which, according to Pliny, is the cause of these deviations by

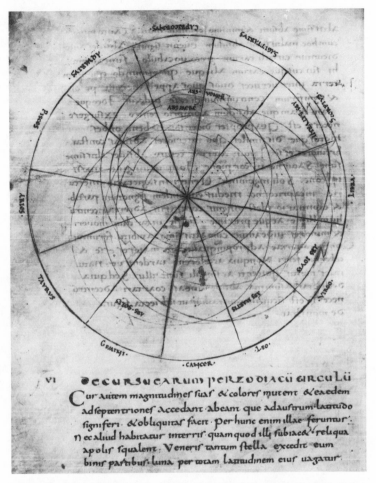

Figure 4. Planetary apsides. (Madrid Bibliotheca Nacional MS lat. 3307, fol. 65v.)

the other planets and does not undergo station or retrograde at all. Further care in tracing the circles will show that not all the planets are moving in the same sense, if the zigzags for stations and retrogrades are taken seriously. Disregarding the orientation of the zodiac, we find that the circle for Jupiter in Figure 4 has stations and retrograde arcs opposed in sense to those of the other planets. If the planets are taken to move clockwise, Jupiter's station and retrograde arc are shown as drops in the planet's position with respect to the center, whereas Saturn and Mars have the reverse situation. With counterclockwise motion

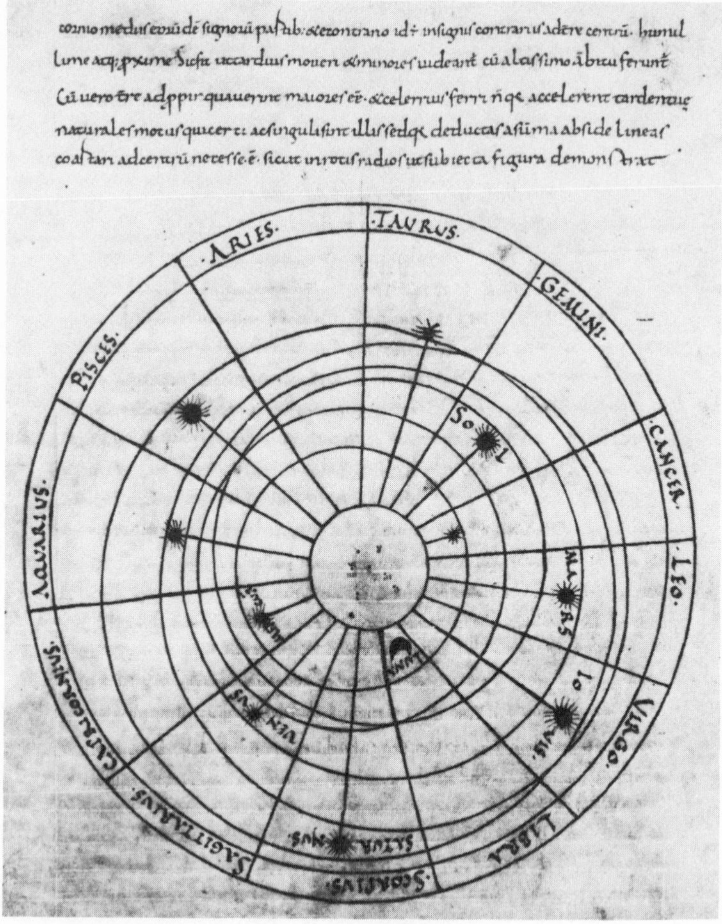

Figure 5.   Planetary apsides. (Paris Bibliothèque Nationale MS lat. 5239, fol. 125v.)

the situation is simply the opposite, neither of which – clockwise or counterclockwise – is acceptable according to Pliny's explanation that the solar rays force the planet outward and away from us at station. The result of careful scrutiny of such a diagram is, therefore, that these details cannot be accepted as correct Plinian doctrine. Whether the errors should be blamed on copying or on the original designer we cannot say.

The third variation noted in these apsidal diagrams is the insertion of planetary bodies with radial lines attached (Figure 5). The labeled planets are placed in the signs of apogee, and the unlabeled planetary bodies are in the opposite signs, or at perigee. Other surviving apsidal diagrams normally have the label alone or only the visually apparent point of apogee to indicate the sign for apogee, unlike Figure 5, with the location of the perigee understood tacitly. But Paris 5239 not only adds the planetary bodies and radii for both apogee and perigee, it modifies the pattern slightly in the interest of visual balance to the detriment of doctrinal completeness. There are no indicators for the perigees of the sun and Venus, since these two have apogees in opposite signs and explicit inclusion of perigees would require two planetary bodies each in Gemini and Sagittarius.

The characteristic, primary elements of the Plinian apsidal diagrams make them recognizably distinct from other astronomical diagrams and certainly from the other Plinian figures. The same can easily be said for the fourth Plinian illustration, designed to accompany the excerpt *De cursu earum per zodiacum circulum*. The diagram appears in only two of the manuscripts of the Seven Book Computus[48] as well as in other early computistical and astronomical collections.[49] The earliest surviving examples show a remarkable uniformity of both conception and realization. The concern of the diagram is to depict planetary latitudes according to the inclinations reported by Pliny near the beginning of the excerpt (Figure 6). This is accomplished with a set of thirteen concentric circles to represent the twelve degrees in width of the zodiacal band plus an eccentric circle for each planet except Saturn and the sun, with the extent of orbital inclination determining the amount of eccentricity of the planet's circle against the background of thirteen concentric circles. Saturn and the sun, in accord with the excerpt, have the same path, which is a serpentine line bounded by the two degree lines at the middle of the zodiac.

The difficulty in reading this circular latitude diagram is inherent in the form. The discernment of the numerical value and latitudinal location of a planetary circle in the zodiac could require visual tracking of the complete circle, especially in the cases of Mars and Jupiter, whose Plinian latitudes are not as remarkable as the rest. The result of this slowness in reading the diagram could be the omission of circles for Mars and Jupiter, as found already in the second half of the ninth century (Figure 7). Later examples simply repeat or even increase this tendency to emphasize the obvious and ignore the less obvious planetary circles.[50]

Each of the four planetary diagrams described above saw a significant modification in form, and the change in each case was dictated by a desire for more rapid communication of information. The most interesting of these transformations took place in the diagram for latitudes. The new form for presenting the data has drawn attention, because it appears superficially to be a graph, but it is not (Figure 8).[51]

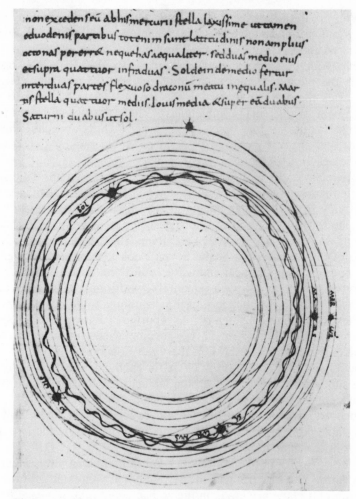

Figure 6. Planetary latitudes. (Monza Biblioteca capitolare MS lat. F.9.176, fol. 73r.)

The earliest extant example of this new, rectangular diagram for latitudes appears in Bern 347, a manuscript which has the circular latitudes diagram on the facing page (Figure 7).[52] This unique coexistence of the two forms together in a ninth-century manuscript suggests strongly that the inventor of the rectangular diagram was its designer for this manuscript, which was probably written at Auxerre in the latter half of the century.[53] The advantages of the invention are striking. If speed of communicating the inclination of orbits is a primary desideratum, the rectangular diagram is vastly superior to the earlier, circular version.

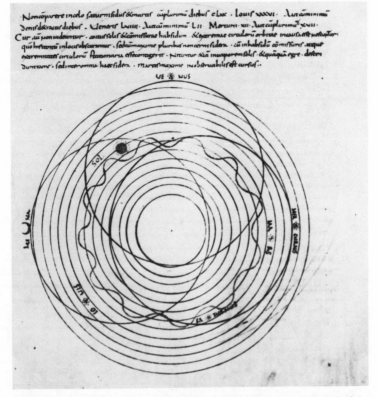

Figure 7.   Planetary latitudes. (Bern Bürgerbibliothek MS 347, fol. 25r.)

On a grid of 12 × 30 squares with the solar serpentine path at the midhorizontal, the most cursory view of the labeled paths in Figure 8 – from top to bottom, Venus, Mercury, sun and Saturn together, Mars, Jupiter, moon – reveals the relative values of the latitudes, and a quick count of vertical squares readily shows the numerical value of any of the latitudes. Almost parenthetically, it should be noticed that this greater ease of reading the diagram has not prevented mistakes. The latitude of Mercury is a full degree too small. Jupiter's latitude is only half the value given in the excerpt and is incorrectly located on the zodiacal band as well; according to the excerpt the latitude is four degrees, covering the middle two and two above that.[54]

The purpose of the grid was clearly limited to showing vertical dimension with nothing other than a sense of symmetry to guide the designer as he proceeded across the grid. Thus there is no graph, be-

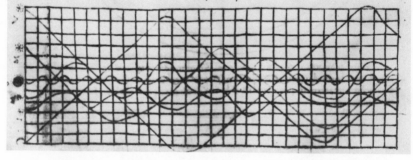

Figure 8.   Planetary latitudes. (Bern Bürgerbibliothek MS 347, fol. 24v.)

cause there is no plotting of rectangularly coordinated positions. The planetary periods are not part of the excerpt, and they are not part of the diagram. As a result we can find a cycle of Venus taking only slightly longer than one for the moon and significantly longer than those for Jupiter and Mars, just the reverse of reality. The basis for horizontal wavelength, as it were, is simply the vertical amplitude. The greater the planetary latitude, the longer the horizontal dimension of a cycle. This rule inclines the diagrams toward a parallelism of planetary lines which is realized and strictly followed in a number of manuscripts of the ninth to eleventh centuries.[55] The rule also means that care in drawing the full horizontal course of the planetary lines is largely a matter of artistic rather than scientific concern.[56] The thorough lack of focus on horizontal, or longitudinal, values becomes clearer when we recognize that the original, circular diagram did not allow even the suggestion of mistaken orbital periods, yet that diagram's scientific superiority in this case was not noticed or else was ignored as irrelevant. Only latitudes are of any significance in the diagrams for the fourth excerpt, and the rectangular grid form of the diagram was immensely successful, utilized in introductory texts on cosmology long after the Plinian excerpts had ceased to be up-to-date science.[57]

The change in the diagram for planetary apsides has been described in detail elsewhere and needs only to be summarized here.[58] The essential concern motivating the change was solely pedagogical, intent on reducing the time for identifying apogee locations, the same sort of reason that led to a revised latitude diagram. The revised apsidal diagram went so far as to reorder the zodiac, thus presenting an element analogous to the grid with latitudes; the unnatural zodiac in one case can be compared with the apparent planetary periods in the other. In each case the counter-natural appearance is insignificant, because it is not the phenomenon to which sole focus is given. The reordered zodiac of the revised apsidal diagram (Figure 9)[59] involved a limited change in which the six signs without apogees were omitted and the six signs for apogee were spread out around the circle, whether at equal intervals or not, but retaining the correct sequence for these six signs.[60] The effect of the reordered zodiac can be seen by comparison of Figures 4 and 9. In Figure 4 most apogees are clear, but those of Mercury and Venus require a pause, and many apsidal diagrams of this form are less immediately clear for other planets as well. In Figure 9 the spreading of the six signs makes every apogee immediately apparent. Even this amount of disturbance to the natural order of the zodiac seems to have occasioned some resistance to use of the revised form of apsidal diagram. While the earliest example comes from the late ninth century and many examples appear in the tenth and eleventh centuries, all but one of these examples is surely or possibly from southern Germany or Switzerland. The innovation seems to have taken place here, perhaps first at St. Gallen, and not to have been copied much at all outside this region.[61] This pedagogical invention was perhaps too questionable for most scholars and teachers of the era.

Even the illustrations accompanying the first two excerpts saw innovation in the ninth century. The diagram for planetary order was changed to a horizontal row of seven small circles labeled with the planetary names in proper order. The diagram for planetary intervals was changed to a vertical list of the intervals, omitting all circles. For these two innovations speed of communication was again the rationale. However, both the earlier and the later forms of each type of diagram

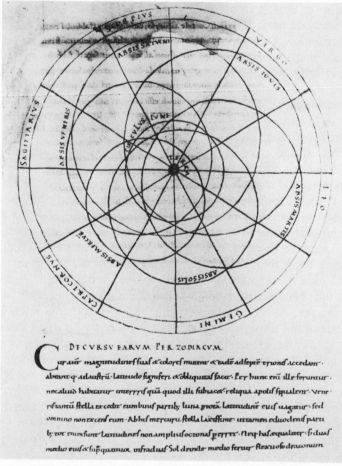

Figure 9. Planetary apsides. (Biblioteca Apostolica Vaticana MS
Palat. lat. 1577, fol. 81v.)

continued in wide use, since the initial forms were as quickly understood as the revised forms.

Conceptual attributes of the Plinian diagrams

Margaret Gibson and John Marenbon have commented on the
simple but fundamental shift they find to have occurred around the
middle of the ninth century, a shift of scholars and their glosses from
precisely limited definition and compilation to a more adventurous approach to their texts wherein they attempted to reason, synthesize, and
make judgments with regard to the materials on which they commented.[62] Evidence for this trend can be drawn from essentially sci-

entific as well as philosophical sources, for example, an important commentary on Bede's *De natura rerum*, made during the third quarter of the ninth century, probably by a scholar of Laon. The commentator shows an eagerness "to find natural and rational explanations," and is willing both to reconcile differences between Bede and Martianus Capella and at times to criticize Bede in favor of Martianus.[63] With respect to the Plinian excerpts and their diagrams a similar conclusion can be made. The excerpts are especially interesting at points where they excise intervening material and even change the sequence of materials presented. Thus the excerpt on apsides displaces the definition and location of apsides and apogees (*Naturalis Historia* II, 63–64) so that, unlike Pliny's sequence, they follow the causal explanation (*N.H.* II, 69–70), which has been placed immediately after the observational data on stations and retrogradations (*N.H.* II, 59–61).[64] The process of partial elimination of information in order to make a more coherent text in the fourth excerpt, on latitudes and related phenomena, exhibits the same tendency. The concern of the excerptor is to create a continuity of parts formerly separated. Not intent on reaching a more fundamental, abstract level of reasoning but rather motivated by a desire to offer an ordered sequence of information, both data and explanation, the creator of these Plinian excerpts attempted to show more coherence in the material than Pliny had. Such a goal is essentially pedagogical, and the composition of the excerpts is better attributed to a *scolasticus* of the early Carolingian era than to a scholar unconnected with teaching.

While less complex than the Plinian excerpts, the planetary diagrams accompanying them offer the same sort of improvement over earlier pedagogy. A common problem in early medieval discussions of geometry will illuminate the situation. As Richard Krautheimer observed some time ago, scholars from Isidore to Gerbert were notably vague in describing geometrical shapes, and yet they were quite precise and emphatic in giving to their readers the number of parts making up a geometrical pattern.[65] The astronomical situation was similar in the late eighth century. Not only were the astronomical materials in Cassiodorus, Isidore, and Bede far less coherent than the Plinian excerpts, there were no diagrams for planetary astronomy except for one Isidorean *rota* (Figure 10). In fact, this *rota* is not an astronomical diagram in the same sense as the Plinian diagrams. Isidore's pattern is a set of concentric bands which do indeed represent the order of the planets from the earth to the firmament. However, the image goes further than this clear and limited purpose by inscribing in each planetary circle a number for the planet's cycle. Not only can the spatial diagram not convey this temporal information directly, but the information given works directly against a straightforward translation of the data into the

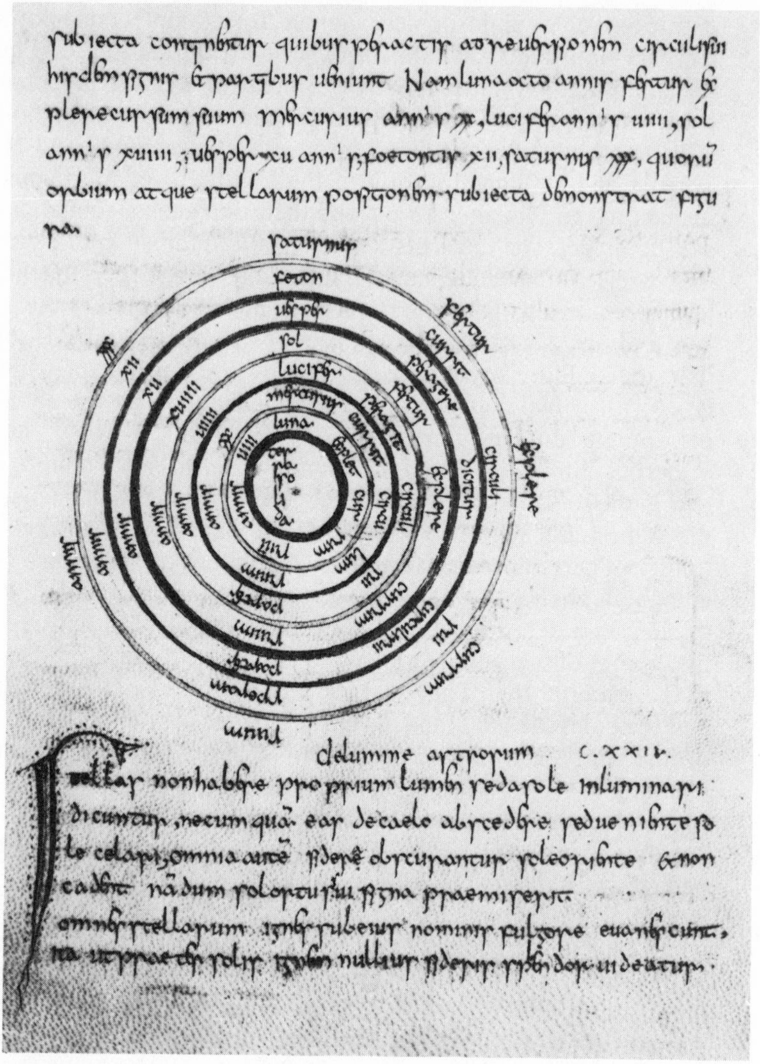

Figure 10. Planetary cycles. (Basel Universitätsbibliothek MS lat. F.III.15(a), fol. 5r.)

terms of the diagram. Some of Isidore's numbers here are years in single circuits, and some numbers are for the planetary circuits in calendrical or other types of compound cycles. Because the information is a focal (and verbal) element in the image, it cannot be ignored. The result is a figure which is not simple, is not integrated, and is not coherent. As with the text of Isidore, we have here a body of discrete information, connected by words but not by a uniform spatial conception.

The diagrams for the Plinian excerpts are witnesses to a definite conceptual advance when viewed in their proper context. In the first two diagrams spatial sequence and spatial intervals are presented directly, unambiguously, and without complication. In the third diagram the location of apogees and perigees and the related phenomena of stations and retrogradations are reduced to functions of the regular variation in distance of points on the periphery of a circle from a point removed from that circle's center. The diagram, in either its original or its revised form, makes the reported apsidal phenomena regular and coherent, even though the circles in the diagram are not to be taken as fully representative of the physical paths of the planets. In fact, the diagram's abstraction from representation of full physical reality is an advantage, which enhances the sense of order in the phenomena represented.[66]

In the fourth latitude diagram both forms show an ability to understand more complex spatial representations. The revised, or grid, form indicates a readiness to transform locations from a circular track to a rectangular framework and vice versa. The original, or circular, form bears witness to at least partial awareness of a means to translate spherical into planar information. The planar circular zodiac with planetary circles eccentric in proportion to their inclination to the ecliptic would seem to be a simplified version of stereographic, or astrolabe, projection. Evidence of knowledge of stereographic projection at the same time as the creation of the excerpts and diagrams has been reported by John North,[67] and there seems no reason to question the ability of someone in the same intellectual orbit to develop the circular latitude diagram. It requires no more than shifting from the plane of the equator to the plane of the ecliptic as plane of projection, and then spreading out the twelve degrees of the zodiac so that its projection allows ample space to perceive the location of the planetary circles, which are inclined to the ecliptic and therefore eccentric to the concentric circles of the zodiac. The simplification involved in this adaptation of stereographic projection does not argue for widespread knowledge of the fundamental technique among students of astronomy, but the diagram does argue for pedagogical ability of a high order for the early Carolingian period. Through the Plinian excerpts and the accompanying Plinian diagrams a virtual breakthrough appeared in the study of astronomy in the Carolingian Revival. Planetary data and definitions were incorporated into a much more analytical and coherent pattern of spatial understanding than had existed in Europe since the end of the ancient world. The character of this new, Carolingian astronomical teaching did not change fundamentally until the eleventh century.

Notes

1 *Martianus Capella*, ed. James Willis (Leipzig: Teubner, 1983), 302–337. Cora E. Lutz, "Martianus Capella," *Catalogus Translationum et Commentariorum* 2 (1971):370–374. The importance of the Capellan texts and glosses is briefly but very intelligently indicated for the ninth and tenth centuries by John Marenbon, *Early Medieval Philosophy (480–1150): An Introduction* (Boston: Routledge & Kegan Paul, 1983), 71–89.

2 These have been studied in detail by Karl Rück, *Auszüge aus der Naturgeschichte des C. Plinius Secundus in einem astronomisch-komputistischen Sammelwerke des achten Jahrhunderts, Programm des Königlichen Ludwigs-Gymnasium* (Munich: Straub, 1888), and by Vernon H. King, "An Investigation of Some Astronomical Excerpts from Pliny's Natural History found in Manuscripts of the Earlier Middle Ages," B. Litt. thesis, Oxford, 1969.

3 A useful introduction, with appropriate emphasis on the importance of manipulation in scientific pedagogy, is that of Guy Beaujouan, "L'enseignement du 'quadrivium,'" *La scuola nell' occidente latino dell' alto medioevo*, Settimane, 19 (Spoleto: CISAM, 1972), 639, 656–662. Similar emphasis appears in André van de Vyver, "L'évolution scientifique du haut moyen âge," *Archeion* 19 (1937):19. On stereographic projection from the eleventh century forward, see Ron B. Thomson, *Jordanus de Nemore and the Mathematics of Astrolabes: "De plana spera"* (Toronto: PIMS, 1978), 53–73. See Ernst Dümmler, "Eine Vorrede Hermanns des Lahmen," *Anzeiger für Kunde der deutschen Vorzeit, 16* (1869):138, for Hermannus Contractus's assertion (s.XIm.) that the science of the astrolabe is the most useful of all for investigating the heavens as well as for either astronomical or geometrical knowledge.

4 *Cassiodori Senatoris Institutiones*, ed. Roger A. B. Mynors (Oxford: Clarendon Press, 1937), II, vii, 2, p. 154.

5 Ibid., 155.

6 This is clear when one realizes that not even the basic diurnal rotation of the celestial sphere is clearly and unambiguously described in Cassiodorus' brief chapter on astronomy. Ibid., 153–157.

7 While Cassiodorus mentions a *minor* and a *maior* work of Ptolemy as well as *canones*, there is no sign of recognition as to what one must know to use any of these works. Ibid., 155–156. Not only does Cassiodorus abhor astrology, he rejects cosmological study as "nimis absurdum," since the divine Scripture teaches us what we need to know in such matters. Ibid., 157.

8 Isidorus Hispalensis, *Etymologiarum sive originum libri XX*, ed. W. M. Lindsay (Oxford: Clarendon Press, 1911), III, xxiv.

9 Ibid., III, xxvii.

10 Ibid., III, xxviiiff.

11 Isidorus Hispalensis, *Differentiarum, sive de proprietate sermonum, libri duo* (J. P. Migne, PL 83, 93D–94A), II, xxxix, 149–150.

12 See, for example, Adolph Berger, *Encyclopedic Dictionary of Roman Law, Transactions of American Philosophical Society*, n.s., 43, pt. 2 (1953), 544. Isidore was, we should remember, a bishop and more than passingly familiar with legal vocabulary and tradition.

13 Isidore, *Differentiae*, II, xxxix, 152 (PL 83, 94B): "Mechanica est quaedam peritia, vel doctrina, ad quam subtiliter fabricas omnium rerum concurrere dicunt."

14 Ibid. "Medicina est scientia curationum, ad temperamentum corporis, vel salutem inventa." This conception of medicine, mixed in with other presumably quantitative disciplines, seems to be in concord with that of the late-sixth-century figure Virgil of Toulouse, whose *Epitomae* gave for geometry the curious definition of "ars disciplinata, quae omnium herbarum graminumque experimentum enuntiat; unde et medicos hac fretos geometres vocamus id est expertos herbarum." Subsequently, Virgil puts *physica* in a distinctly lower position than the purely verbal arts when he says, "Omnis igitur humana industria, omnis ad hoc spectat sapientia, ut de inferioribus ad superiora conscendat, quo scilicet naturalem omnium rerum notitiam hoc est fisicam disputans, ethicam quoque quae ad morum emolumenta pertenditur legitime transcendens, logicam ipsam hoc est rationabilem supernarum rerum attinguat disputationem." *Les Epitomae de Virgile de Toulouse*, ed. Dominique Tardi (Paris: Boivin, 1928), 57–59.

15 Bernhard Bischoff, "Eine verschollene Einteilung der Wissenschaften," in *Mittelalterliche Studien* (Stuttgart: Hiersemann, 1966), I, 276–280.

16 Ibid., 282–284.

17 Ibid., 274–275, 285.

18 As I have described above, this was so as much for astronomy in the Cassiodoran tradition, which preserved the Boethian quadrivium, as it was in the Isidorean tradition, which joined the quadrivial subjects with astrology, mechanics, and medicine – disciplines of practical experience rather than abstract theoretical doctrine at the time.

19 Both Isidore's *De natura rerum* and Bede's treatise of the same name used Pliny for bits of information and doctrine. Isidore seems generally to have preferred others to Pliny for information found mutually, for example, Isidore's use of Lucan rather than Pliny for the effects of solar rays on planetary motions; see Isidore de Seville, *Traité de la nature*, ed. Jacques Fontaine (Paris: CNRS, 1960), 255–257 (chap. 22). In the whole book Isidore shows knowledge or verbatim use of Pliny at eighteen points, especially concerning weather and winds. Bede, in contrast to Isidore, found Pliny a congenial and useful source on astronomical matters. In the fifty-one chapters of his *De natura rerum*, Bede used Pliny for much or most of the material in thirty-two chapters. Even so, Bede did not include all that is found in the Plinian excerpts of the Carolingian era, nor were there any astronomical diagrams in Bede, either originally or in the Carolingian manuscripts.

20 This judgment on the influence of Plinio-Capellan astronomy has only suggestive evidence to date, although I am preparing a study which will establish it more clearly. For partial evidence in support, see my arguments and information in "Notes on the Planetary Configuration in Aberystwyth N.L.W. MS. 735C, f. 4v," *National Library of Wales Journal* 22 (1981):132–134; "The Chaster Path of Venus (*orbis Veneris castior*) in the Astronomy of Martianus Capella," *Archives Internationales d'Histoire des Sciences* 32 (1982):153–154; "Origins and Contents of the Leiden Planetary Configuration (MS Voss. Q. 79, fol. 93v) . . . ," *Viator* 14 (1983):10–14, 19–23; and my "Plinian Astronomy in the Middle Ages and Renaissance," in *Science in the Early Roman Empire: Pliny the Elder, His Sources and His Influence*, eds. Roger French and Frank Greenaway (Beckenham, England: Croom Helm, 1986), pp. 207–212.

21 Rück, *Auszüge* (see above, note 2), 34–50, edits the texts for the six excerpts, giving in the texts the section numbers in parentheses, showing

exactly where in Book II (excerpts 1–4) and book XVIII (excerpts 5–6) of Pliny the elements are drawn from.

22 Ibid., 5–25: Bern 347 (s.IX), Bern 265 (s.XI), Montpellier H334 (s.IX), München CLM 210 (s.IX), CLM 6362 (s.XI), CLM 6364 (s.X), CLM 14436 (s.X), Paris BN 8663 (ca. 1000), Paris BN 12117 (s.XI), Wien NB 387 (s.IX).

23 Ibid., 81–89.

24 Ibid., 87–88.

25 On these two copies, CLM 210 and Wien 387, see the recent article by Patrick McGurk, "Carolingian Astrological Manuscripts," in *Charles the Bold: Court and Kingdom*, ed. M. Gibson and J. Nelson (Oxford: BAR, 1981), 321, and the literature cited there in note 36. They are also discussed extensively by King, "Investigation" [see above, note 2], 3–27.

26 See King, "Investigation," 54–79, for a full discussion of elements relating both to date and to origin.

27 Isidore, *Traité*, 257–261. Compare this succinct account with that in Isidore, *Etymologiae*, III, xlvii–lxx, which gives a fuller account for sun and moon but in diffuse form, and offers less precision and more ambiguity for the other five planets.

28 On the Plinian determination of interplanetary distances, see the commentary of Jean Beaujeu to Pline l'Ancien, *Histoire Naturelle* II (Paris: Belles Lettres, 1950), 172–175. On Pliny's neo-Pythagoreanism here, see E. Bickel, "Neupythagoreische Kosmologie bei den Römern," *Philologus* 79 (1924):358–364; and Walter Burkert, "Hellenistische Pseudopythagorica," *Philologus* 105 (1961):34, 36, 42. Lambert of St. Omer, *Liber floridus* (ca. 1120), used the data given in the medieval Plinian excerpt for his computation of interplanetary distances; see Gand UB MS 92, fol. 94v, and Paris BN MS 8865, fol. 57r. Lambert omitted the superscript for 1,000 in recording *stadia*.

29 Bede, *De natura rerum* 14, ed. Charles W. Jones, CCSL 123A (Turnhout,: Brepols, 1975), 205–207. Bede's chapter is essentially the same as the last one-third of the Plinian excerpt; Rück, *Auszüge*, 39, lines 10–40, line 7.

30 Bede, *De natura rerum* 16, ed. Jones, 207–208. Cf. Rück, *Auszüge*, 40, lines 11–41, line 7; that is, ten lines out of the total sixty-one in the excerpt.

31 Here there is a remarkable difference between the Plinian excerpts and the unbroken text of the *Naturalis Historia*. The excerpt reads "Tantumque haec una totidem diebus quot subierat descendit"; all the early manuscripts of the excerpts support this reading (ed. Rück, 42, 4–5). The continuous text contains the reading "tantumque haec et luna totidem diebus, quot subiere, descendunt" (ed. Beaujeu, 32, 19–20: II, 76). The earliest manuscript with continuous text for the first half of Pliny contains the reference to the moon but introduces another difficulty in the reading "tantumque haec et luna totidem diebus quod subire non descendunt" (New York Pierpont Morgan MS M871, fol. 22 rb, 7–9, s.IX[1]). This manuscript, formerly Phillipps 8297, may have originated at Lorsch, or perhaps Arras. Bischoff refers to it as "Saint-Vaast-Stil aus Lorsch"; see Bernhard Bischoff, *Lorsch im Spiegel seiner Handschriften* (Munich: Arbeo-Gesellschaft, 1974), 32–33, pl. 11.

32 Pliny, *Histoire Naturelle*, ed. Beaujeu, II, 78 (p. 33), has 69 here; Rück, *Auszüge*, p. 43, has 79 but notes the reading 69 in Paris 12117. The value 69 is also given in Madrid 3307, fol. 67r; Vat. lat. 645, fol. 71v; Monza F.9.176, fol. 74r; as well as the early continuous text in Morgan M871, fol. 22rb; all of which manuscripts are ninth century.

33 The reading in Beaujeu's edition (p. 33) is 13, while Rück, *Auszüge*, p. 43, and all the manuscripts mentioned in note 32 read 12.
34 This is a circular diagram, taking the full page, on fol. 123r in both codices, CLM 210 and Wien 387.
35 This is the same design in each codex, on fol. 123v in each.
36 Details on the Seven Book Computus as well as description and reproductions of Madrid BN MS 3307 (ca. 820–840, perhaps at Metz) can be found in Wilhelm R. W. Koehler, *Die karolingischen Miniaturen* (Berlin: Deutscher Verein für Kunstwissenschaft, 1960), III, 119–127, and pl. 60. In Madrid 3307 the diagrams are located at fols. 63v (planetary order), 64r (intervals), 65v (apsides), 66r (latitudes); in each case the diagram follows immediately the relevant excerpt.
37 Koehler, *Karol. Miniaturen,* 121. See Patrick McGurk, *Catalogue of Astrological and Mythological Illuminated Manuscripts of the Latin Middle Ages* (London: Warburg Institute, 1966), IV, 52–61, for further details on this manuscript. Monza F.9.176 has the first three planetary diagrams, following their respective Plinian excerpts, at fols. 70v, 71r, 72v. The fourth diagram, for latitudes, divides its related excerpt in appearing at fol. 73r.
38 Fritz Saxl, *Verzeichnis astrologischer und mythologischer illustrierter Handschriften des lateinischen Mittelalters in römischen Bibliotheken,* Heidelberg Akademie der Wissenschaften, Sitzungsberichte, Philos.–hist. Kl. (1915), 71–76. Koehler, *Karol. Miniaturen,* 120. Claudio Leonardi, "I Codici di Marziano Capella," *Aevum* 34 (1960):475–476, with further bibliography. In Vat. lat. 645 the diagram for planetary order (fol. 67v) follows that for intervals (fol. 66v), and the four planetary excerpts are on fols. 68r–72r.
39 For Vat. Regin. lat. 309, a manuscript from St. Denis, written in the third quarter of the ninth century, see McGurk, "Carolingian Astrological Manuscripts," 320; Saxl, *Verziechnis,* 59–66. For Paris BN n.a.l. 456, a manuscript dated variously from s.IX–X (Van de Vijver) to s.XIIin. (Delisle), with provenance from St. Orient d'Auch, see Leonardi, "I Codici," 446, and further sources cited there.
40 Rück, *Auszüge,* 26. These manuscripts, with letters to indicate which of the four Plinian planetary excerpts have accompanying diagrams in the manuscript, are: Bern 265 (s.XI) a b c d; Bern 347 (s.IX) a b c d; CLM 6362 (s.XI) b c; CLM 6364 (s.X) a b c d; CLM 14436 (s.X) a b c d; Montpellier H 334 (s.IX) none; Paris 8663 (ca. 1000) a; Paris 12117 (s.XI) none.
41 King, "Investigation," 80–117, discusses such manuscripts up to ca. 1100, and (pp. 117–124) during the twelfth century. Because my focus is on the diagrams, I shall ignore those he covers without diagrams and shall introduce others not described by King.
42 Figure 1 shows the Vienna manuscript diagram. CLM 210, fol. 123r, has only one difference, an addition clearly made sometime later to the diagram. The addition is the insertion of abbreviations for the twelve signs of the zodiac in the appropriate planetary circles, connecting the sun and the moon with Leo and Cancer respectively and each of the other planets with two signs in accord with the prescription of Macrobius for the arrangement of the planets at the birth of the world; see *Commentarii in somnium Scipionis,* ed. James Willis (Leipzig: Teubner, 1970), 89 (I.xxi. 24–26). This insertion in the intervals diagram is not found in other manuscripts of s.IX–X.
43 That is, the same diagram is found on fol. 123v of both Wien 387 and CLM 210 and nowhere else – at least, nowhere else in conjunction with a known Plinian apsidal excerpt.

44 While it helps only to disqualify Figure 2 as an apsidal diagram, I have found one source which might be responsible for the seven equally spaced radial lines. In some, but by no means a majority, of the *rotae* for planetary orbits in manuscripts of Isidore's *De natura rerum*, c. 23, the texts on the planetary bands are divided into seven segments, either equally or unequally distributed, so that the concentric bands show seven radiating lines of text from the center outward. Examples are Basel UB lat. F.III. 15(a), fol. 5r, ca. 800; Laon 422 fol. 15v, s.IX^1/3; Bamberg Nat. 1, fol. 27r, s.IX: Vat. Regin. lat. 1573, fol. 106r, s.XI. See my Figure 10 for the first of these examples. I am not convinced that this is the source of the seven radial lines in the incomplete diagram under consideration.

45 For the locations of the diagrams in this and other manuscripts of the Seven Book Computus, see above, notes 36–38.

46 London BL Harl. 647, fol. 19r, s.IX^2/4, is exactly the same excerpt for boxes around the planetary names and a doubling of the circles to make narrow bands rather than simple circles.

47 Monza F.9.176, fol. 70v, and Madrid 3307, fol. 63v, are quite the same. Vat. lat. 645, fol. 67v, varies only by adding a figure, holding a plant and a serpent, for *terra* in the central circle. The details of the heads for the planets also vary a bit. No illustration accompanies this excerpt in either Vat. Regin. lat. 309 or Paris BN n.a.l. 456. As late as Wien 12600, fol. 27r, s.XII ex., provenance Prüfening, the illustration remains very much the same, and intervening variants are minor.

48 See above, notes 36–37.

49 Examples are Paris BN n.a.l. 1615, fol. 161r, s.IX^1, and Bern 347, fol. 25r, s.IX^2.

50 Examples are Bern 265, fol. 59r; CLM 14836, fol. 124r; Paris 11130, fol. 74v; Vat. Palat. 1577, fol. 83r; Vat. Regin, 123, fol. 169v. An exception is the rather nicely drawn example in Vat. Ross. 247, fol. 200v, A.D. 1018, which lacks only the circle for Jupiter; however, this diagram has a zodiac of 14° rather than 12°.

51 Harriet P. Lattin, "The Eleventh Century MS Munich 14436: Its Contribution to the History of Coordinates, of Logic, of German Studies in France," *Isis* 3 (1947–48):205–225, and earlier studies cited there.

52 On this manuscript, which is a segment of a much larger compilation, see Otto S. Homburger, *Die illustrierten Handschriften der Bürgerbibliothek Bern. Die vorkarolingischen und karolingischen Handschriften* (Bern: Bürgerbibliothek, 1962), 134–136, and R. W. Hunt et al., *The Survival of Ancient Literature* (Oxford: Bodleian Library, 1975), 53–54.

53 Detailed study of the varying and evolving form of the rectangular latitudes diagram can be found in my two recent articles, "Characteristics of the Plinian Astronomical Diagrams in a Bodleian Palimpsest, Ms. D'Orville 95, ff. 25–38," *Sudhoffs Archiv* 67 (1983):8–11, and "Mss. Madrid 9605, Munich 6364, and the evolution of two Plinian astronomical diagrams in the tenth century," *Dynamis. Acta Hispanica* 3 (1983):272–276.

54 Just as with scribal errors in copying texts, the tradition of rectangular latitude diagrams shows many errors, often missing by one or two the correct number of horizontal squares for the grid and often laying out latitude lines on the wrong number of squares vertically.

55 Examples with rounded extremes and thereby a bit less obvious are: Oxford Canon. class. 279, fol. 34r, s.IX^2–X^1; Paris 5239, fol. 39r, s.X; Strasbourg 326, fol. 123r, s.X. Examples with angular extremes and strictly rectilinear

ascent and descent are: London BL Add. 11943, fol. 49v, s.XI; Madrid 9605, fol. 12v, s.XI; CLM 6364, fol. 24v, s.X^2; Oxf. Lyell 54, fol. 26v, s.XI; St. Gallen 250, p. 2, s.IX^2; Zurich Car. C 176, fol. 193v, s.X–XI.

56 The best example of the degeneration thereby allowed in the rectangular latitude diagram is in Vat. Palat. 1577, fol. 82v, s.X–XI, admittedly extreme in its carelessness, where the planetary lines begin at likely positions on the left vertical, fail to show even the initial amplitudes for three of the planets, and dissipate into vaguely wavy lines on the right except for the Venus and lunar lines.

57 In addition to the examples listed above [notes 55 and 56], some further locations are: s.X – Genève (Cologny) Bodmer 111, fol. 41r, CLM 14436, fol. 61r; s.XI – Bern 265, fol. 59r, London BL Cott. Vit. A. XII, fol. 9r, Vat. Regin. 1573, fol. 52r, Zurich Car. C 122, fol. 42r; s.XII – Baltimore Walters W. 73, fol. 5v, Cambridge St. John's I. 15 (221), pp. 287, 353, Erfurt Amplon. Q 351, fol. 1v, London BL Roy. 13.A.XI. fol. 143v, Oxford St. John's 17, fol. 38r; s.XIII – London BL Cott. Tib. E. IV, fol. 142r, London BL Eg. 3088, fol. 83v. Surely the most curious of all the latitude diagrams I have discovered is that in Wroclaw UB IV.0.11, fol. 59r, s.XII; it actually combines the two forms by laying out the planetary zigzags from a rectangular diagram bent around upon the thirteen concentric circles from a circular diagram. No Plinian excerpt accompanies this hybrid diagram in the manuscript.

58 See above, note 53: Eastwood, "Characteristics," *Sudhoffs Archiv*, 2–8; and "Mss. Madrid 9605," *Dynamis*, 267–72.

59 There is some difference of modern opinion on the date of this manuscript, Vat. Palat. 1577. Bernhard Bischoff, *Lorsch*, 63, 87 n.90, gives s.XIin. Valerie M. Lagorio, "Excerpts from Pliny's *Natural History* in Codex Pal. Lat. 1577," *American Journal of Philology* 94 (1973):290, gives s.X. I incline toward the earlier dating.

60 There is one and, so far as I know, only one contradiction to this rule: Madrid 9605, fol. 12r. I consider the reversal of Capricorn and Sagittarius there to be an error, not just a further modification. A reproduction can be found in *Dynamis* [see above, note 53], p. 277.

61 The earliest example is that of St. Gallen 250, p. 23. While the codex has been dated to the later ninth century, I have some question about this page, which contains two diagrams and no text; the other diagram is from Macrobius. With regard to an additional peculiarity of this diagram, the incorrect addition of an apogee for the moon, see *Dynamis*, 269, and n. 20.

62 Margaret Gibson, "The Continuity of Learning circa 850–circa 1050," *Viator* 6 (1975):2–5; John Marenbon, *From the Circle of Alcuin to the School of Auxerre* (Cambridge: Cambridge University Press, 1981), 140–141.

63 Frances R. Lipp, "The Carolingian Commentaries on Bede's *De natura rerum*," Ph.D. dissertation, Yale University, 1961, pp. 63–67. The text of the commentary in question is edited in the notes to Bede, *De natura rerum*, ed., Jones, 192–234. The commentary appears in Berlin DSB Phillipps 1832 (130), fols. 1r–9r. Extensive discussion of the glosses to Bede's *De natura rerum* and *De temporum ratione* in this manuscript can be found in John Contreni, *The Cathedral School of Laon from 850 to 930* (Munich: Arbeo-Gesellschaft, 1978), 39 n.34, 98, 125, 137, et al. Contreni's provisional judgment (p. 125 n. 35) is that the two sets of glosses to these two works in Phillipps 1832 are independent. The *De natura rerum* glosses are not by a Laon hand.

64 Rück, *Auszüge*, 37–40.
65 Richard Krautheimer, "Introduction to an 'Iconography of Mediaeval Architecture,'" *Journal of the Warburg and Courtauld Institutes* 5 (1942):7–8. John Murdoch, *Album of Science: Antiquity and the Middle Ages* (New York: Scribner, 1984), 126–127, has remarked on this problem.
66 This effect is analogous to the uses of the exploded view and the orthographic projection in preference to a naturalistic perspectival view for conveying spatial information in the clearest and most coherent way.
67 John D. North, "Monasticism and the First Mechanical Clocks," *The Study of Time, II*, ed. J. T. Fraser and N. Lawrence (New York: Springer-Verlag, 1975), 386–387. Most notably, the diagram cited by North is CLM 210, fol. 113v, a page in one of the two surviving witnesses of the Three Book Computus! The illustration is not to be found in Wien 387, nor has a space been left for it there. Knowledge of stereographic projection in Carolingian Europe has many possible sources, and transmission from Byzantium is not to be discounted.

6

Coordinates and categories: the graphical representation of functions in medieval astronomy

J. D. NORTH

"Dixit Ptolomeus . . ." There is something to be said for this medieval way with intellectual history, where the main concern is simply to repeat what has been said, rather than provide a wider context for it. Historical interpretations are always dependent to some extent on current orthodoxy, and the greatest danger occurs when we try to look across a significant historical divide, perhaps even one so great that later scientific groups have not been aware of it, and have revived an earlier style of thought without realizing it. In such instances, we are prone to use the wrong categories to describe what we find. What we cannot categorize we cannot compare. The tradition to be considered here – from the late Middle Ages – is a case in point. It is a case of "coordinate geometry" from the domain of astronomical computation that cannot, without distortion, be assimilated either with Greek analytical geometry or with later "Cartesian" forms. It will be shown to have much more in common with what was later called "nomography" – itself a byway in mathematics, and one that presents its own problems of historical analysis. Those who wrote on nomography seem to have been quite oblivious of their mathematical forebears.

There have been many attempts to show the existence of "Cartesian geometry" before Descartes. That this has been so is a simple consequence of the important status that has always been assigned to his work. It has more to do with historians' yearning for continuity than with the excitement to be had from furthering a priority dispute. Of course, the analysis of the Cartesian revolution in geometry has not gone unqualified. Descartes has been seen by some as marking a stage in the evolution of deeper trends in intellectual history, these having supposedly been revealed in the light of later developments in mathematics; and the question again arises as to just how far it is right to allow later trends to dictate historical categories. How far, in other

173

words, may historians be allowed to stray from the "Dixit Ptolomeus" mentality?

Some idea of this problem, as it relates to our present subject, can be had as soon as we have drafted a rough definition of analytical (or better, "algebraic") geometry in general terms. It may be described as the correlation of what had earlier been regarded as "geometric" concepts and objects – no matter what was meant by the appellation "geometric" – with certain algebraic concepts and objects, which are then proved, by algebraic methods, to have such and such properties. These properties will now therefore be describable in geometric as well as in algebraic terms. As it is often more briefly put, the essence of the analytical method in geometry is the study of geometrical loci by means of their equations. This is a narrower definition, and the primary concern of those who use it is with *geometry*. Modern mathematicians have tended to lose this perspective, but the two-way traffic between algebra and geometry is not new. Fermat, for example, in studying the intersections of the circle and the parabola as a problem equivalent to that of finding the solutions to a quartic equation in a single variable, seems to have been as interested in the one problem as the other, and one might take many similar examples from him, or Descartes, or even – with a few qualifications as to the meaning of "algebra" – from Greek precursors such as Menaechmus or Apollonius.

Of course, analytical geometry has taken many new forms since the seventeenth century, and some of the more recent revisions of mathematical attitudes have left historians unsure of how best to describe events. With the advent of a better appreciation of the axiomatic methods that geometry had bred from the time of the Greeks, geometers were able first to shed the metrical element. The effect of this move was to make some historical writers feel that projective geometry was what really mattered, and that they should be casting history into categories dictated by the projective approach. So much the worse for the fashionable historian, then, when mathematicians unashamedly reintroduced axioms for metrical bases.

There have been many such changes in conceptual signposting, by which mathematicians have left historians over the last century wondering where fashion was leading. There have been numerous new analytic constructions of geometry, for example, developed from linear algebra and the theory of vector spaces over a field, that have suggested new ways of relating the axiomatic structures with the algebraic. Thus, as a now conventional basis for the geometry of vector spaces, there is the theorem that the Pappus planes are simply the affine-coordinate planes defined by a two-dimensional vector space over a field (a skew field).[1] This insight immediately lends itself to a new way of looking at the history of Cartesian geometry, and at what Descartes was "really" doing; but then, in much the way I have suggested already, the temptation is checked somewhat by the realization that there are geometers who favor other approaches. The systems of axioms for the Euclidean plane that lead to metrical theorems, or that are themselves overtly metrical, are somehow closer not only to the world of common experience but to the way history has always – at least, to the superficial gaze – seemed to have developed. The feeling is reinforced that perhaps after all the greatest sin is anachronism, and the best way of avoiding it is only to look backward in time. The ideal is at best dull and at worst unattainable. There is something to be learned from the modern tendency to look at algebra and geometry in an evenhanded way, when considering algebraic geometry, and we should not allow ourselves to slide back into the old ways of introducing the history of the subject through the Roman *agrimensores*, Oresme's "latitudes of forms," and the like.

Speaking generally, there have been three influential attitudes to the historiography of early analytical geometry. The first has emphasized the heavy dependence of seventeenth-century mathematics on Fermat and Descartes, to the exclusion of anything medieval. It has been held that medieval contenders for the title to the invention of their methods were helpless, ignorant as they were of the use of free variables. The second approach finds coordinate geometry in innumerable places – for example, in the Egyptian practice of dividing the land into rectangular districts, in Hipparchus' use of stellar coordinates, and in early medieval graphs of planetary movement in latitude. Such examples and more were produced more than a century ago by Sigmund Günther, who was in turn answered by H. G. Zeuthen.[2] If priorities are in dispute, Zeuthen's position was surely the most satisfactory. In broad terms, this was that analytical geometry was an invention of the Greeks.[3] Fermat and Descartes had, of course, a far better understanding of the geometrical equivalents of the six basic algebraic operations, and obviously a great deal of latitude is required in our understanding of what the word "algebra" means in a Greek context.[4]

The neat modern division between algebra and geometry, the two branches of mathematics that are to be put into correspondence with each other, is historically misleading, inasmuch as it obscures the way in which specific correspondences were first usually established. In the first place, there is more to the subject than simply setting up a correspondence between elements. As a minimum requirement, a transference of rules amounting to an algebra of geometrical line segments is called for, so that addition, subtraction, multiplication, division, raising to a power, and extracting a root can all be given geometrical equivalents – if possible with homogeneous results. Descartes was particularly clear on the order in which he should proceed, but in the great majority of cases, mathematicians dipped into Euclid without any regard for conceptual tidiness, and the seventeenth-century situation was that much of the algebra was tacitly borrowed along with such calculational rules of standard Euclidean geometry as Pythagoras' theorem. The fact that so much of the "algebraic" work had already been done by "Euclid," means that it has not always been easy to see the nature of the correspondence between the *structural* elements of the two subjects. There is some truth in the view often expressed, that much algebraic geometry amounted to a trivial rewriting of old geometry with a new terminology. It is this intermingling of elements, in fact, that has obscured a fundamental difference between most of what was done then and in the Middle Ages.

An important stage in the evolution of algebraic geometry is reached – and was reached, as Zeuthen showed, by Menaechmus – when the new algebraic relationship is looked upon as standing proxy for the geometrical object – say a curve – so that the connections between it and other geometrical objects may be deduced by the manipulation of their algebraic equivalents. A third stage comes when the algebraic relationship is considered to be, as it were, that in which is encapsulated *all* that can be said about the curve, once it is properly interpreted. When Fermat and Descartes showed, for example – in their different ways – that the equation to a curve determines the direction of the normal at any point, there was a tacit assumption of this principle.

For something to merit a historical description as "algebraic geometry" it is clearly not enough that it involve a formal correspondence between algebra and geometry. There is also the question of motive, and of the wish to further the one subject by the help of the other. This does not mean that assistance had to be offered in only one direction. Descartes' third and crucial book of his *Géométrie* contains a novel algebraic theory of equations, and is therefore quite in keeping with his ambition, as expressed in a letter to Isaac Beeckman in 1619, to formulate an algebra by which all questions may be resolved as regards any sort of quantity, continuous or discrete. It is in keeping, too, with Descartes's general concern to *construct* points satisfying a certain relationship, and later to construct the roots of equations of various degrees. This is not the Descartes so many have imagined, creating a royal road to geometry that anyone could tread, but, on the contrary, a Descartes who seems almost to have meant geometry as a substructure for his algebra. His characteristic philosophical priorities and his great confidence in geometry as a basis for truth were certainly not those of the modern mathematician. At all events, the graphical representations of preexisting algorithms that we shall be considering were meant to assist not in the establishment of general mathematical truths but in numerical computation. One might say of these cases that "algebra was being assisted by geometry," but certainly not in any strong epistemological sense.

The algebraic or quasi-algebraic relationship whose graphs we shall consider have as their main purpose the correlation of sets of numbers – such as the temperature of an object and its weight, or the coordinates of a planet and the time. These coordinates do not have to be empirical, of course, nor do they have to be positional. If they represent physical magnitudes, then we require only that their functional relationship is known. As for positional quantity, as soon as one mentions position there is a potential confusion, for we wish to go a stage farther and introduce a correspondence between our algebraic relationship and a geometrical locus, or other geometrical entity. In short, we wish to consider the notion of a geometrical curve (or whatever) corresponding to an algebraic relation that is of central interest and is regarded as having been antecedently established. This "graph of such and such a function," in short, is to be thought of not as coming before the function but as something derived from it. The sequence is the reverse of that with which we began our catalogue of approaches to algebraic geometry. The medieval episode to be considered stands quite outside the history of algebraic geometry in the "Fermat–Descartes" sense.

This last type of correspondence between algebra and geometry, which has so often been treated as a precursor of what is more usually meant by "algebraic geometry," fails to live up to this claim twice over. As explained, it lacks the necessary generality, and it is pointed, so to speak, in the wrong direction. In the conventional phrase, it does not "provide a knowledge of geometrical loci through their equations," but rather a knowledge of equations, and the correlation of the variables they contain, through geometrical loci. In most cases, the word "knowledge" is misleading, although graphing might certainly help us with visualization, empirical understanding, or calculation. It was precisely with these things in view, though, that the medieval astronomer developed what – on account of the direction in which it is aimed – might reasonably be called his "geometrical algebra." The confusion between this and "algebraic geometry" in the usual sense stems to a great extent from the fact that the graphed quantity has so often been a positional coordinate.

Consider such typical astronomical tables as are found in single- or double-entry form. (Many tables ostensibly in the latter form are only trivially describable as double-entry, such as when days are entered down the left column and months across the top.) The chances are that such a table – at least if it is not Babylonian – was arrived at on the basis of a geometrical algorithm. Consider the simplest of all single-entry tables, those for mean motion. These give a steadily increasing angle as a function of the time, and to anyone with the barest knowledge of rectangular graphing techniques they beg to be turned into straight-line graphs (broken, of course, at 360-degree intervals). At a more advanced level of conceptualization, the resulting planetary positions recorded, after much calculation, in an almanac, would have been very usefully graphed, whether in rectangular or polar coordinates. At the very least, interpolation problems would have been better understood and more easily executed. These steps, however, were not taken, and while there are medieval survivals that look very much like our "squared paper," not to mention polar coordinate grids, we must resist the temptation to suppose that they were always viewed in the way we find most natural, namely, as the bearers of some previously evaluated function, the contents, so to speak, of a table of values. This way of looking at things – of which we shall find evidence only in the later Middle Ages – is quite alien to an earlier period, as may be very well illustrated by the most notorious of medieval graphical representations, those belonging to what might be called "Plinian" astronomy, which had a revival between roughly the ninth and the twelfth centuries.[5]

The diagrams in question simply show the variation of planetary latitude (the ordinate) with time, or rather with zodiacal position – and this in a very loose sense, as will be explained later. That the theory behind the representation was ill understood may be guessed from the very notion of capturing within a single circuit of the zodiac all potential latitude variations, in a situation where the nodes of the planetary orbit are not stationary. What matters more to us is that the genesis of the graphs that are now best known, and that are seemingly based on rectangular coordinates, was not at all along the route that we, with our expectations based on the modern graphical tradition, are inclined to suppose. The route becomes plain as soon as we consider the diagrams that had gone before. These were *circular* in form. They comprised a background of thirteen evenly spaced concentric circles, with each planetary path an eccentric circle, the eccentricity being settled by the maximum and minimum potential latitudes attained by the planet in question. At this stage, although the diagrams might be grandiosely described as being in polar coordinate form, all that has really been done is to put a three-dimensional picture on the two-dimensional page. (To speak of a projection would be too strong. There is no sense of perspective here.) At the later, rectangular, stage, the band of latitudinal degrees is straightened out, and, of course, inevitably broken in the process. What is more, at this later stage something very odd happened: The planetary paths, while having the roughly sinusoidal forms that are to be expected, and while having the right amplitudes in latitude, no longer follow a single cycle, as the original diagrams suggest they ought to have done. Instead, they show a completed number of longitudinal waves that are usually inversely proportional to the latitudinal amplitude, the moon defining the basic unit.[6] It would be impossible to deny that these diagrams carry information, but it is in a form so garbled, and so far removed from the geometrical algebra we are seeking, that there is little point in discussing it further here.

There is something to be learned, even so, from the Plinian latitude graphs, namely, that at an early historical stage a coordinate mesh is more likely to have been superimposed on the pre-existing figure – a picture of reality, so to speak – than to have been the route to that figure. This is the case in a yet older and much more important cluster of traditions, those centered on the astrolabe.

The astrolabe may be characterized in many different ways, but for our purposes there is much to be said for describing it as an analogue computer. It provides the user with numbers – times, altitudes, azimuths, longitudes, and the like – as functions of other numbers – altitudes, ascendant ecliptic longitudes, and whatever. Its two main parts are the fretwork rete that represents the sphere of the fixed stars, and the plate of local coordinates over which it turns. Their relative disposition is analogous to that of the two main sets of circles as conceptualized in elementary spherical astronomy.[7] As is well known, they are produced on the instrument by stereographic projection, and this precise geometrical way of turning one figure into another is a far cry from the crudely impressionistic transformation found in the former example.

The astrolabe illustrates several of the different elements that ought to be distinguished in any account of the evolution of geometric algebra. These include the notions of (1) coordinate assignment; (2) second-order graphical representation; (3) coordinate transformation; (4) the graphing of quasi-algebraic functions, that is, over and above any geometrical substructure that might have led to them; (5) contour lines; and (6) the use of these graphs as substitutes for the algorithms on which they are based.

The first of these conceptual elements, *coordinate assignment*, is a minimal requirement, of course. In the broadest sense, coordinates are a means to solving the problem of saying where things are, so as to be able to locate them on another occasion, or have others do so. The discovery of the historical use of simple coordinate systems does not seem to me to be an occasion for great excitement. It is a matter of great importance that in astronomy, where the most significant developments in the use of coordinates took place, coordinates were most commonly angular, and therefore often cyclical. The most common ways of assigning coordinates to points of the celestial sphere were through ecliptic latitude and longitude, right ascension and declination, or mediation (the degree of the ecliptic that culminates with the object) and declination.

By *second-order graphical representation* nothing more is meant here than a mapping, a turning of one figure or space into another, not necessarily as a self-consciously mathematical action. It is what was done by the "Plinian" astronomers who turned the zodiacal belt on the sphere into a belt of thirteen concentric circles on the plane surface of their books. It is what a painter does; and if painters had needed to wait for mathematical rules of perspective, which arrived on the scene only in the late Middle Ages, their art might never have begun. This is not to say that perspective painting is without interest: Those who practiced it were, after all, generally doing a better job than the "Plinian" astronomers. It leaves something to be desired, even so, from a mathematical point of view. When a painter turns three dimensions into two, there is a loss of information along the way. Each point in the painting may correspond to any or all points of a visual ray in the three-dimensional original. Information in this case is tacitly restored by the topological sense of the onlooker, who knows that eyes belong with heads, feet belong with the ground, and so on. In the astronomical case, where only directions and not radial distances are held to be significant, the two-dimensional representation used on an astrolabe loses nothing of importance. We shall later come across ways in which three independent variables (and more) were represented in a plane without loss, in the context of astronomy.

In effecting the sort of graphical representation found on an astrolabe, and doing so in a consciously correct geometrical way, one is practicing a sort of second-order geometry, somewhat analogous to what is done in the geometrical algebra and algebraic geometry of which we have already spoken. It is a stage on the road to the former, in fact, as will be shown later, from a number of medieval examples.

Coordinate transformation is one of the principal functions of an astrolabe. A typical example is where the rete is positioned for a particular moment in the day, allowing us to deduce the altitude and azimuth of a star whose latitude and longitude (or mediation and declination, perhaps) were needed to position it on the rete in the first place. Astrolabe techniques provided a basis for certain instruments used for conversion between one system of celestial coordinates and another. Such instruments were used, for instance, to turn one type of star table into another – and, of course, they could not be very precise, by comparison with the usual long trigonometrical methods of calculation. Oddly enough, the reason for effecting the conversion, in many cases, was to have coordinates (mediation and declination) that were more convenient for placing stars on an astrolabe. Some of the more sophisticated techniques of coordinate transformation were developed in connection with the *universal* astrolabe projection, usually known in the West as the *saphea*.[8] There were many other forms of coordinate transformation used by astronomers, however. Angle-measuring devices (such as the triquetrum, Jacob's staff, even most of the numerous forms of sundial) that use a linear scale to yield an angular coordinate are clearly being used to effect a transformation of coordinates, and although this might seem to be a pretentious way of describing what is done in these cases, familiarity with them had at the very least a psychological importance for the evolution of stage (4) in our progression. The message was simply that measures are not always what they seem, that a distance is not always the measure of a distance, and that even when it is so, the one distance need not be directly proportional to the other.

The *graphing of quasi-algebraic functions* is something best approached in stages, for it is easily confused with second-order graphical representation. Consider the case of the lines on an astrolabe that show the unequal hours. It is not important to know here how they are used, but only to realize that they are arcs of circles that the point marking the sun (on the rete) crosses in the course of the daily rotation, its crossing marking the traditional divisions of day and night. (A twelfth part of the day, a "seasonal" or "unequal" hour of the day, will only be equal to a twelfth part of the night at the equinoxes.) It is how these arcs are obtained that matters to us. For every position of the sun on the ecliptic, its movement from setting to rising – to take the case of the night hours – around the lower part of the astrolabe plate describes an arc, which we may suppose to be circular. (Since the sun's longitude changes by a trifle, in the course of the night, there is a slight approximation involved.) A point a twelfth part of the way along that arc will measure the position of the sun at the end of the first unequal hour of the night. The totality of all such points, for all possible polar distances of the sun, will be the form of the hour-line we are seeking. Each point on the hour-line is such that an algorithm was known for calculating it. It would have been possible in principle to have drawn that hour-line on the basis of the known algorithm, in which case we should have had what I have called the graphing of a quasi-algebraic function. In fact it was drawn otherwise, that is, as an arc of a circle passing through three points of symmetry that are easily constructed.

There are many other cases in medieval astronomy that should be classified in much the same way as the hour-lines example, although they are often instances where the construction is an exact one. The lines on an astrolabe by which the astrological houses are computed using what I have elsewhere called the "fixed boundaries methods," lines that are often confounded with the unequal-hour lines, are of the same general sort.[9] They are *contour lines*, moreover, in a way that should be obvious: the hour-lines, for example, are labeled from 1 to 12 with the hour variable, if we may call it that. In a sense, many other lines on an astrolabe are analogous to them, and of most the same could be said: There are algorithms, and there are tables embodying those algorithms, to which they correspond, *but there is no immediate genetic relationship between algorithm and graph.*

The astrolabe was used for problems in spherical astronomy, and was not as such connected with the problem of computing planetary positions. Here there was a large class of instruments available, under the generic name "equatorium." Despite their being described by this one name – the name stems from their function of evaluating the "equations" to be added to the mean positions of the planets to give their true positions – they may be classified in many different ways, and one of the most enlightening of classifications concerns the degree of simulation of the fundamental planetary (usually Ptolemaic) model. Whether or not the Ptolemaic planetary models are themselves meant to simulate reality is beside the point here: We shall suppose that they are agreed as the means to the end of calculating planetary positions. With each geometrical model goes a whole series of trigonometrical procedures. The geometrical model, with its deferent circles, epicycles, apselines, equants, and the rest, may be simulated precisely, in principle, by graduated circles of metal or parchment or whatever, rods, pivots, and so forth, the whole arrangement being set for a particular moment of time in accordance with the mean position-angles that the tables of mean motions give for that moment. Most equatoria fall into this general class.

There is another sort, however, that follows more closely the sequence of algorithms gone through in an ordinary Ptolemaic calculation of planetary position. Perhaps the most intricate of these equatoria was the so-called Albion of Richard of Wallingford, dating from around 1326, but his was not by any means an isolated example.[10] Speaking very generally, there is an element of simulation, but only in the component computations. Thus, if an adjustment term, an "equation," depends on a sine factor, one might find a procedure for using certain scales of the instrument to perform a construction that amounts to the evaluation of that sine, just as one might do it with a ruler and protractor on paper; and so for more complex algorithms. There is simulation here, but not of models in their entirety, and certainly not of anything the astronomer might have been tempted to describe as reality.

Once this sort of procedure had been established, once the spell had been broken that held astronomers to a procedure simply picturing the Ptolemaic picture, the way was clear to a new phase in experiment with more powerful graphical techniques. It is certainly not possible to say that the inventors of equatoria in this second style were alone responsible for this new phase, but it does seem to be true that the most significant combination of graphical (subsidiary) instruments came along with Richard of Wallingford's Albion, which was in an unmistakably "trigonometrical" form. I shall here give attention to just two of the graphical devices incorporated in it, one for conjunctions and oppositions of the sun and moon, and the other for eclipses of the sun and moon.[11]

The use of both instruments is extremely involved, and it is impossible here to give more than the barest sketch of the general principles behind them. At first sight, the instrument for conjunctions (which also gives lunar velocity) is a series of arcs arranged spirally, each with a certain number attached, exactly as on the unequal-hour lines of an astrolabe. The idea for these contour lines undoubtedly came from that ubiquitous instrument. The moon's argument settles the radial distance of a bead on a thread through the center of the instrument. At this distance, we may imagine a scale whose graduations are where the imaginary scale crosses the numbered contours. The thread is then set to a certain longitude difference (that between the sun and the moon at mean syzygy, in fact) on a peripheral scale, and the bead will then indicate a certain "equated (corrected) time" on the imaginary scale graduated by the contours. In fact, as with all contours, as every map user knows, the method of interpolation is intuitively obvious, but carrying out the task is only easy when the contours are closely packed.

The difference in principle between this instrument and the hour-lines on an astrolabe, with which it is functionally related, is one of intuitive appeal. The hour-lines mark out quite simply the divisions over which the sun, in its daily course, may be considered to pass. The syzygy instrument lends itself to no such interpretation. It simulates nothing that can be readily pictured as happening in the heavens. It is a series of (polar) graphs that together allow one variable to be evaluated as a function of two others, and that have their justification in a fairly complicated algorithm due to Ptolemy. The *form* of those graphs is still not very complex. They are still arcs of circles through three points. There is good reason for suggesting, however, that here we have a case of what I called earlier the graphing of quasi-algebraic functions.

The eclipse instruments that form a part of the Albion are no less complex in their entirety, but briefly they may be said to contain a series of eccentric (semicircular) arcs that are polar graphs of different functions of certain dimensions (lunar, solar, and earth's shadow at the lunar distance) that matter to anyone calculating an eclipse. The radial scale is a linear scale of those (angular) dimensions, and the angular argument is the lunar argument during eclipse. The loci in question are not quite what they should be according to Ptolemaic eclipse theory, but the important thing is that Richard of Wallingford was trying to make them so. He was working within the limits set by circular arcs. (And lest anyone be tempted to weave a story of the hidebound medieval astronomer, constrained by a Greek philosophical obsession with the perfection of the circle, let it be said that no one used ovals more intelligently than Richard of Wallingford, whether on his Albion or in his astronomical clock.)

The Albion was an instrument of great complexity, with its many subsidiary parts, and over sixty scales – its very name meant "all by one." It is hardly surprising that later astronomers tended to abstract its subsidiary instruments from it. The only set of surviving fragments of an Albion made in metal is now in Rome.[12] Richard of Wallingford's treatise on the instrument was extremely influential, however, and some of those writers whom it influenced were themselves influential, notably in central Europe. There are at least a dozen editions or treatises that derive immediately from it, not to mention several minor tractates that make use of its parts, and these last usually concentrate on the instrument for conjunctions and the double eclipse instrument. The first new edition was done by Simon Tunsted, before 1369. The Viennese astronomer John of Gmunden produced what was perhaps the most widely copied, around the year 1430, not altering the instrument significantly but adding explanations where the original was too cryptic. He added a thousand words or so in connection with the instruments singled out for attention here. Regiomontanus (1436–76) produced a rather careless version. John Schöner published an equatorium treatise in 1522 that included Richard of Wallingford's eclipse instruments. The most striking printed work to make use of the treatise *Albion*, however, was the lavishly illustrated *Astronomicum Caesareum* of Peter Apian, published at Ingolstadt in 1540.

The writers named here not only were intrigued by the instruments mentioned but show distinct signs of having tried to extend the underlying principles to the solution of other problems. Apian was not alone in this, nor was he the first, but his book was a rich source of new ideas, and will serve to illustrate another stage in the evolution of the graphing of functional dependences.

In two *enunciata* (20 and 21), he made use of the principle that Richard of Wallingford had used in his conjunctions instrument. The problems are somewhat similar: They are to determine small changes in the aspects of the moon with the planets and the planets with other planets. As before, one variable is evaluated as a function of two others, the dependent variable being given by spiraling contour-lines, the independent by polar coordinates. *Enunciatum 26* concerns an instrument of the same general character, to relate the time to the hourly changes in lunar and solar argument at syzygy, and this begins a re-

markable series of instruments that develop the theme. All are of the
same general form, that is, they express a variable as a function of two
others, and all are used in the same way as those already explained.
On them, however, the loci which on the Albion and immediately derivative instruments are circular arcs are now curves of a much more
intricate form. They are contour lines such as might have charted the
form of a sand dune whipped up by a tornado. They are contour lines
of constant time difference, that might have been found a "Ptolemaic"
algorithm, although that would have been extremely complex, for reasons that will shortly emerge. In fact they were *plotted from previously
calculated tables*. The data in the worked examples included in the
text are there far too precise to have been found from the instruments.
On the other hand, there is no doubt that when he was designing his
latitude instruments, he had in front of him the latitude instruments
designed by Sebastian Münster, who had not proceeded in this way.
Münster's work was published in his *Organum Uranicum* (Basle,
1536).[13] The change of style marks an important step in the evolution
of graphical methods.

We need not enter into the details of Münster's solution to the problem of latitudes. It followed the same general principle as that adopted
by Richard of Wallingford in his conjunctions instrument, and there
are clear links between the two. The radial polar coordinate (set by a
seed pearl on a radial thread) was to be set, now, with the help of an
ancillary circular scale of degrees, a sort of epicycle. This first variable
(the radius vector) was thus effectively itself a function of another (the
"true argument" set in the epicycle). A peripheral scale gave the variable known in astronomy as the "true center," while, as a function of
these two, the latitude was given by the contour on which the seed
pearl fell – or, rather, the imaginary scale graduated by the contours.
The important thing to notice is that the contours were still *arcs of
circles*. What Apian did to Münster's design was highly significant: he
created a *linear* radial scale of the (true) epicyclic argument. As explained, there had not previously been a radial scale at all, since the
length of the radius vector had been found as a function of an epicyclic
argument, but the effective radial scale was nonlinear. By simplifying
the instrument in this respect, Apian had made it necessary to produce
noncircular contours. Their form was complicated by yet another fact:
he set the 360 degrees of the peripheral scale, that of the true center,
into about 340 degrees of the instrument, leaving space in the other 20
degrees for his radial scale. His were therefore not polar coordinates
as we know them, but they were none the worse for that. Apian had
realized that what others might have seen as the distortions of his
coordinate scales did not matter in the slightest, if he had at his disposal
tabulated information to plot on the plane of his instrument. It was not

what we usually mean by "empirical" information, but the same principle would have applied, had it been so.

Apian's latitude instruments had a more influential career than their publication in the sumptuous "Imperial Astronomy" alone would have guaranteed them. They were adapted, for example, by Jacques Bassantin in his work *Astronomique discours* of 1557, having already been given a more public form on the face of the extraordinary astronomical clock by Philip Imser of Tübingen, dated 1555.[14] It is hard to believe that many of those who contemplated with the intricately engraved discs of contour lines on Imser's clock knew how to interpret them.

Apian's *Astronomicum* has the eclipse instruments of the Albion, in a heavily disguised form, but here there is nothing essentially new.[15] In some versions of the original eclipse instruments, it should be added, the semicircles are backed by a square mesh, giving the appearance of our modern squared paper. Its function here is quite different, however, for it is simply to assist the user in dropping perpendiculars to the scales, as the canons require. As so often in the history of "coordinate geometry," appearances are deceptive.

It is not suggested that the examples given here are unique, or the first of their kind. There is a very different type of instrument for planetary latitudes, for instance, as drawn by Johann Werner in his *Organa* of 1521, which may be generously interpreted as yielding the product of two numbers m and x as the ordinate to the straight line $y = mx$.[16] The technique of "geometric multiplication" was, of course, as old as Euclid, but the graphical form of Werner's figures, with their rectangular mesh, certainly enriched the "nomographic" tradition being discussed here. I have said nothing of the contour lines on sundials – for example, on cylinder dials – or of influences from outside Europe. The tradition of engraving the so-called sinecal quadrants on Indian astrolabes, for example, although seemingly late, would repay investigation. Doing so would not, however, affect significantly the thesis being put forward, that by the early sixteenth century the path to a "geometric algebra" was all but complete, if we interpret the word "algebra" liberally enough. The algorithms mentioned have been essentially trigonometrical, and the great notational advantages of the sixteenth and seventeenth centuries were still in the future, but the astronomers we have been discussing were undoubtedly conscious that they were using the graphs as substitutes for the underlying algorithms, and that the graphs were very *immediate* representations of the working-out of those algorithms. This might even be argued in the case of Richard of Wallingford, although he seems to have made no use of the representation of tabulated data.

The works of Richard of Wallingford, Apian, and their intermediaries, did not, in any direct way, concern the graphical representation of empirical data. One does not find this conspicuously in contour-form before the eighteenth century. It seems that the first important example might have been Philippe Buache's use of contours in his charting of submarine channels.[17] The older tradition is much more reminiscent of the graphical scales introduced by Louis Pouchet in 1790, as substitutes for the double-entry tables used in metric conversion. Pouchet's work in turn seems to have given rise to an entire subject, and one that was thought to be new – nomography. (The name is due to Maurice d'Ocagne, and dates only from 1891.) It is a subject that seems to have owed nothing to Fermat, Descartes, and their immediate followers. It made use of graphical procedures representing three and more coordinates in a plane. As it happens, Apian – or whoever was his source of inspiration – had already shown the way to introducing any number of variables into the plane, by the successive use of different "instruments" on the same page: c is a function of a and b; e is a function of c and d; g of e and f; and so forth. In short, the technique exploited by Richard of Wallingford lent itself to enormous generality. Nomography in its later manifestation came into its own in a technological milieu, where much rapid calculation was needed in any one of a class of essentially similar problems. This is very much the situation in which medieval astronomers had found themselves. But what is more to the point: Medieval astronomy had yielded techniques very similar to those found in the eighteenth and nineteenth centuries, and had done so in a way that many historians, aware of the chronology of developments in algebraic geometry, would have judged impossible a priori. Such a mistaken judgment can only arise from a poor historical categorization of "coordinate geometry." To this there are clearly more sides than one.

Notes

1 The phrase "Pappus planes" is a conventional expression having only a tenuous link with Pappus, and I use it simply because it is now a fairly standard way of referring to planes defined by the affine form of the famous theorem of Pappus and Pascal concerning collineations in connection with a hexagon. They are sometimes called "Pappus–Desargues planes," since in an affine plane Desargues's theorem follows from the limited form of that of Pappus and Pascal.

2 S. Günther, "Die Anfänge und Entwicklungsstadien des Coordinatenprinciples," *Abhandlungen der naturforschenden Gesellschaft zu Nürnberg* 6 (1877):3–50; H. G. Zeuthen, "Note sur l'usage des coordonneés dans l'antiquité, et sur l'invention de cet instrument," *Oversigt over Danske Videnskabernes Selskabs Forhandlinger* 1 (1888):127–44. The

literature spawned by these two classic papers is very considerable but not very relevant here.

3 For Zeuthen's most important statement of this, see his *Die Lehre von den Kegelschnitten im Altertum* (Copenhagen, 1886).

4 Descartes has an especially clearheaded statement of the equivalence, at the beginning of the first book of the *Géométrie*. For a convenient edition, with the French edition of 1637 facing an English translation (from the French and Latin), see D. E. Smith and M. L. Latham, *The Geometry of René Descartes* (repr. New York: Dover, 1954).

5 For more information about the background to these and other "Plinian" illustrations, and reproductions of many of them, see B. S. Eastwood, "Plinian Astronomy in the Middle Ages and Renaissance." *Science in the Early Roman Empire: Pliny the Elder, His Sources and His Influences,* eds. Roger French and Frank Greenaway (Beckenham, England: Croom Helm, 1986), pp. 197–251.

6 This far from obvious interpretation was offered by B. S. Eastwood (Ibid.).

7 For an elementary introduction to the instrument, see my "The Astrolabe," *Scientific American* (January 1974):96–106.

8 The universal astrolabe is so described because it is valid for all geographical latitudes. A conventional astrolabe needs a separate plate for every latitude for which it will be used.

For four tracts concerning instruments for conversion, all of them possibly associated with Richard of Wallingford (ca. 1292–1336), see my *Richard of Wallingford: An Edition of his Writings, with Introductions, English Translation and Commentary,* 3 vols (Oxford: Clarendon Press), vol. 2, 302–8. For an introduction to the *saphea*, and its various Western forms – it was incorporated in many Renaissance instruments, for instance – see vol. 1, 331–3; vol. 2, 187–92; and vol. 3, 36.

9 See my *Horoscopes and History* (London: The Warburg Institute, forthcoming). The "standard method" may actually be operated with the hour-lines themselves.

10 The instrument was the subject of a treatise edited in the edition cited in note 8. For a general study of equatoria, see E. Poulle, *Equatoires et horlogerie planétaire du XIIIe au XVIe siècle,* 2 vols. (Geneva: Droz, 1980).

11 See the edition of *Albion* and the commentary on it (as cited in note 8; above), sections III. 18–19 (for the first instrument's use) and III.20–34 (for the use of the second).

12 See plates XXIII and XXIV of my edition (note 8 above). There is another part, not illustrated there. For amplification of the remainder of this paragraph, see especially vol. 2, 127–36, 270–86, and appropriate places in the text, commentary, and illustrations (vol. 3).

13 There are many manuscript versions of Münster's work on equatoria. For a general survey, see Poulle, *Equatoires et horlogerie planétaire* [note 10 above], especially chapter B1. For the latitude instrument mentioned below, see Poulle, figure 123 and plate L.

14 See E. Poulle, *Equatoires et horlogerie planétaire,* plates LII–LIII.

15 For a survey of Apian's work, see *Richard of Wallingford* (note 8, above), vol. 2, 278–285.

16 The remarks about nomography in this paragraph are largely based on a paper read by Dr. H. A. Evesham to the British Society for the History of Mathematics in September 1983.

17 These are illustrated from Bodleian Library MS Digby 132 in Poulle, *Equatoires et horlogerie planétaire* (see note 10, above), plates XLVII–XLIX.

7

Eccentrics and epicycles in medieval cosmology

EDWARD GRANT

Between approximately 1160 and 1250, two rival cosmological systems entered Western Europe and vied for acceptance. In one, derived from the works of Aristotle, it was assumed that the stars and planets were carried around on concentric, or homocentric, spheres; in the other, derived ultimately from Claudius Ptolemy's *Hypotheses of the Planets,* the planets were assumed to be carried around by a system of material eccentric and epicyclic spheres.[1] Despite its momentous role in medieval cosmology, the *Hypotheses of the Planets* was not translated into Latin during the Middle Ages. In some as yet unknown manner, however, its fundamental ideas reached Western Europe, probably in works translated from Arabic, although the precise treatise, or treatises, in which these ideas were embedded has yet to be identified. Although works attributed to Alhazen (Ibn al-Haytham), Alfraganus (al-Farghani), and perhaps Thabit ibn Qurra included descriptions of material eccentric and epicyclic spheres, the basic scholastic version of eccentrics and epicycles described in this chapter finds no counterpart in the Latin translations of these Arabic treatises.[2]

Although references to epicycles and eccentrics appeared in such widely used thirteenth-century works as Sacrobosco's *Treatise on the Sphere* and in the anonymous *Theorica planetarum,* neither author indicates any knowledge of, or interest in, eccentrics and epicycles as real, material, solid orbs.[3] If Roger Bacon (ca. 1219–ca. 1292) was not the first to mention material eccentrics and epicycles in the Latin West, he may well have been the first scholastic natural philosopher to present a serious evaluation of the cosmological utility of eccentrics and epicycles.[4] Although hesitant and somewhat ambiguous, Bacon appears to have inclined toward Aristotle and rejected physical eccentrics and epicycles. Ironically, his description of the system of material eccentrics and epicycles was the one most widely adopted during the Middle

189

Ages and which still found defenders well into the seventeenth century. Because most discussions assumed three eccentric orbs for each planet, we shall frequently refer to the "modern" theory as the "three-orb system," and even on occasion as the "Aristotelian–Ptolemaic system," since, as we shall see, it also assigned a significant role to concentric spheres.

The literature on eccentrics and epicycles

Before describing the arguments for and against the three-orb system, a brief statement about the sources of those arguments is in order. Medieval cosmological discussions about the existence and nature of eccentrics and epicycles are found most frequently in commentaries or questions on Aristotle's *De caelo* and, to a lesser extent, in commentaries on Sacrobosco's *Treatise on the Sphere, Commentaries on the Sentences of Peter Lombard,* Book 2, distinction 14, which was concerned with the second to fourth days of creation, and commentaries or questions on Aristotle's *Metaphysics*. They are rarely found in straightforward astronomical treatises. Although many scholastic authors discussed the problem of the existence of material eccentrics and/or epicycles, few devoted much space to it. Roger Bacon and Pierre d'Ailly (ca. 1350–1420) were notable exceptions, the former writing perhaps the most extensive section in the Middle Ages and the latter devoting a lengthy question to it in his *Commentary on the Sphere of Sacrobosco*. Others who considered the problem of eccentrics in more than cursory fashion were Albertus Magnus (ca. 1200–1280), John Duns Scotus (1265–1308), John of Jandun (ca. 1289–1328), John Buridan (ca. 1300–ca. 1358), Albert of Saxony (d. 1390), Johannes Versor (d. after 1482), and Johannes de Magistris (fl. second half of the fifteenth century). Except for Duns Scotus, who treated eccentrics in his *Commentary on the Sentences,* and Jandun, who did so in his *Questions on the Metaphysics,* the other five took up the problem of eccentrics and epicycles in *Questions on De caelo* or, as with Albertus Magnus, in a commentary on *De caelo*.

The system of eccentrics and epicycles

What was the system of eccentrics and epicycles that Bacon described and that was destined to replace Aristotle's system of concentric spheres? After demonstrating the impossibility of any system constituted of eccentric orbs in which the eccentricity is due to an eccentric convex surface, an eccentric concave surface, or the eccentricity of both surfaces, Bacon introduces another interpretation – "a certain conception of the moderns," as he put it[5] – in which the external surfaces of each planetary orb are concentric but contain at least three eccentric orbs. To illustrate the system, Bacon describes the motions

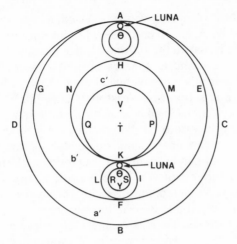

Figure 1

of sun and moon. Because the sun has only an eccentric, it will be useful to follow his account of the moon, which has both an eccentric and an epicycle.

In the diagram (Figure 1),[6] let *T* be the center of the earth and world and also the center of the lunar orb. The entire sphere of the moon lies between the convex circumference *ADBC* and the concave circumference *OQKP,* which are both concentric to *T*. Between these two circumferences, three orbs are distinguished (namely *a'*, *b'*, and *c'*) by assigning another center *V*, toward the moon's aux, or apogee. Around *V* as center are two circumferences, *AGFE* and *HNKM,* which signify the surfaces that enclose the lunar deferent and form the eccentric orb, *b'*. Surrounding the eccentric orb or deferent is the outermost orb, *a'* lying between surfaces *ADBC* and *AGFE;* and surrounded by the eccentric orb is the innermost orb, *c'*, lying between the concave surface *HNKM* and the convex surface *OQKP*. Between the surfaces of the middle, or eccentric, orb is a concavity that contains a spherical epicycle. The latter may be conceived in two ways: either as a solid globe, which Bacon calls a "convex sphere" (*spericum convexum*) like a ball (*pila*) because it lacks a concave surface; or as a ring with two surfaces, one convex (*KLFI*), the other concave (*RYSθ*), where the central core belongs wholly to the eccentric orb and forms no part of the epicyclic sphere itself. The moon, or planet, is a solid spherical figure which has only a convex surface and is located within a concavity of the epicyclic orb. The eccentric sphere is assumed to move around its center, *V*, carrying the epicycle with it; the epicyclic sphere, in turn, has its own simultaneous motion and carries the planet with it.

When extended to all the planets, it was this system that was widely adopted during the Middle Ages. Even those who did not accept the three-orb system believed it saved the astronomical appearances better than did the systems of concentric spheres proposed in Aristotle's *Metaphysics*[7] and in al-Bitruji's more technical *De motibus celorum*.[8] Not even Averroës' strong support for Aristotle's concentric astronomy and cosmology[9] could deter medieval natural philosophers from conceding to Ptolemaic eccentric orbs the greater ability to save the phenomena. Indeed only a few unambiguous defenders of Aristotle's purely concentric cosmology can be identified.[10] During the thirteenth, and even into the fourteenth, century, other scholastic natural philosophers, although not explicitly defending Aristotle, refused to concede that the world was constructed almost wholly of eccentric orbs. In a number of places in his writings, Thomas Aquinas was either noncommittal (*De trinitate*) or rejected eccentric orbs (*Commentary on the Metaphysics*). In his last treatise (*De caelo*) he argued that existence of eccentrics and epicycles was undemonstrated because even a system that saved the astronomical appearances might not be physically real.[11] John of Jandun was convinced that the three-orb system could save the astronomical appearances. Indeed, he proclaimed that he knew of no argument that could repudiate it. But in the end he rested content to proclaim it as merely "possible."[12] But even as Bacon rejected the three-orb system, his contemporary Albertus Magnus accepted the existence of eccentrics, although their arrangement differed from Bacon's description. By the end of the thirteenth and the early fourteenth centuries, this hesitancy was abandoned and more and more scholastics came to assume the truth of material eccentrics, among whom we may mention Duns Scotus, Aegidius Romanus (ca. 1245–1316), and Durandus de Saint–Pourçain (d.1334).

The three-orb system at the end of the fourteenth century: Pierre D'Ailly

By the time Pierre D'Ailly enters the picture near the end of the fourteenth century and some 130 years after Bacon's account, the three-orb system had received its definitive scholastic form. Indeed, no one expressed it better than Pierre D'Ailly, who presented as detailed an account as could be expected from a natural philosopher who was not a technical astronomer. Keeping in mind the diagram of the moon's spheres as described by Bacon, it will be useful to sketch D'Ailly's description of the three-orb system as he presented it in his *14 Questions on the Sphere of Sacrobosco*.[13] The objections he raised, and the solutions he proposed, were representative of the way material eccentrics and epicycles were interpreted by most natural philosophers from the late Middle Ages to the end of the sixteenth century.

According to D'Ailly, the heaven is made up of a combination of concentric and eccentric orbs. The totality of every orb (*orbis totalis*) is concentric and includes within it all other orbs necessary to produce the position of the planet. Within the concentric surfaces of each planetary orb are the eccentric orbs. Each eccentric orb or sphere, usually described as a "partial orb" (*orbis partialis*),[14] contains the center of the world as well as its own proper center outside the center of the world.

In a manner similar to Bacon, D'Ailly distinguished three types of eccentric orb but divided them into two classes: one, called *eccentricus simpliciter,* has the same center for both its concave and convex surfaces; the other, designated *eccentricus secundum quid*, has the center of the world as the center of one surface and a point outside the center of the world as the center of the other surface. The former surface is concentric; the latter, eccentric. Thus the eccentric surface of an eccentric orb *secundum quid* may be either convex or concave, yielding two different types of eccentric orbs for a total of three. Because it has two eccentric surfaces, an *eccentricus simpliciter* will always be of uniform thickness, whereas an *eccentricus secundum quid,* with one surface concentric and the other eccentric, is thicker in one part and thinner in another. When eccentric orbs are moved, the thin part of one moves with the thick part of another and conversely.[15]

The three-orb scheme is described next in a manner that differs little from Bacon's earlier description of the theory of the "moderns." According to D'Ailly, astronomers imagine three eccentric orbs as constituting the whole sphere of a planet. Two of these orbs are eccentrics *secundum quid,* that is, eccentric with respect to one surface only. One of them is the outermost orb and the other the innermost orb. As eccentrics *secundum quid,* the outermost orb is eccentric only with respect to its concave surface, while the innermost orb is eccentric only with respect to its convex surface. Between these two orbs lies the third, which is *eccentricus simpliciter* and therefore has both surfaces eccentric. Indeed the middle eccentric is constituted of the concave surface of the outermost eccentric sphere and the convex surface of the innermost sphere. The middle orb is called the deferent orb and carries the planet itself. Although D'Ailley's three-orb system incorporates the three types of eccentric orbs that he, and Bacon, distinguished, the surface of the outermost eccentric sphere must be concentric, as also the concave surface of the innermost eccentric orb. This was, of course, a primary feature of the Aristotelian–Ptolemaic compromise that saved the geocentric system. Because a concentric orb has the geometric center of the world as its center, D'Ailly explains that "the first moveable sphere [the *primum mobile,* sometimes equated with the sphere of the fixed stars] is a concentric orb, and generally every total orb is concentric, where 'total orb' (*orbis totalis*) is taken as the aggregate of all the orbs required to save the total motion of a planet."[16] In this manner, Aristotle's cosmology of concentric spheres was "saved," even though, in violation of Aristotle's physical principles, eccentric orbs with centers other than the center of the world formed the basis of the compromise system.

As we saw, the middle orb, with its two eccentric surfaces and uniform thickness, is called the deferent because it carries the planet. The deferent orb is divided into four equidistant points: the aux, which is most distant from the center of the world; the opposite of the aux, which is the point on the deferent nearest to the center of the world; and the two opposite points located between the aux and opposite of the aux, which are called the mean distances (*longitudines mediae*). Although the deferent orb carries the planet, it does so by an epicycle, which, as D'Ailly describes it, is "a small circle on the surface of the deferent orb that does not contain within itself the center of the world; and the body of the planet is imagined to be in it. And this epicycle is assumed to be contiguous, and not continuous, with the eccentric deferent because it is moved with a motion other than the motion of the eccentric deferent."[17] Like the eccentric deferent, the epicycle has four equidistant points, the aux of the epicycle; the opposite of the aux of the epicycle; and two points equidistant from the aux and the opposite of the aux called stations (*stationes*)[18]

Eccentric and epicyclic orbs explain the various observed motions of the planets, namely direct motion, retrogradations, and stations. When a planet is in the aux of its epicycle, its motion is said to be direct and quickest because the direction of its motion on the epicycle is the same as that of the eccentric deferent. But when the planet is in the opposite of the aux of the epicycle its motion is retrograde and slower because it now moves in a direction opposite to that of the eccentric deferent. Should the planet arrive at one of the points of station, it would move neither with the deferent nor contrary to it, so that its speed will seem neither to increase nor decrease. The planet will actually appear stationary.

Although D'Ailly mentions rival conceptions that reject epicycles and eccentrics, he regards the Ptolemaic theory of eccentrics and epicycles as "more common" (*est magis communis*) and unambiguously adopts it.

Cosmological problems with eccentrics and epicycles

By enclosing each set of eccentric planetary orbs within concave and convex surfaces that were themselves concentric with respect to the earth's center, Ptolemy himself had seemingly made a strong gesture toward reconciling his own cosmology with that of Aristotle's.

In both systems, the earth's center was the center of motion for each total planetary orb. Natural philosophers could thus continue to believe that the most basic structure of Aristotle's system was preserved – the external surfaces of all planetary spheres were concentric with the earth. Ptolemy sharply diverged from Aristotle, however, by his assumption that within the external concentric surfaces that constituted each total planetary sphere or orb were three or more partial eccentric orbs with centers other than that of the earth. In violation of Aristotle's dictum that all celestial spheres move with uniform motion around the earth as center, Ptolemy assumed the motion of all his eccentric spheres to be around points other than the earth's center. Although most scholastics recognized that eccentric orbs and epicycles explained planetary variations in distance and latitude that went unaccounted for in Aristotle's system of concentric spheres, they were also made aware of a number of significant problems that directly threatened Aristotelian cosmology and physics.

The medieval conflict between the Aristotelian and Ptolemaic systems centered on efforts to demonstrate that eccentric and epicyclic orbs did not imply consequences that were subversive and destructive of Aristotelian cosmology and physics. In what follows, we shall describe the manner in which medieval natural philosophers coped with material eccentrics and epicycles within the framework of Aristotelian cosmology and physics. Many of the cosmological objections that were raised against the three-orb system of material eccentrics and epicycles can already be found, though not always clearly expressed, in Roger Bacon's treatises of the 1260s. These arguments, and others, would be repeated and occasionally altered over the centuries.

1. Vacua and condensation and rarefaction in the heavens

In one of the most important arguments against the existence of material eccentrics, Roger Bacon declared that "it is impossible to assume an eccentric orb of any planet because then it would be necessary that the celestial body be divisible; or that two bodies be in the same place; or that a vacuum exist."[19] In D'Ailly's language, this possibility applies only to eccentrics that are *secundum quid,* that is, to eccentrics that have, with respect to the center of the world, one of their two surfaces eccentric and the other concentric. As with Bacon, D'Ailly and the scholastics were well aware that such eccentrics were of unequal thickness because the points of apogee and perigee were unequally distant from the center of the world (for example, in the figure, a' is thickest between points FB, where F is the point of perigee, and thinnest at point A, the point of apogee; whereas eccentric sphere c' is thickest at OH, where H is the point of apogee, and thinnest at point K, the point of perigee). The absurdities described by Bacon arise from the rotations of such orbs. For example, if we assume that eccentric orb c' has rotated 180 degrees, it follows that the thickest and thinnest parts of it will have exchanged places. Because the thickest part of orb c' will occupy more space than the thinnest part, it must make a space for itself by pushing away some of the surrounding matter that now occupies the place it must enter; or it must occupy the same place with that matter. If it replaces the matter that was previously there, the displaced matter, in turn, must find a place for itself and therefore must divide adjacent celestial matter. Within the set of planetary orbs that contain c', matter must condense somewhere so that the thicker part of orb c' can occupy a greater place. To accommodate the rotations of such orbs, celestial matter must be conceived as divisible and condensible, both of which possibilities were ruled out by Aristotle and most of his followers. At the other side of the eccentric orb, the thinnest part will be unable to fill the space formerly occupied by the thickest part. In these circumstances, either a vacuum will form or, in order to prevent a vacuum, matter adjacent to orb c' would instantaneously fill any empty spaces. In this situation, either void spaces exist in the heavens, or celestial matter is divisible and capable to rarefaction to fill a potential vacuum.

A factor that should have been of great importance in treating of the celestial orbs, but which was usually ignored or left vague, was the relationship that allegedly obtained between the external surfaces of any two successive orbs – that is, between the convex surface of a contained sphere and the concave surface of its containing sphere. In Aristotelian terms, three possibilities could be envisioned: (1) the surfaces are continuous, that is, they coincide; (2) the surfaces are contiguous, that is, they are distinct but in direct contact at every point; or (3) they are wholly or partially distinct and without contact. For Aristotle, two kinds of contact were possible. In one way, things that are in succession and touch are said to be "contiguous." Thus each of two distinct surfaces in contact at all points would be successive and contiguous. But if those two surfaces became one and the same, they would be said to be "continuous." Indeed "continuity would be impossible if these extremities are two." To emphasize his point, Aristotle explained further that "if there is continuity there is necessarily contact, but if there is contact, that alone does not imply continuity; for the extremities of things may be together without necessarily being one."[20] Aristotle employed the definition of continuity in his conception of the place of a thing, which he defined as "the boundary [or inner surface] of the containing body at which it is in contact with the contained body." But the two bounding surfaces of container and contained formed only one surface and were therefore continuous. Or, as

Aristotle expressed it, "place is coincident with the thing, for boundaries are coincident with the bounded."[21]

With this in mind, how did medieval natural philosophers interpret the relationship between the extreme surfaces of two successive celestial orbs? The manner in which they measured the distances of planetary spheres strongly suggests an assumption of continuity. In his widely used *Theorica planetarum,* Campanus of Novara declared that the convex surface of one planetary sphere was exactly equal to the distance of the concave surface of the next upper celestial sphere. Indeed, they formed one continuous surface; or, as Campanus expressed it, "the highest point of the lower [sphere] coincides with the lowest point of the higher."[22] He was apparently convinced that only by a fusion of these two celestial surfaces could waste space be avoided – whether in the form of a vacuum, or as some kind of separate matter distinct from the orbs themselves. Because most natural philosophers assumed with Campanus that the convex and concave surfaces of two successive orbs were equidistant from the center of the world, it is likely that they, too, thought of those surfaces as one and continuous.

Continuity of successive celestial surfaces posed insurmountable problems, however, as Roger Bacon recognized when he declared that continuous surfaces would cause those orbs to "be moved with equal velocity, even with the same motions, which is contrary to experience."[23] Although Bacon's contemporary Robertus Anglicus thought he could reconcile continuity of celestial surfaces with diversity of celestial motions,[24] Bacon, and probably most other natural philosophers, believed that the obvious facts of astronomy required a denial of continuity and the assumption of contiguity. The different directions in which planets were carried by their orbs or the diverse speeds at which they rotated made it all too plain that successive celestial surfaces could not form a single unified, continuous surface but had to be distinct entities in contact at every point.

However, if those successive surfaces were contiguous and in contact at all points so that they could move with independent circular motions, could something lie between any two such surfaces? For example, could a vacuum lie between them? Or perhaps some other kind of matter? To my knowledge, this question received only an indirect answer. By assuming continuity of the relevant surfaces and believing that this prevented the formation of waste space, Campanus and others like him appear to imply that contiguous surfaces were rejected because they would permit the dreaded vacuum or some kind of strange matter between the surfaces of two successive orbs. Continuity also triumphed because it fit perfectly with the prevailing ideas about the distances of the planetary orbs.

And yet, Campanus the astronomer, and medieval natural philosophers as a whole, might have assumed the contiguity of the successive
celestial surfaces and made it fit as well as, and very likely better than,
the assumption of continuity. Why is this? The strength of the continuity argument lies solely in its ability to explain why extraneous spaces
could not occur between celestial orbs. It utterly failed to explain how
successive orbs could move in different directions and at different
speeds. Rightly interpreted, the contiguity thesis, with its two distinct
successive surfaces, was better and more comprehensive than the continuity thesis. For not only did it allow for successive orbs to move in
different directions and at different speeds, there is no obvious reason
why it could not also have avoided the unacceptable existence of waste
space between its distinct surfaces. If the convex surface of an orb is
contiguous to the concave surface of the next higher celestial orb and
they are assumed in contact at all points along their respective surfaces,
how could any extraneous space or matter fit between them?

Instead of adopting contiguous spheres, however, many medieval
natural philosophers seem to have unwittingly held two incompatible
views simultaneously: When they considered the distances of the planetary spheres and the impossibility of waste space they assumed that
the proximate surfaces of two successive orbs were continuous and
one; but when they focused attention on the observed fact that successive planetary orbs moved in different directions and at different
speeds, they assumed, perhaps unknowingly, the contiguity of those
same planetary orbs.

In fact, the dire consequences that Bacon and others imagined could
not have occurred on the assumption of contiguous spheres. For on
the assumption of contiguity – that is, on the assumption that all orbs
are nested and their external surfaces are in contact at every point –
all surfaces would simply retain their respective locations as each
sphere rotated in a fixed position. Thus, no gaps could occur by loss
of contact at the points where the thinnest part of an orb enters places
formerly occupied by the thickest part; and similarly two bodies would
not come to occupy the same place, nor would any condensation of
celestial matter be required when the thickest part of an eccentric orb
enters places previously occupied by the thinnest part. If I have understood him rightly, Bernard of Verdun, in the thirteenth century, is at
least one scholastic who seems to have understood this important feature of the eccentric system.[25]

The potential impossible consequences that Bacon and others described do not seem to have been deduced as the result of an analysis
or comparison between a world conceived in terms of continuous or
contiguous celestial spheres. Those consequences may have derived
from a quite different concept, namely from the idea that an eccentric
material orb that possessed one concentric surface and one eccentric

surface not only was of unequal thickness but was actually ovoid in shape. If so, then Aristotle may have furnished the basis for this mistaken notion when, in demonstrating the sphericity of the heaven, he declared that if the world were not spherical, but "lentiform, or oviform, in every case we should have to admit space and void outside the moving body, because the whole body would not always occupy the same room."[26] In commenting on this passage, Nicole Oresme distinguished different circumstances under which oval-shaped planetary orbs would or would not produce the impossible consequences described above. He argues that "if the planetary spheres were oval in shape, being moved in a manner different from the sovereign [or last] heaven and on different axes . . . , either there would have to be an empty place or penetration in the heavens – that is, one heaven would pierce through the other – or there would have to be condensation or compression, all of which are impossible in nature."[27]

Judging from some of the responses, many scholastics seem to have believed that eccentric orbs of uneven thickness would indeed produce the alleged impossibilities were it not for an otherwise unexplained synchronization of motions. On the assumption that eccentric orbs move uniformly, Pierre D'Ailly, for example, held that when the thickest part of one eccentric is moved toward its opposite side, another eccentric orb moves uniformly in the opposite direction. When the two eccentric orbs – say, *a* and *b* – have simultaneously moved 180 degrees, the thickest part of orb *a* will have come to occupy the place formerly occupied by the thickest part of orb *b;* and similarly, the thinnest part of orb *a* will also have come to occupy the place formerly occupied by the thinnest part of orb *b*.[28] In this manner, a balance is always maintained and the dreaded impossibilities are perpetually avoided. How and why such synchronization of orbs should occur is nowhere explained or even considered.

The existence of material eccentrics was made to seem viable in yet another way: by the assumption of intervening matter between the orbs, which, as a consequence, implied the rejection of continuous or contiguous spheres. Albertus Magnus adopted this position in his *Commentary on De caelo*.[29] Like Bacon, Albertus seems to have considered eccentrics as oval-shaped and was therefore convinced that if eccentrics are nested one within the other, their motions would cause gaps between their surfaces.[30] Because, in Albertus' judgment, these gaps or spaces could not be void, one of two alternatives must occur: either (1) the various eccentric spheres would rarefy and condense to prevent formation of a vacuum; or (2) another body must intervene between any two successive eccentric spheres.

To refute the first alternative, Albertus argues that the surfaces of two successive spheres cannot be perfectly contiguous. At some points, and over stretches of their circumferences, loss of contact will occur.

Since the planetary orbs cannot themselves expand and contract, gaps that would inevitably develop between the surfaces of the successive orbs might perhaps be filled with a corporeal medium capable of rarefaction and condensation. But Albertus rejects this suggestion because when contiguous surfaces are in perfect point-by-point contact, the contiguous spheres and the intervening medium would occupy the same place. This follows because when the contiguous surfaces would separate at those points, the intervening corporeal medium would have to expand to prevent formation of the vacuum. Consequently, it must have been there all the time.

"Because of this," Albertus concludes, "I say that they [the successive spheres] never touch but that intervals [or gaps between the spheres] in some particular place are sometimes greater and sometimes smaller, and the rare or dense body existing between the circles [or spheres] fills them." Albertus adopted this interpretation and identified it with "the opinion of Thebit, a wise philosopher, in a book which he composed on the motion of the spheres."[31] Thus every sphere is separated from its immediate neighbors by a certain kind of celestial matter that is capable of rarefaction and condensation.

That Albertus adopted such an opinion – and it seems that he did – is quite astonishing. It marked a radical departure from Aristotle's cosmology. Albertus dramatically distinguished the matter that lies between orbs from the ether, or fifth element, that composes the rest of the celestial region. Unlike ether, which is incorruptible, indivisible, and therefore suffers no rarefaction or condensation, the intervening matter can rarefy and condense and must therefore be divisible. Although the creator of the celestial orbs made some parts permanently rarer and some parts permanently denser than other parts, he made "the intervening body contractable and extendable so that it should fill what lies between the spheres."[32] Albertus had thus abandoned the important Aristotelian concept of celestial homogeneity and assumed instead the existence of two different kinds of eternal celestial substances: one divisible and therefore changeable; the other indivisible and unchangeable. Moreover, the celestial orbs were no longer in direct contact.

That few chose to follow Albertus comes as no surprise.[33] His was a radical theory. Convinced that eccentrics were essential to account for the astronomical phenomena, Albertus was obviously prepared to abandon certain important Aristotelian concepts in favor of a system that would save the phenomena and also preserve a viable cosmology. As we shall see, others would proceed in a similar fashion.

2. If eccentrics exist, can the earth lie at the center of the world?

Although the potential impossibilities just described were probably considered the most serious cosmological difficulties for scholastic authors, a number of other objections appeared rather regularly. Pierre D'Ailly reports that some questioned whether, if eccentrics existed, the earth could lie at the center of the universe.[34] This objection was apparently based on the assumption that all celestial orbs are eccentric. It would therefore follow that the earth could not be their center. D'Ailly, and others, replied that the earth lies at the center of the "total orb," that is, it lies at the center of all the concentric surfaces that serve as boundaries for each set of planetary orbs.[35] As Albert of Saxony explained, the absence of the earth from the center of eccentric orbs posed no problems because eccentrics are included within the totality of planetary orbs, the external surfaces of which are concentric.[36] Here, of course, was the fundamental feature of the compromise between Ptolemaic astronomy and Aristotelian cosmology. With the extreme surfaces of every planetary sphere assumed concentric with respect to the earth, the eccentric orbs contained within those concentric surfaces could possess their own centers without any adverse effect on Aristotelian cosmology.

3. Eccentrics and the problem of a plurality of centers

But even if the earth could serve as a center of the universe, the existence of at least one other center for the eccentric orbs would involve at least two different centers for celestial bodies. If two such centers existed, could a heavy body move naturally downward to its natural place, when the latter is defined in terms of a unique center that functions as a *terminus ad quem?* According to D'Ailly, some denied that a heavy body could reach its natural place at the center of the world. They argued that a heavy falling body would either have to move to both centers simultaneously; or, because it could not choose between them, it would not move at all.[37]

D'Ailly responded that despite the different centers, every heavy body would nonetheless move toward the center of the world because the latter is the center of the "total orb," that is, the center of all the concentric surfaces which enclose all the eccentric orbs.[38] In a similar vein, Johannes Versor argued that a plurality of eccentric and concentric centers would not render meaningless the idea of a unique, absolute "down" location. "Down" in the universe, Versor explains, was usually taken "in relation to the whole heaven, or in relation to the whole orb, but not with respect to partial circles [or orbs]."[39] But the heaven "is concentric with respect to each of its extremal surfaces; and the same holds for any orb, even though a partial surface of one part of the orb has a center distinct from the center of the world."

Although the concentric surfaces of each planetary sphere enabled the earth to retain its cosmic centrality, and although the earth remained the natural place of heavy bodies, the defenders of solid eccentrics and epicycles had made a significant departure from Aristotelian cosmology: They allowed celestial bodies to move around more than one center. Eccentric orbs were assumed to move around their own centers rather than around the earth as center. To accept the three-orb system as truly representative of the physical cosmos was to admit that, contrary to Aristotle, Averroës, and Maimonides, planetary spheres could rotate around geometric points other than the center of the earth, that is, other than the geometric center of the universe. Most scholastics passed over this significant shift with little or no comment, but a few, like Nicole Oresme and John Buridan, met the issue head on. Oresme flatly declared that "whether Averroes likes it or not, we must admit that they [the heavenly bodies] move around various centers, as stated many times before, and this is the truth."[40] Buridan insisted that "the Commentator [Averroës] speaks improperly when he says that the spheres are located by a [common] center; . . . modern astronomers (*astrologi*) do not concede that all celestial spheres have the same center; indeed they assume eccentrics and epicycles."[41]

4. Would planets move with rectilinear motion if eccentrics and epicycles existed?

Because a major purpose of eccentrics and epicycles was to account for changes in planetary distances from the earth, it was alleged that eccentrics and epicycles would cause planets to ascend and descend rectilinearly as they alternately approached and withdrew from the earth. Following Albumasar, Roger Bacon held that motion is three-fold, namely, from the center (*media*), toward the center, and around the center of the world.[42] Celestial bodies only move around the center of the world, that is around the earth. For if a planet moved around another center, it would sometimes be nearer the earth and sometimes farther away. But if a celestial body varied its distance from the earth in this manner, it could only do so by a rectilinear motion, or by a motion compounded of rectilinear and circular. To move toward or away from the earth rectilinearly, a celestial body would have to be either heavy or light, or compounded of both; or it might have an entirely different nature. But rectilinear motion toward or away from the earth and out of a circular orbit would involve a celestial body in violent action, which was contrary to the nature of the celestial ether.[43]

The usual response was to deny that variations in planetary distances involved rectilinear motion. Pierre D'Ailly argued[44] that up-and-down motions could only happen where generation and corruption occurred, namely in the terrestrial region.[45] Paul of Venice met the same objection by a different argument. To qualify as rectilinear ascent and descent, motions must be measured along a radius of the world. Such measurements were therefore not applicable to circular motions,[46] from which it followed that the motion of planets on eccentrics and epicycles did not qualify as rectilinear.

5. The problem with epicycles

Although it would have appeared that the acceptance of eccentric orbs committed one to acceptance of epicyclic orbs, at least one scholastic, John Buridan, accepted the former but not the latter. Epicycles posed a special problem because of the moon's observed behavior. Aristotle had argued that because the moon always shows the same face to us, it cannot be said to rotate or revolve. On the assumption that all planets were alike in their basic properties, he inferred from the moon's behavior that no planets rotated around their own axes.[47] Aristotle's denial of rotation to the moon and other planets would play a significant role in arguments about the reality of material epicycles.

If the fundamental problem about epicycles can be traced as far back as Roger Bacon in the thirteenth century,[48] it was John Buridan and Albert of Saxony in the fourteenth century who clearly reveal the two most basic approaches that were open to natural philosophers. In Book 12, question 10, of his *Questions on the Metaphysics,* Buridan inquired "whether epicycles are to be assumed in celestial bodies."[49] His opinion was based on the behavior of the "man in the moon," that is, the spot on the lunar surface that had the appearance of a man whose feet always point toward – or lie at – the bottom of the moon. If the moon had an epicycle, the man's feet should sometimes appear in, or point toward, the upper part of the lunar disk. Thus, if the man's feet are at the bottom of the lunar disk when the moon is in the aux, or apogee, of the epicycle, the feet ought to be in the upper part of the lunar disk when the moon reaches the opposite of the aux, or perigee, of the epicycle. Buridan argues, however, that this is never observed. The feet always remain at the bottom of the lunar disk. Buridan suggests a way to account for this phenomenon and retain the epicycle. We would have to assume that "just as this epicycle is moved around its proper center, so also is the body of the moon moved around its proper center in a motion contrary to that of the epicycle and with an equal speed."[50] Only in this way will the upper part of the man always appear in the upper part of the lunar disk.

Assuming with Aristotle that all planets possess the same basic properties, Buridan infers that if the moon has a proper rotatory motion, all the other planets should also possess that same motion. Like Aristotle, however, he was convinced that no planet could rotate around its own center. Planets not only move from one position to another, they also cause transmutations in sublunar bodies. Consequently, if planets rotated around their own centers, the rotations ought to affect the way in which they cause sublunar effects. That is, each planet ought to produce differential effects; otherwise its rotatory motion would be superfluous. Taking the sun as exemplar, Buridan argues that it does not produce such differential effects, probably because it is a uniform, homogeneous body whose upper and lower parts are identical. Any rotatory motion by the sun around its own center would therefore be superfluous because no sublunar changes would result. "But if the sun does not have such a motion, it does not seem reasonable that the moon should have it, since the sun is much nobler than the moon."[51]

Buridan now derives the following consequence: If the moon does not have a proper motion around its own center, it cannot have an epicycle. For if the moon had an epicycle but lacked a proper motion, the head and feet of the man in the moon would change positions every time the epicycle's apogee and perigee rotated 180 degrees. Because no such change is observed, Buridan concludes that the moon could have no epicycle, from which it seemed to follow that "if an epicycle is not posited in the orb of the moon it ought not to be posited in the orb of the other planets, since all reasons which apply to the other planets should also apply to the moon."[52] Thus does Buridan conclude that "all appearances can be saved by eccentrics [alone] without epicycles."

One response to Buridan's tactic was to allow that a particular planet might indeed behave differently from its sister planets. Albert of Saxony adopted just such a strategy. After describing the problem much as Buridan had,[53] Albert assumes that the moon possesses a proper motion around its own center in a direction contrary to the motion of its epicycle. As for those who say that "other planets do not have proper motions around their proper centers, therefore the moon does not," Albert counters, without elaboration and perhaps with Buridan in mind, that the moon's nature differs from that of the other planets because the moon's upper and lower parts can affect sublunar things differentially. Its proper motion around its own center is, therefore, not superfluous but brings the lower part of the moon to the upper part and the upper to the lower. Because the moon's proper motion is contrary to the motion of its epicycle, we do not observe these continuous and regular turnings of the spot in the moon.[54]

Albert's interpretation prevailed.[55] Although both men sought to save the observed behavior of the spot in the moon, they did so in radically different ways. Where Buridan insisted on the uniformity of planetary behavior and properties, Albert permitted divergence. Buridan sought for consistency: either all planets rotated around proper centers, or none did; either all planets moved on epicycles, or none did. The astronomical appearances could only be saved if planetary homogeneity and uniformity were preserved. By contrast, Albert of

Saxony thought it more important to save the appearances than to preserve the uniformity of planetary behavior. In Albert's scheme, it was not necessary that all planets should move on epicycles (the sun did not). Nor, as we saw, was it necessary that either *all* planets or *no* planets move around their own centers. If the phenomena could be saved by assuming that some planets really moved around their own centers and others did not, Albert was satisfied.

Conclusion

Although a few other arguments were sometimes cited for and against eccentric and epicyclic spheres,[56] those mentioned here were unquestionably the most basic for cosmology. Despite the conviction of most scholastic natural philosophers that eccentrics and epicycles saved the astronomical phenomena and that Aristotelian concentric astronomy did not, they were also aware that those same eccentrics and epicycles appeared to violate important aspects of Aristotelian cosmology. In order to save the astronomical phenomena and avoid alleged cosmological impossibilities, some, and in a number of instances many, natural philosophers made major departures from Aristotelian cosmological principles. Among the most significant were the assumptions that (1) eccentric celestial orbs move with circular motion around centers other than the earth; (2) that the moon and all other planets have proper motions around their own centers in a direction opposite to that of their epicycles; (3) that successive orbs are not in direct contact and the space between those orbs is occupied by a celestial substance that is divisible, though incorruptible; and, finally (4), that celestial bodies, and therefore the celestial substance, need not be homogeneous.[57]

These were seemingly significant changes and raise momentous questions about the very meaning of the term "Aristotelian cosmology" during the period 1200 to 1700. To the departures mentioned here, others equally, and perhaps even more, significant would emerge in the controversies of the sixteenth and seventeenth centuries. Most, though not all, of the ideas on eccentrics and epicycles described here became a part of Aristotelian cosmology. By contrast, other ideas that were deemed essential features of that cosmology during the period 1200 to 1500 lost that status in the seventeenth century.[58] The complex of ideas and concepts that constituted the core of Aristotelian cosmology not only varied with time but was also often enough at odds with Aristotle himself. For these reasons, any attempt to describe a definitive sense of Aristotelian cosmology between 1200 and 1700 is fraught with difficulty.

Despite significant departures from Aristotle's cosmology, Aristotle remained the supreme authority on matters cosmological. This was no doubt due in part to the compromise nature of the three-orb system, which retained the basic Aristotelian concentricity of every total planetary orb while assuming Ptolemy's eccentric orbs and epicycles within the concentric surfaces of each planetary sphere. The compromise nature of medieval cosmology may have worked to obscure the significance of the departures. Their seemingly dangerous consequences were explained away to the apparent satisfaction of most natural philosophers. The departures were apparently deemed a small price to pay for a system that saved the astronomical phenomena and also retained the earth as the physical center of each total planetary sphere.

From discussions of the three-orb system, it is obvious that those who accepted the new system believed in the physical reality of material eccentrics and epicycles. The controversy in the Latin West was not between those who argued for a system that merely saved the appearances regardless of physical reality and those who insisted that any astronomical system must not only save the phenomena but also represent physical reality.[59] Rather, the dispute involved a decision as to which system of cosmic spheres best represented physical reality – a purely concentric system or a mixture of concentric and eccentric spheres. Repeated invocations of dire physical consequences that might or might not follow from one or the other of the two rival systems serve only to confirm that medieval natural philosophers were arguing about the structure of cosmic reality and not about convenient and arbitrary arrangements of geometrical figures that might save the astronomical appearances.

Although verbal and diagrammatic representations of the medieval cosmos from the Middle Ages to the present frequently depict it as a system of concentric spheres, those depictions were sufficiently ambiguous to embrace both the purely concentric cosmos of Aristotle and the concentric shells of the Ptolemaic system with its eccentrics and epicycles omitted.[60] A purely concentric system thus became identified with the medieval conception of the cosmos. But, as we have seen, such a system never took root in the Middle Ages and was rejected by the great majority of astronomers and natural philosophers. The compromise three-orb system that derived ultimately from Ptolemy's *Hypotheses of the Planets* was by 1300 the dominant interpretation of celestial motion. It would remain so until the seventeenth century, when, in the aftermath of the new Copernican astronomy and cosmology, the concept of real celestial orbs vanished forever.

Notes

1 In his monumental *Le système du monde. Histoire des doctrines cosmologiques de Platon à Copernic* (10 vols.; Paris: Hermann, 1913–59),

Pierre Duhem described the controversy between the defenders of concentric orbs and those who sided with Ptolemy and assumed eccentric and epicyclic orbs. Duhem, however, did not present this important topic in an independent and integrated manner but treated individual authors, usually in chronological order. Sections of varying lengths are thus isolated within descriptions of individual authors and are nowhere adequately summarized or synthesized. Because seventy or more years have passed since Duhem's heroic effort, it is time to reexamine this significant issue in medieval natural philosophy.

2 Although it has been said (Olaf Pedersen, "Astronomy," in David C. Lindberg, *Science in the Middle Ages* [Chicago: University of Chicago Press, 1978], pp. 321 and then 319) that the machinery of the material spheres suggested by Ptolemy's *Hypotheses of the Planets* was presented to Western astronomers in two brief cosmological treatises by Thabit ibn Qurra, namely Thabit's *De hiis quae indigent antequam legatur Almagesti* and *De quantitatibus stellarum et planetarum et proportio terrae,* an examination of these treatises reveals nothing relevant for the problem of material eccentric and epicyclic spheres (for the texts, see Francis J. Carmody, *The Astronomical Works of Thabit b. Qurra* [Berkeley: University of California Press, 1960], pp. 131–139, 145–148). What Thabit may have passed on to the West on the subject of material eccentric spheres will be seen later in the chapter.

In his *Theoricae novae planetarum,* Georg Peurbach (1423–1461) provided detailed descriptions of the eccentric and epicyclic machinery for the different planets. Willy Hartner believes that early Renaissance astronomers drew on Arabic sources (at least on Alhazen and al-Jaghmini) but has been unable to identify the channels of transmission. See Willy Hartner, "The Mercury Horoscope of Marcantonio Michiel of Venice, A Study in the History of Renaissance Astrology and Astronomy," *Collectanea III: Willy Hartner, Oriens-Occidens, Ausgewählte Schriften zur Wissenschafts- und kulturgeschichte Festschrift zum 60. Geburtstag* (Hildesheim, W. Germany: Georg Olms, 1968), p. 480. Although Alhazen's *Liber de mundo et coelo,* which is the Latin version of a treatise bearing an Arabic title akin to *Configuration of the World,* is a possible source of influence on Peurbach, it does not qualify as a direct source for the Middle Ages. Indeed, if the Latin translation of Alhazen's treatise was actually made in the late thirteenth century (see Noel M. Swerdlow, "Pseudodoxia Copernicana: or, Enquiries into very many received tenents and commonly presumed truths, mostly concerning spheres," *Archives internationales d'histoire des sciences* 26 [1976]:117 n.14), then it probably became available sometime after Roger Bacon's description of material celestial spheres made in the 1260s (see below, note 4).

3 For Sacrobosco's discussion, see Lynn Thorndike, ed. and tr., *The "Sphere" of Sacrobosco and Its Commentators* (Chicago: University of Chicago Press, 1949), pp. 113–114 (Latin), pp. 140–141 (English). For the *Theorica,* see Olaf Pedersen's translation in Edward Grant, ed., *A Source Book in Medieval Science* (Cambridge, Mass.: Harvard University Press, 1974), pp. 452–465 (eccentrics are defined on p. 452).

In speaking of "real, material, solid orbs," I deliberately refrain from signifying whether those orbs were conceived as hard or soft. Few natural philosophers took up the problem, and it is one that I shall consider separately.

4 Sometime in the 1260s, Bacon presented almost identical accounts in his
Opus tertium and *Communia naturalium*. But in the latter, Bacon added a
significant chapter on whether eccentrics and epicycles were consistent with
Aristotelian cosmology and followed this with a lengthy description of the
Ptolemaic system. For the *Opus tertium*, see Pierre Duhem, ed., *Un
fragment inédit de l'Opus tertium de Roger Bacon précédé d'une étude sur
ce fragment* (Ad Claras Aquas [Quaracchi] prope Florentiam: ex
typographia Collegii S. Bonaventurae, 1909). Bacon's discussion on
eccentrics and epicycles extends over pp. 99–137. For the *Communia
naturalium*, see *Opera hactenus inedita Rogeri Baconi*, fasc. 4: *Liber
secundus Communium naturalium Fratris Rogeri De celestibus*, partes
quinque, ed. Robert Steele (Oxford: Clarendon Press, 1913), pp. 419–456.
    Crombie and North believe the *Opus tertium* was written sometime
between 1266 and 1268 (see A. C. Crombie and J. D. North, "Bacon,
Roger," *Dictionary of Scientific Biography* [16 vols.; New York: Scribner,
1970–80], vol. 1, p. 378) and suggest that Bacon may have written his
Communia naturalium in this same period. By contrast, David C. Lindberg
(*Roger Bacon's Philosophy of Nature*. A Critical edition, with English
translation, introduction, and notes, of *De multiplicatione specierum* and *De
speculis comburentibus* [Oxford: Clarendon Press, 1983], p. xxxii), who
accepts 1267 as a date of composition for the *Opus tertium*, believes that no
firm date can be attached to the *Communia naturalium* and that "all that
can be said is that it represents the early stages of the broadening of
Bacon's outlook, the usual guess placing it in the early 1260s."

5 "de quadam ymaginatione modernorum." *Opus tertium*, p. 125 (and pp.
125–134 for the exposition and critique of the modern theory), and
Communia naturalium, p. 438 (and pp. 437–443 for the modern theory).
Although the "modern" theory seems ultimately derived from Ptolemy's
Hypotheses of the Planets, Bacon's immediately preceding discussion
(*Opus tertium*, pp. 119–125; *Communia naturalium*, pp. 433–437),
concerning the impossibility of a total planetary orb being composed of
eccentric orbs where one or both of the external surfaces is eccentric, may
have derived from an earlier attempt to materialize eccentrics on the basis
of Ptolemy's description in the *Almagest*.

6 The figure appears on p. 129 of Duhem's edition of the *Opus tertium;* the
relevant text on pp. 128–131. The description follows my account in
"Cosmology," in Lindberg, ed., *Science in the Middle Ages*, pp. 281–282.
To identify the three distinct orbs, I have added the letters a', b', c'.

7 *Metaphysics* 12.8.1073b.11–1074a.14.

8 The Latin text appears in Francis J. Carmody, ed., *Al-Bitrûjî De motibus
celorum*, Critical Edition of the Latin Translation of Michael Scot
(Berkeley: University of California Press, 1952).

9 Averroës' defense of Aristotle was made in his middle commentaries on *De
caelo* and the *Metaphysics*. The relevant passages have been collected and
analyzed by Francis J. Carmody, "The Planetary Theory of Ibn Rushd,"
Osiris 10 (1952):556–586.

10 For example, William of Auvergne, Alexander of Hales, and St.
Bonaventure. See Duhem, *Le système du monde*, vol. 3, p. 404.

11 For these passages, see Thomas Litt, *Les corps célestes dans l'univers de
Saint Thomas d'Aquin* (Louvain: Publications Universitaires; Paris:
Beatrice-Nauwelaerts, 1963), pp. 348, 350–352. Aquinas' position is rather
akin to that of Moses Maimonides (see Moses Maimonides, *The Guide of*

the Perplexed, translated with an Introduction and Notes by Shlomo Pines with an introductory essay by Leo Strauss [2 vols.; Chicago: University of Chicago Press, 1963], 2, bk. 2, chap. 24, pp. 322–327).

12 *Ioannis de Ianduno in duodecim libros Metaphysicae iuxta Aristotelis et magni Commentatoris intentionem* . . . (Venice: apud Hieronymum Scotum, 1553; rep. Frankfurt: Minerva, 1966), bk. 12, question 20 ("whether a plurality of eccentric orbs and epicycles are really in celestial bodies"), fols. 141r, col. 1–142r, col. 1.

13 Pierre D'Ailly devoted the thirteenth of his fourteen questions on the *Sphere* of Sacrobosco to the question of "whether it is necessary to assume eccentric and epicyclic circles to save the appearances in planetary motions." In the edition I used, D'Ailly's treatise is one of sixteen on astronomy that appeared in Venice, 1531, with the title *Spherae Tractatus Ioannis de Sacro Busto Anglici viri clariss.; Gerardi Cremonensis Theoricae planetarum veteres; Georgii Purbachii Theoricae planetarum novae; Prosdocimo de Beldomando Patavini Super tractatu sphaerico commentaria, nuper in lucem diducta per L.GA. nunquam amplius impressa. . . . Petri Cardin. de Aliaco episcopi Camaracensis 14 Quaestiones. . . . Alpetragii Arabi Theorica planetarum nuperrime Latinis mandata literis a Calo Calonymos Hebreo Neapolitano, ubi nititur salvare apparentias in motibus planetarum absque eccentricis et epicyclis.* For question 13, see fols. 163v–164v.

14 Although the expression "partial orb" was rather common, D'Ailly did not use it.

15 See D'Ailly, *14 Questions,* fol. 163v. Without employing the terms *eccentricus simpliciter* and *eccentricus secundum quid,* Albert of Saxony divided eccentric orbs into the same three types as did Bacon and D'Ailly. See Albert's *Questions on De celo,* in *Questiones et decisiones physicales insignium virorum: Alberti de Saxonia in octo libros Physicorum; tres libros De celo et mundo; duos lib. De generatione et corruptione; Thimonis in quatuor libros Meteororum; Buridani tres lib. De anima; lib. De sensu et sensato . . . Aristotelis. Recognitae rursus et emendatae summa accuratione et iudicio Magistri Georgii Lokert Scotia quo sunt Tractatus proportionum additi* (Paris, 1518), *De celo,* bk. 2, question 7 ("whether for saving the appearances of the planetary motions, it is necessary to assume eccentric orbs and epicycles"), fol. 106v, col. 1. Because D'Ailly's arguments are similar to, and even follow the order of, Albert's, it is not unreasonable to suppose that D'Ailly may have used Albert of Saxony's question as one of his major sources.

16 "Unde orbis concentricus dicitur orbis sub utraque eius superficie continens centrum mundi et habens eius centrum cum centro mundi. Isto modo primum mobile est orbis concentricus et generaliter quilibet orbis totalis est concentricus et ibi capitur orbis totalis pro aggregato ex omnibus orbibus requisitis ad salvandum motum totalem unius planetae." D'Ailly, *14 Questions,* fol. 163v.

17 Ibid., fols. 163v–164r.

18 Ibid., fol. 164r. The four points on the epicycle are in fact not equidistant, as D'Ailly asserts. Although the first and second stations must always be equidistant from the true apogee (aux) of the epicycle, the points of station always fall nearer to the perigee of the epicycle (opposite of the aux) than to the apogee. It follows, therefore, that the four points on the epicycle mentioned by D'Ailly cannot be equidistant. See Campanus of Novara's

Theorica planetarum, pp. 225–227, 231, 313, of the edition cited below in note 22. For bringing this to my attention, I am grateful to my colleague Victor E. Thoren.

19 *Opus tertium,* p. 119; *Communia naturalium,* pp. 433–434. Sometime between 1322 and 1327, Cecco d'Ascoli (1269–1327), in a treatise *De eccentricis et epicyclis,* specified the various impossible consequences that would result from the existence of eccentrics and epicycles. He explains that "if there were eccentrics and epicycles, then rarefaction or condensation would occur, which is impossible by the first [book] of [*De*] *celo et mundo;* or a vacuum would occur, which is impossible, as is said in the fourth [book] of the *Physics;* or there would be a separation of the spheres, which is impossible, as is obvious in the second [book] of *De celo et mundo;* or there would be a penetration of bodies, which is false, as is obvious in the fourth [book] of the *Physics.*" See G. Boffito, "Il 'De eccentricis et epicyclis' di Cecco d'Ascoli novamente scoperto e illustrato," *La bibliofilia rivista dell'arte antica in libri, stampe, manoscritti, autografi e legature,* diretta da Leo S. Olschki, vol. 7 (1905–6), pp. 161–162.

20 Aristotle *Physics* 5.3.227a.9–12, 21–23. The translation is by R. P. Hardie and R. K. Gaye, in *The Complete Works of Aristotle: The Revised Oxford Translation,* ed. Jonathan Barnes (2 vols; Princeton: Princeton University Press, 1984), vol. 1.

21 *Physics* 4.4.212a.5–6 and 212a.30. The bracketed addition is mine. Aristotle's doctrine of place was as applicable to the nested celestial spheres as it was to terrestrial objects, as is evident from the fact that he denied the existence of places beyond the world (*De caelo* 1.9.279a.8–15), thereby implying that places existed throughout the cosmos.

22 *Campanus of Novara and Medieval Planetary Theory: Theorica Planetarum,* edited with an Introduction, English translation, and commentary by Francis S. Benjamin, Jr., and G. J. Toomer (Madison: University of Wisconsin Press, 1971), p. 331. See also pp. 331–337 and 53–55. It was usually assumed in medieval Islamic and Latin astronomy that the distance of the convex surface of one planetary sphere was equal to the distance of the concave surface of the next sphere (see the tables of distances and dimensions in *Campanus of Novara,* pp. 356–363). In effect, since the two distances were identical so were the surfaces. Although the distances of the innermost and outermost circular surfaces were fixed, the distances of the planets varied within their respective epicycles.

23 *Opus tertium,* p. 123; *Communia naturalium,* p. 436.

24 In his *Commentary on the Sphere of Sacrobosco,* written around 1271, Robertus avoided the major problem with continuous orbs by assuming that the outer surfaces of celestial orbs were immobile with only their middle parts, which he likened to a fluid, being capable of motion. Under these conditions, each orb could move independently of the others. See Thorndike, *The "Sphere" of Sacrobosco and Its Commentators,* p. 147 (Latin) and pp. 202–203 (English).

25 After mentioning the usual charge that eccentrics would produce the dreaded impossibilities, Bernard explains that the "different parts" of eccentric orbs "succeed themselves continually in the points or places of the farther and nearer distance" – that is, in the points of apogee and perigee – "that are imagined in the convexity of the surrounding orb." The translation is mine from *Source Book,* p. 523 (I have here replaced "longitude" with "distance").

26 *De caelo* 2.4.287a.12–24 (tr. J. L. Stocks, in *The Complete Works of Aristotle*).

27 *Nicole Oresme Le Livre du ciel et du monde,* ed. Albert D. Menut and Alexander J. Denomy, C.S.B.; translated with an introduction by Albert D. Menut (Madison: University of Wisconsin Press, 1968), bk. 2, chap. 10, p. 391. I have added the words in square brackets.

28 See D'Ailly, *14 Questions,* fol. 163v, for the argument, and fol. 164v for D'Ailly's brief response. Between 1322 and 1327, Cecco d'Ascoli described the same mechanism for synchronizing the motions of eccentric orbs by what he called "proportional motions" (*proportionales motus;* "De eccentricis," *La Bibliofilia,* vol. 7, 166). Although Cecco, who defended the existence of eccentrics and epicycles, thought the idea of "proportional motions" was a good idea, he denied that Ptolemy had it in mind. Some two hundred years after D'Ailly, Christopher Clavius, the famous Jesuit astronomer, would present, and approve of, the same explanation. See *Christophori Clavii Bambergensis ex Societate Iesu In Sphaeram Ioannis de Sacro Bosco Commentarius nunc quarto ab ipso auctore recognitus et plerisque in locis locupletatus* (Lyon: sumptibus fratrum de Gabiano, 1593), p. 521 (for description of the impossibilities), and p. 523 (for the response).

29 *Alberti Magni Opera Omnia,* vol. 5, part 1: *De caelo et mundo,* ed. Paul Hossfeld (Monasterii Westfalorum, Münster: Aschendorff, 1971), bk. 1, tract 1, chap. 11, pp. 29–30.

30 Albertus was not reporting Bacon's "modern" three-orb system but rather assigned only one eccentric to each planet.

31 Albertus Magnus, *De caelo,* p. 30. Hossfeld, the editor of this work identified Albertus' source as Thabit's (or Thebit's) *De motu octavae sphaerae* (see Hossfeld's note to line 29 on p. 30). In Carmody's edition of this treatise (*The Astronomical Works of Thabit b. Qurra,* pp. 102–107), there is no such passage. Hossfeld's claim, however, is based on a single manuscript, MS Paris, Bibliothèque Nationale, fond latin, 7195, fols. 140 vb–143va, which is but one of a number of manuscripts that Carmody used for his edition. If such a passage existed in Thabit's *De motu octavae spherae,* it is unlikely that all traces of it would have disappeared from Carmody's edition. Indeed, the theory of intervening matter between the celestial orbs does not appear in any other of Thabit's works that were translated into Latin and that have been edited by Carmody. Yet even before Albertus Magnus, Moses Maimonides also cited Thabit as one who demonstrated that a body must lie between successive celestial spheres (see Maimonides, *The Guide of the Perplexed,* vol. 2, p. 325). What Thabit may have had in mind is left vague.

32 Albertus Magnus, *De caelo,* p. 30, where Albertus adds that this is also the opinion of Avicenna (in the *De caelo et mundo* of the latter's *Sufficientia*) and Averroës (in the *De substantia orbis,* which Albertus cites as *Liber de essentia orbis*).

33 One who did was Cecco d'Ascoli, who declared that "orbs are neither continuous nor contiguous, but there is an intervening body between them, which, according to Thebit and Albertus, is capable of being compressed." For the Latin text, see Cecco's *Commentary on the Sphere of Sacrobsoco,* in Thorndike, *The "Sphere" of Sacrobosco and Its Commentators,* p. 353. In the seventeenth century, Giovanni Baptista Riccioli cited this very passage as evidence that D'Ascoli believed in the fluidity of the heavens (see Riccioli's *Almagestum novum astronomiam veterem novamque*

complectens observationibus aliorum, et propriis novisque theorematibus,
problematibus ac tabulis promotam . . . (Bologna, 1651), p. 239, col. 2. It is
worth noting that in his *De eccentricis et epicyclis,* Cecco d'Ascoli makes
no mention of bodies intervening between successive orbs.

    Although he seems not to have adopted it, Thomas Aquinas mentioned
the theory without reference to Albertus when he explained (in his
Commentary on Boethius' *De trinitate*) that supporters of eccentrics and
epicycles believed that this opinion avoided the dilemma that two bodies
might have to occupy the same place and that the substance of the spheres
could be divided. Thomas describes the intervening matter as "another
substance, which lies between the spheres and which, like air, is divisible
and without thickness, although [unlike air] it is incorruptible." For the
Latin text, see Litt, *Les corps célestes,* p. 348.

34 D'Ailly, *14 Questions,* fol. 163v.

35 Ibid., fol. 164v. Paul of Venice accepted the same argument and also used
the expression "total orb of the planets" (*totalis orbis planetarum*); see
Paul of Venice, *Summa naturalium* (Venice, 1476): *Liber celi et mundi,* p.
31, col. 2 (the last two lines; because the work is unfoliated and provided
with few signatures, the page numbers have been determined by counting
from the beginning of the *Liber celi et mundi*).

36 Albert of Saxony, *Questions on De celo,* bk. 2, question 7, fol. 106r
(mistakenly foliated 107r), col. 2, for the objection; fol. 106v, col. 2, for the
reply.

37 D'Ailly, who reports this argument (*14 Questions,* fol. 163v) speaks of only
two centers, one for the world and the other for all eccentric orbs. But
eccentric orbs had many centers because differences in their planetary
eccentricities precluded a common center. Paul of Venice also spoke of two
centers (*Summa naturalium,* p. 31), but Albert of Saxony spoke of "many
centers" (*plura centra*) (*Questions on De celo,* fol. 106r, col. 2, for the
statement of the objection, and fol. 106v for Albert's reply).

38 D'Ailly, *14 Questions,* fol. 164v. Paul of Venice offered the same solution
(*Summa naturalium,* p. 31), explaining that the earth is "in the middle of
the total orb of the planets because the orb is totally concentric to the
world" ("tamen est [i.e., the earth] in medio totalis orbis planetarum eo
quod orbis totaliter est concentricus mundo"). Albert of Saxony presented
the same objection with much the same response (see his *Questions on De
celo,* fol. 106r, col. 2, for the objection, and fol. 106v, col. 2, for the
response).

39 *Questiones subtilissime in via sancti Thome magistri Iohannis Versoris*
super libros De celo et mundo Aristotelis . . . Questiones Versoris super
Parva Naturalia cum textu Arestotelis [sic] . . .Tractatus compendiosus
sancti Thome De ente et essentia. . .(Cologne: H. Quentell, ca. 1493), *De*
celo et mundo, bk. 2, question 9 ("Whether eccentric, concentric, and
epicyclic circles [i.e., orbs] are to be assumed in the heaven to save the
appearances of the planetary motions"), fol. 22v, col. 1, for the objection,
and fol. 23r, col. 1, for Versor's reply. Here we see the common distinction
that most natural philosophers drew between "the whole orb" (*orbis totalis*
or, as Versor put it, *orbis integer*), which embraces three or more eccentric
orbs, and a "partial circle" or "orb" (*circuli* [or *orbes*] *partiales*), which
refers only to one of the constituent eccentric orbs of a "whole orb."

40 *Nicole Oresme Le Livre du ciel et du monde,* ed. Menut and Denomy, bk.
2, chap. 16, p. 463.

41 *Iohannis Buridani Quaestiones super libris quattuor De caelo et mundo,* ed.
Ernest A. Moody (Cambridge, Mass.: The Mediaeval Academy of America,
1942), bk. 2, question 14, p. 191, lines 19–23.

42 Bacon, *Communia naturalium,* p. 444, cites Albumasar's *De
conjunccionibus* as his source.

43 "But there is no violence in the heavens, as Aristotle says in the book *On
the Heaven and the World* and in the eighth [book] of the *Physics;* and it is
obvious that nothing perpetual is violent." Bacon, *Communia naturalium,*
p. 444, lines 26–29.

44 D'Ailly, *14 Questions,* fol. 163v.

45 Ibid., fol. 164v. Albert of Saxony had earlier presented the same objection
and resolution (see his *De celo,* bk. 2, question 7, fols. 106r, col. 2
[objection], and 106v, col. 2 [reply]).

46 *Summa naturalium,* p. 31, col. 2.

47 Aristotle *De caelo* 2.8.290a.25–27.

48 Bacon, *Opus tertium,* pp. 130–131; the *Communia naturalium* omits this
section. Duhem (*Le système du monde,* vol. 3, p. 436), conjectures that
Bacon may have been the first to propose this objection to the existence of
solid epicycles. Since Bacon speaks as if others had already proposed the
criticism, this seems unlikely.

49 *In Metaphysicen Aristotelis Questiones argutissimae Magistri Ioannis
Buridani in ultima praelectione ab ipso recognitae et emissae* . . . (Paris,
1518; rep. Frankfurt: Minerva, 1964), fols. 73–74r. The quotations are from
my translation of this question in Grant, *Source Book,* pp. 524–526.

50 Grant, *Source Book,* p. 526.

51 Buridan believed that planets were also unlikely to have proper motions
because each such motion would require a special mover. We would then
have to assume "as many intelligences as there are stars in the sky because
each star would require a special mover for its special motion." But
"Aristotle did not assign [or concede] such a multitude [of motions and
intelligences]." Ibid.

52 Ibid., p. 525.

53 Albert of Saxony, *De celo,* bk. 2, question 7, fol. 106r, col. 2 (the fifth
principal argument).

54 Although Albert of Saxony admitted that he had often seen a black spot in
the moon, he denied that it resembled a man.

55 In agreement with Albert's position were Pierre D'Ailly (*14 Quaestiones,*
fol. 163v [for the objection] and 164v [for the response]) and Paul of Venice
(*Summa naturalium,* p. 31, col. 2 [for the objection] and p. 32, col. 1 [for
the response]). Paul argued that because the moon has "diversity in its
parts," it requires a proper motion, whereas the other planets lack diversity
and need no proper motions. Without invoking diversity, Bernard of Verdun
(Grant, *Source Book,* pp. 523–524) retained the lunar epicycle and also
assumed that the moon somehow turns, or is turned, so that "the spot
always appears to us in the same shape [or form]."

56 For example, Bacon argued that although the surface of an eccentric sphere
is spherical, the sphere itself is nonuniform, as is evident from its varying
thickness. Natural philosophers, however, insist that celestial bodies must
be simple and homogeneous and therefore unable to vary in thickness. This
is but another aspect of the homogeneity argument. See Bacon, *Opus
tertium,* p. 133 (for a few additional arguments, see pp. 132–137);
Communia naturalium, p. 440 (and pp. 439–443 for the same additional
arguments).

57 As is evident by the assumption that the substance between orbs differs from that of the orbs themselves; that the planets have different basic properties; and, as we saw in the preceding note, that one and the same sphere may vary in thickness.

58 For example, celestial incorruptibility, which was assumed an essential feature of Aristotelian cosmology during the thirteenth and fourteenth centuries, lost that status in the seventeenth century, as numerous scholastic natural philosophers assumed some degree of corruptibility in the heavens.

59 Maimonides and Aquinas appear to have adopted the first alternative. Sympathetic to Aristotelian cosmology but aware that it could not save certain crucial astronomical phenomena and also disturbed by the cosmological dilemmas inherent in any system of solid eccentrics, they argued that the phenomena might perhaps be saved in ways that had not yet been understood or, as Maimonides put it [for Aquinas, see note 11 above], "the deity alone fully knows the true reality, the nature, the substance, the form, the motions, and the causes of the heavens." Although what Aristotle says about the sublunar region "is in accord with reason," Maimonides believes that the heavens are too far away and too noble for us to grasp anything "but a small measure of what is mathematical." *The Guide of the Perplexed,* vol. 2, p. 327. Few in the Middle Ages shared the cosmological uncertainty exhibited by Aquinas and Maimonides.

60 For further elucidation, see Grant, "Cosmology," pp. 278–280. For a typical diagram of eleven concentric spheres representing the celestial region, see Gregor Reisch, *Margarita Philosophica* (4th ed.; Basel, 1517; repr. Stern-Verlag Janssen, Düsseldorf, 1973), p. 244.

PART III. OPTICS

8

Psychology versus mathematics: Ptolemy and Alhazen on the moon illusion

A. I. SABRA

That celestial magnitudes appear larger at the horizon than at higher altitudes is a commonly known phenomenon that has been recorded and investigated since antiquity. Because the phenomenon is particularly noticeable in the case of the moon, it has sometimes been referred to in recent times as the "moon illusion," a designation also reflecting the accepted understanding of the apparent enlargement as a psychological effect. But that is not how the phenomenon was always understood. A traditionally held belief in antiquity found expression in a brief passage in Ptolemy's *Almagest*, which compared the enlargement to the apparent magnification of objects immersed in water, thus proposing an explanation in physical rather than psychological terms. And it was this explanation that enjoyed wide acceptance in late antiquity and in the Islamic Middle Ages up to the end of the thirteenth century, which testified both to the great authority of the *Almagest* and to a remarkable lack of understanding of the mathematical theory of optical refraction. Already in the first half of the eleventh century, however, the mathematician Alhazen (Ibn al-Haytham) had freed himself from the erroneous view in the *Almagest* and, setting off from a new level of understanding some of the elements of which he found in Ptolemy's *Optica*, he had offered in his *Book of Optics (Kitāb al-Manāẓir)* a psychological explanation in terms of what modern psychologists have called, with some exaggeration, the size–distance constancy principle. The principle itself is clearly stated in Ptolemy's *Optica* and, until recently, it had been more or less taken for granted that a passage in the same work also contains an application of it to the phenomenon in question. This assumption was, however, challenged in 1976 when two psychologists showed that it was due to a confusion introduced by Della Porta in 1593 as a result of misreading a passage in Roger Bacon's *Perspectiva*.[1]

The fact is that the enormous literature concerned with explaining the phenomenon itself is in glaring contrast to the extreme paucity, almost the nonexistence, of significant studies devoted to the historical development of its investigation, especially in the ancient and medieval periods. The present chapter is a step toward improving this situation. Concentrating on the works of Ptolemy and the closely related writings of Alhazen, I shall follow a method favored by Marshall Clagett, namely, the method of presenting and elucidating texts, which seems particularly appropriate in this case. In analyzing the relevant texts in those two authors I shall be mainly concerned to describe the transition from a mathematical to a psychological explanation of the moon illusion. A secondary aim of my analysis will be to dispel one or two historical illusions.

I. Ptolemy

We shall examine three relevant passages that occur in three different works of Ptolemy: *Almagest*, *Planetary Hypotheses*, and *Optica*. It will be convenient to refer to these passages as the *A*-passage, the *PH*-passage, and the *O*-passage, respectively. It is generally accepted that the *Almagest* was the earliest of the three works, and some believe that the *Planetary Hypotheses* was probably the last of Ptolemy's compositions.

The A-*passage*

TEXT

The text of the *A*-passage reads as follows in G. J. Toomer's English translation of the *Almagest*:

> To sum up, if one assumes any motion whatever, except
> spherical, for the heavenly bodies, it necessarily follows that
> their distances, measured from the earth upwards, must
> vary, wherever and however one supposes the earth itself to
> be situated. Hence the sizes and mutual distances of the
> stars must appear to vary for the same observers during the
> course of each revolution, since at one time they must be at
> a greater distance, at another at a lesser. Yet we see that no
> such variation occurs. For the apparent increase in their
> sizes at the horizons is caused, not by a decrease in their
> distances, but by the exhalations of moisture surrounding
> the earth being interposed between the place from which we
> observe and the heavenly bodies, just as objects placed in
> water appear bigger than they are, and the lower they sink,
> the bigger they appear.[2]

COMMENTARY

Later we shall see what Alhazen thought of this passage in several of his writings. Here we shall note that most commentators of the *Almagest*, from Theon of Alexandria (fourth century A.D.) to Otto Neugebauer, have understood it to involve a strict analogy with optical refraction. But since, in viewing the apparently enlarged heavenly body, the eye is located in the denser medium of moist air, whereas the opposite is the case when observing an object immersed in water, the analogy seems to be false. (Or, as Neugebauer has put it, the optical situation invoked by Ptolemy is "not relevant" and the analogy itself "incorrect.")[3] How, then, could Ptolemy have committed such an error while his *Optica* clearly asserts that objects in a rare medium appear smaller, not larger, than they are to an eye placed in a denser medium? Several conjectures have occurred to those who pondered over this puzzle: Ptolemy cannot be the author of the *Optica* (a conjecture no longer seriously entertained by historians); or, Ptolemy did indeed write the *Optica*, but only after the *Almagest*, and after he had gained a better understanding of refraction and of optical matters generally; or, the analogy in the *Almagest* is so brief as to suggest that Ptolemy was simply helping himself to a current explanation which he had not carefully examined and which, therefore, should not be taken seriously. The suggestion that Ptolemy was referring not to refraction but to aerial perspective has little to support it; and it does not seem to have occurred to his ancient and medieval commentators.

Alhazen, at one time a subscriber to a totally erroneous conception of refraction that he tried to apply to the phenomenon in question, continued to see merit in the *A*-passage even after he later realized its basic irrelevance, and even after he had come upon his own psychological explanation of the phenomenon as an optical illusion. It will be interesting, when we consider his views in the second part of this chapter, to have before us a notion of what is contained in the most articulate interpretation of the *A*-passage that has come down to us from antiquity – that of Theon of Alexandria.

Relying on an account of refraction which he found (as he tells us) in a work of Archimedes on *Catoptrics*, Theon, in his *Commentary on the Almagest*, formulates an argument that incorporates the following statements:[4]

1. Two unequal magnitudes such as *AB*, *GD* (Figure 1), seen from *E* by means of the rectilinear rays *EAG*, *EBD*, will appear to be of equal size because they are viewed through one and the same angle *GED*.

2. *AB*, when placed below the water surface *ZH*, and seen by means of refracted rays such as *ETA*, *EKB*, will appear larger than it is because it is now viewed through an angle *TEK* > *AEB*.

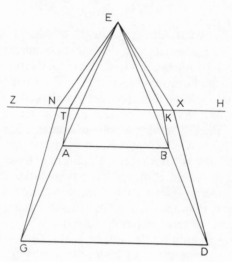

Figure 1

3. In the same manner, *GD* will look bigger when viewed through refracted rays *ENG*, *EXD* than when it is seen through the rectilinear rays *EG*, *ED*.

4. "Therefore the unequal objects *AB*, *GD*, which in pure air appear to be equal, will appear unequal when placed in water, and the lower one will look bigger, because they will be viewed through unequal angles."

5. Theon's construction of the enlarged images shows them in the same plane as the two magnitudes, that is, on the extensions of *AB* and *GD*.

The following comments may now be made:

First, statement (1) is untrue, and it implies that Theon either did not know or did not understand Ptolemy's *Optica* when he wrote his *Commentary on the Almagest*. The *Optica* clearly recognized the fact that apparent size is a function of the distance of the seen object from the eye as well as of the visual angle (a third variable also considered by Ptolemy is the inclination of the object's surface to the axis of the visual cone).[5]

Second, the explanation in (2) and (3) is correct in terms of a theory of monocular vision and a correct account of refraction stipulating that refraction into a denser medium takes place *toward* the normal to the surface of separation. However, (3) does not take us beyond (2), both being concerned with the same situation. Theon expressly attributes the explanation in (2) to Archimedes.

Third, the conclusion in (4) is irrelevant since *GD* is already *assumed* to be larger than *AB*. Theon's argument fails to show that one and the same magnitude appears to grow in size when it sinks deeper in water.

Fourth, (5) is another piece of evidence of Theon's ignorance of the *Optica*, in which work he would have noted that images of immersed objects appear closer to the water surface than the objects themselves.[6]

Fifth, Theon's argument simply consists in applying the irrelevant conclusion (4) to the phenomenon under discussion by observing that in horizontal viewing of a heavenly object, sight has to cover a longer distance through the moist air than in vertical viewing.

The PH-*passage*

Ptolemy wrote the *Planetary Hypotheses* in two books, of which only part 1 of Book I has survived in the original Greek. A ninth-century Arabic translation (revised by Thābit ibn Qurra, d. 901) of the complete Books I and II is extant in two manuscripts, one at Leiden University Library, the other at the British Library. In 1967, Bernard Goldstein published a facsimile of the BL MS together with variants from the Leiden MS, and his edition included a translation of the final part 2 of Book I and a commentary.[7] As is now well known, part 2, on planetary distances and sizes, is an account of "the Ptolemaic system of the world," the system representing the universe as a sequence of nested spheres which together occupy the entire celestial region between the sublunar sphere of fire and the sphere of the fixed stars.

Ptolemy first computes planetary distances from the earth on the basis of data derived from the *Almagest*: the geocentric extremal distances of the moon, the mean solar distance, and the ratios of extremal distances for every one of the planets. Following this, and using the value in the *Almagest* for the solar diameter, he computes the true diameters of the planets from an estimation of their apparent diameters and their distances.[8] Then, at the very end of this final part of Book I, and as if to answer a possible objection, Ptolemy adds what we have called the *PH*-passage. In the Appendix at the end of this chapter, I have provided an edition of the Arabic text of the *PH*-passage based on the two manuscripts mentioned and a quotation of the same passage in the *Nihāyat al-suʾl* of the Damascene astronomer Ibn al-Shāṭir (fourteenth century). Following is my translation of this edited text.

TRANSLATION OF THE *PH*-PASSAGE

{1} As for the reason why what appears to sight and is imagined in it regarding the size of the planets is not in proportion to their distances, we must recognize that this is the error that occurs to sight on account of the difference of perspective[s]. {2} This is manifest in all that appears and is seen from large distances. {3} For just as the magnitude of the distances themselves cannot be recognized in what appears [in this case] to the eye, so also the difference between objects of unequal magnitudes cannot be known at those distances according to their proportion. {4} For by condensing and contracting that difference, sight reduces it to what is more familiar to it. {5} And, for this reason, every one of the stars is seen closer to us than it is in fact, because sight falls back to the distances to which it is accustomed and with which it is familiar. {6} Thus the case with the increases and decreases that size undergoes according to the increase and decreases in distances – which [increases and decreases] are less than the proportion [of the distances] – is like the case with the distances [themselves]. {7} For sight, as we have said, is unable to discern and perceive the amount of difference between all the kinds [of magnitude] we have mentioned.

NOTES ON THE TRANSLATION OF THE *PH*-PASSAGE

{1} 1. "sight" renders *naẓar* (first occurrence) and *baṣar* (second occurrence). 2. "the size of the planets"] The Arabic has "the size of their [or "its"] body": *ʿiẓami jirmihā*. Grammatically and contextually the reference could be to the planets generally, or to Venus, which is mentioned in the sentence immediately preceding the *PH*-passage (see the commentary following these notes). 3. "proportion" here renders *nisab* (ratios); in {3} *tanāsub* (proportion) is used. 4. "the difference of perspective(s)": *ikhtilāf al-manẓar* (or *al-manāẓir*)] A standard expression which refers generally to the variable appearance of objects in regard to magnitude and spatial relations. The Leiden MS and the quote in Ibn al-Shāṭir (in five of six manuscripts consulted) have the singular *al-manẓar*; the BL MS has the plural *al-manāẓir*.

{2} 1. "This is manifest"] *wa yatabayyanu dhālika*. I read *yatabayynu*, as in the quote in Ibn al-Shāṭir, rather than *nubayyinu*, as in the BL MS. To adopt the latter reading, meaning "we show this" or "we will show this," would be puzzling, since the sentence stops short of telling us where the showing is to take place. The word in the Leiden MS lacks the diacritical point(s) for the first letter, and can therefore be read *tabayyana* (it has been shown) as well as *nubayyinu*. *Yatabayyanu* is certainly more natural than any of these other possibilities.

{3} 1. "recognized" and "known" here translate one and the same Arabic verb, *ʿalima*. 2. "according to their proportion"] *ʿalā al-tanāsub allatī (sic) hiya ʿalayhi*. The pronoun *hiya* could refer to *al-ashyāʾ* (the objects) or to *al-abʿād* (the distances) – see the following commentary.

{4} 1. "condensing" and "contracting"] The Arabic words are *jam*ᶜ
and *qabḍ*, respectively. The verb *jamaᶜa* means to draw or bring together, or to collect. The second verb, *qabaḍa*, indicates the narrowing
or shrinking effect of this collecting; hence my choice of the English
words. 2. "that difference" replaces a singular masculine pronoun
which, grammatically, can only refer back to the *tafāḍul* (difference
or inequality), or the *tanāsub* (proportion), or to "what appears to the
eye," all mentioned earlier in the same sentence.

{7} 1. "all the kinds [of magnitude] we have mentioned"] The two
kinds mentioned are size and distance.

COMMENTARY

Neugebauer has interpreted this passage as indicating an aban-
donment of the explanation in the *Almagest* and as recognizing the
phenomenon described in the *A*-passage as an optical illusion.[9] This is
surprising since neither the context of the *PH*-passage nor the arguments in it would suggest such an interpretation. The *PH*-passage is
clearly concerned with how the size of a celestial magnitude appears
to vary with distance, not with how it appears to vary with the direction
of sight while assuming the distance to be constant. Ptolemy is now
thinking in psychological terms, but whether he now has a psychological explanation of the situation examined in the *Almagest* remains an
open question.

Let us look at the statements in the *PH*-passage one by one. The
passage begins by stating in {1}, as a matter of observation, that the
size of a planet does not appear to vary in proportion to its distance
from us, and, in {2}, that this is an error of sight which occurs whenever
we look at very far objects. This means either that no reduction of size
is observed when the distance of the far object increases, or that the
reduction appears to be less than is required by the geometry of the
visual cone. Subsequent statements in the passage imply that the latter
is the case. The remark is then made in {3}, with reference to unequal
objects at far distances from us, that we fail to perceive their inequality
according to the proportion of their (true) sizes, just as we are not able
to obtain a correct perception of the distances themselves. The Arabic
is ambiguous (see {3} 2 of preceding "Notes on the Translation . . ."),
allowing that the proportion intended might be that of the objects' distances, not their sizes; but then, on this alternative, we would be left
with the question why the objects are supposed to be unequal. Thus
while {1} is concerned with judging the size of what could be taken to
be one and the same object at different distances, {3} seems to be
concerned with estimating the apparent relative sizes of unequal objects at equal distances.

It is to be noted that while {3} asserts that we misjudge both distances and relative sizes of far objects, nothing is said about whether misjudging one of these magnitudes depends on misjudging the other. But the nature of the "error" involved in both cases is made clear in {4}: It consists in reducing the perceived magnitude, be it distance or size, to a more familiar and smaller scale; and this is stated explicitly in {5} with regard to distance.

Finally, {6} returns to considering the apparent variation of size with distance: Such variation does occur, we are told, but the apparent increase or decrease in size is less than the ratio of the distances because of sight's inability to estimate the amount of difference in this case – for the reason stated in {4}.

Apparently because the difference between extremal distances is (according to Ptolemy) appreciably greater for Venus and Mars than for any other planet, Ibn al-Shāṭir takes the *PH*-passage to be especially addressed to a question suggested by observation of these two planets. He wrote in a note at the end of Book I of his *Nihāya*: "You must know that the diameters of Venus and Mars are not seen in accordance with (ʿalā muqtaḍā) their distances. For their maximum distances being many times their minimum distances, their diameters should appear according to (tabaʿ) the distances: I mean that the diameter of each of them should appear to be many times greater at perigee than at apogee; but this is not so."[10] After quoting Ptolemy's passage, Ibn al-Shāṭir gives two reasons of his own: the diminution of the visible part of a sphere as it approaches the eye, and the disturbing effect of rays emanating directly from luminous bodies. (He assumes that all planets, except the moon, are self-luminous.)[11]

The O-passage

We come now to the passage in Ptolemy's *Optica* that has been generally understood to contain an explanation of the moon illusion in terms of the size–distance invariance principle. The *Optica* survives in a twelfth-century Latin translation from an earlier Arabic version of Books II to IV and part of Book V; neither the Greek original nor the Arabic translation is extant.

LATIN TEXT OF THE *O*-PASSAGE
***erit distantiis, eo quod debilitas sensum plus fit penes coniunctionem. {1} Vniuersaliter enim, cum uisibilis radius, quando cadit super res uidendas aliter quam inest ei de natura et consuetudine, minus sensit omnes diuersitates que in eis sunt, similiter etiam erit sensibilitas eius de distantiis quas comprehendit, minor. {2} Videtur autem hac de causa quod de rebus que sunt in celo et subtendunt equales

angulos inter radios uisibiles, ille que propinque sunt puncto qui super capud nostrum est, apparent minores; que uero sunt prope orizontem, uidentur diuerso modo et secundum consuetudinem. {3} Res autem sublimes uidentur parue extra consuetudinem et cum difficultate actionis.[12]

TRANSLATION OF THE *O*-PASSAGE
. . . {1} For, generally, just as the visual ray, when it strikes visible objects in [circumstances] other than what is natural and familiar to it, senses all their differences less, so also its sensation of the distances it perceives [in those circumstances] is less. {2} And this is seen to be the reason why, of the celestial objects that subtend equal angles between the visual rays, those near the point above our head look smaller, whereas those near the horizon are seen in a different manner and in accordance with what is customary. {3} But objects high above are seen as small because of the extraordinary circumstances and the difficulty [involved] in the act [of seeing].

NOTES ON THE TRANSLATION OF THE *O*-PASSAGE
{1} 1. "generally"] *uniuersaliter*. My guess is that the Latin renders a word like *bi-al-jumla* (Greek *holōs*). Ross and Ross translate: "it is a universal law that." The word occurs many times in Ptolemy's *Optica*.
2. "senses/*sensit*"] If the phrase *uisibilis radius* translates *shuʿāʿ al-baṣar* (the ray of sight), then the subject of *sensit* could be "sight," or the faculty of sight, rather than the "visual ray" itself, which would make a difference to an argument by Ross and Ross.[13]
3. "all their differences"] *omnes diuersitates que in eis sunt*. "Diuersitates" suggests the Arabic *ikhtilāfāt*, which it would not have been objectionable to translate as "inequalities," had it not been for the presence of *"omnes." "Omnes"* clearly implies that all kinds of visual differences between objects, including differences of size, are meant. Ross and Ross have "characteristics," which also preserves the general meaning. In the *PH*-passage, difference of size is expressed by *tafāḍul*.
4. "perceives"] *comprehendit*. The Latin word is used in this sense many times in Ptolemy's *Optica* and in the Latin version of Alhazen's *Optics*. Ross and Ross translate it here as "covers." The corresponding Arabic word is, most probably, *adraka*.
{2} 1. "in accordance with what is customary"] *secundum consuetudinem*. That is, according to what sight is accustomed to in familiar situations. The corresponding Arabic word may well be *iʿtāda* or *alifa*, both of which are used in the *PH*-passage.

COMMENTARY

The *O*-passage raises so many questions that a paraphrase of it is not possible without inserting a number of more or less crucial interpretations of the Latin text, in addition to those inevitably involved in translating it. One question already raised by Ross and Ross concerns the meaning of "less" (*minus, minor*) in sentence {1}. What is it that is here said to be reduced when objects are looked at in unusual circumstances: Is it the sensation itself or the seen object? Or, as Ross and Ross have put it, does Ptolemy's expression refer to "reduced ability to discriminate" or to "perception of something as smaller?"[14] (An Arabic word such as *aqall*, a probable equivalent of the Latin *minus*, would have been equally ambiguous. *Aṣghar*/smaller is less probable as it could hardly apply to *sensibilitas* = *?iḥsās*.) The application of *minor* to *sensibilitas* would seem to suggest that the former alternative is intended. But however we understand the words in {1} it is definitely asserted in {2} that an apparent *reduction of size* takes place, at least as a consequence.

It would seem to me that a reduction of distance is also intended. Ross and Ross cite results of experimental psychology to support the hypothesis that Ptolemy may well have associated apparently reduced size with *increased* distance.[15] But we have seen that Ptolemy, in the *PH*-passage, clearly asserts that the distances as well as the sizes of heavenly bodies do appear to be *smaller* than they really are, giving as a reason sight's tendency to compress *all* magnitudes perceived from afar. True, he makes this assertion there as a general statement, applicable in all situations. But the *PH*-passage shows at least that Ptolemy's beliefs may not have always been those of modern psychology. The real difficulty lies in making an inference with regard to the *O*-passage from a statement in another passage which may have been written later. And it remains true that neither in the *O*-passage nor in the *PH*-passage does Ptolemy assert a *dependence* of reduction of size upon a reduction of distance.

We are not told in {1} what an example of an unusual situation is, but such an example is provided by the contrast in {2} and {3} between viewing a heavenly object at a point near the zenith and viewing the same object near the horizon "in accordance with what is customary": Vision is said to be more difficult in the former, unusual case, with the result that the object looks smaller. Adding to this our conjecture that Ptolemy considered the distance, too, to appear smaller in the unusual situation, one might be tempted to offer the following interpretation of all these statements put together: (1) In the customary, horizontal viewing, a heavenly object appears to be farther away from us than when it is viewed with difficulty by looking upward; (2) objects that subtend the same angle at the eye appear to be smaller when they seem to be nearer; (3) therefore a zenith-star will appear to us smaller than a horizon-star. It may be argued that such an interpretation would be consistent with the *O*-passage as we have understood it, but the trouble with this interpretation is that neither (1) nor (2) can be found in that passage. (1) and (2) would have to be brought in and developed from other parts of the *Optica*. Would Ptolemy have only cited the difficulty of looking upward had he intended an explanation like the one just given?

II. Alhazen

Not surprisingly, it was the passage from the *Almagest*, the *A*-passage, that attracted the attention of Islamic mathematicians. In the Islamic world the *Optica* never achieved anything like the wide circulation enjoyed by the *Almagest*, and neither the *O*-passage nor the *PH*-passage was likely by itself to compel a comparison with the situation described in the *Almagest*. Almost all Arabic commentaries on the *Almagest* and all parallel discussions of its contents included remarks on the *A*-passage. At first these remarks were neither extensive nor illuminating, being for the most part limited to paraphrasing Ptolemy's text; but they grew in variety and sophistication after Alhazen's *Optics* became widely known through the "Revision" prepared by Kamāl al-Dîn at the beginning of the fourteenth century A.D. Alhazen's *Optics* thus marked a turning point in the Arabic discussions of the moon-illusion problem.

From his writings we gather that the problem claimed Alhazen's attention at various times that ranged over his entire scientific career.[16] Thus his full, mature explanation of the moon illusion is contained in Book VII of his *Optics*, a work composed relatively late in his career. His first comments on the *A*-passage are, however, to be found in a *Commentary on the Almagest*, undoubtedly an early composition which Alhazen wrote before he had access to Ptolemy's *Optica*. In a short treatise "On the Appearance of the Stars" (*Fī Ru'yat al-kawākib*), written probably before the *Optics*, Alhazen made an attempt to reconcile the *A*-passage with the doctrine of refraction as expounded in Ptolemy's *Optica*. This treatise was thus written after Alhazen became acquainted with the *Optica*. He returned to the same subject in two treatises that dealt with doubts or difficulties (*shukūk*) in Ptolemy's works. One of these treatises has reached us in the form of a series of comments bearing the title "Solution of Difficulties in the *Almagest* Which a Certain Scholar Has Raised" (*Ḥall shukūk fī al-Majisṭī yushakkiku fīhā ba'ḍu ahl al-'ilm*). The other is his critique of Ptolemy known as *Dubitationes in Ptolemaeum* (*al-Shukūk 'alā Baṭlamyūs*), very likely the last of all these works in order of composition and, as far as our problem is concerned, the least informative.[17] In what follows

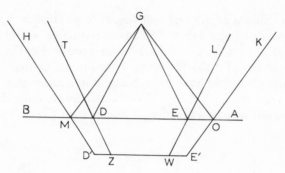

Figure 2

I shall deal first with Alazen's writings before the *Optics*, then with his views in the *Solution*, leaving his mature theory in the *Optics* to the end.

Alhazen's Commentary on the Almagest *and his*
Treatise on the Appearance of the Stars
It will be instructive to look briefly at the primitive argument in the *Commentary on the Almagest*,[18] the sole purpose of which was to explain the apparent magnification of objects immersed in water and the increase of magnification with deeper immersion. The argument can be found already in an early-ninth-century compilation *On Optics and on Burning Mirrors* by a certain Aḥmad ibn ʿIsā, from which Alhazen may have obtained it.[19] The "explanation" begins with this curious statement:[20] "It has been shown in books of optics that rays of sight are reflected (*tanʿakisu*) from the surfaces of visible objects at equal angles and in straight lines, such as *EL, DT* [Figure 2], and that these lines enter into (*tanfudhu fī*) the transparent bodies and reach the object immersed in those bodies so that vision would occur by means of the reflected rays (*al-shuʿāʿāt al-munʿakisa*)."
Alhazen then argues as follows: The object *DE*, placed upon the water surface *AB* is seen by the eye at *G* through angle *DGE*. When sunk into the water to position *D'E'* the same object must now be seen through an angle *MGO* such that the extensions of the reflected rays *MH, OK* will reach *D', E'*. "Thus it is evident that *DE* is seen on the water surface through the smaller angle *EGD*, and, when immersed in water, through the larger angle *OGM*; but that which is viewed through a larger angle appears larger, for we have shown previously that an object is seen according to the angle of vision; and for this reason, the deeper the object sinks the larger it will appear."
Such was the retarded understanding of refraction before Ptolemy's *Optica* became known.

With the treatise *On the Appearance of the Stars*[21] Alhazen reaches
an entirely different level of understanding. He has now read and understood Ptolemy's theory of refraction in the *Optica* and he has become
aware of new problems which, he feels, have not been fully dealt with
by anyone, including Ptolemy. Most people think (he says at the beginning of the treatise) that the heavenly bodies are seen by rectilinear
rays, while "experts in optics and mathematics" believe that the visual
rays by means of which we perceive a star diverge upon striking the
concave surface of the ether, thus causing the star to be seen as smaller
than it is because the angle produced in the eye by the refracted rays
will be smaller than that contained by the straight lines directly drawn
to the extremities of the star's diameter. Both views are wrong, says
Alhazen. The first ignores refraction altogether, and the second (he
explains later in the same treatise) is true only in the special case in
which the observed star is near the zenith.

> As for the excellent Ptolemy, he did not say in his *Optics*
> how sight perceives the bodies of the stars,[22] nor did he say
> how the ray [conceived as a solid volume] encompasses the
> star's body. He neither mentioned the angle through which
> the star is seen nor the mutual positions of the two rays that
> contain that angle and are refracted to the extremities of the
> star's diameter. [Ptolemy in the *Optics*] said [only] that a
> visible object is seen as smaller than its true magnitude
> when the eye is located in a medium denser than that in
> which the object is placed and when the two rays are
> refracted to the object's extremities away from the normal
> drawn from the eye to the separating surface. This is the
> case frequently described by writers on optics and repeated
> in their books.

To correct the generally received opinion Alhazen devotes the first
part of his treatise to showing geometrically that (1) the apparent size
of a star is diminished, regardless of altitude, as a result of being viewed
by refraction; and (2) that the size is diminished less when the star is
near the zenith than when it is near the horizon, because in the former
case the amount of refraction is less and the difference between the
angles of incidence and of refraction is less. The reduction in size
should therefore in principle be more noticeable in horizontal viewing
than in zenith viewing – which gives rise to the problem formulated
by Alhazen as follows:

> A difficulty may occur to many mathematicians when they
> put Ptolemy's discourse in the *Optics*, where he shows that
> the magnitude of an object placed in a rarer medium than
> that in which the eye is located, appears to be smaller, and
> the greater the density of the latter medium the greater the

diminution, and also says that the body in which the star is located is rarer than air, side by side with his discourse in the *Almagest* (*al-Ta'ālīm*), where he says that the stars appear larger at the horizons because the vapor from the moisture surrounding the earth is interposed between the eye and the stars – for these two discourses are in appearance opposed to one another.

But only "in appearance," for, according to Alhazen, it is "possible" (*qad yumkinu*) that the interposition of thick vapor may cause the star to appear larger than it would when viewed in the absence of the vapor. He considers two cases (illustrated by Figures 3a and b) both yielding the same result. In both cases, let *ABE* be the plane of the azimuthal circle passing through the star *AT*. *AE* is the intersection of the circle with the horizontal plane through *E* where the observer is located. In Figure 3a let the vapor continuously fill the space between *E* and the concave surface of a rarer layer of air contained between *LN* and the ether surface *GD*. In Figure 3b a thin air lies on either side of the thick vapor contained between *OZ* and *LN*.

In both figures, let angle *GEK*, smaller than *AET*, be the angle of viewing *AT* in the absence of any dense moisture in the air between *E* and *GD* (I have added the broken lines *GA* and *KT*).

In the first case, in Figure 3a, the star *AT* will be seen by rays *EL* and *EM* that are refracted, first away from the normals at *L* and *M* into *LC* and *MH*, then again away from the normals at *C* and *H* into *CA* and *HT*, respectively.

In the second case, in Figure 3b, rays *EO* and *EF* are first refracted toward the normals at *O* and *F* into *OL* and *FM*, then refracted away from the normals twice at *L*, *M*, and at *C*, *H*, respectively.

After drawing the perpendiculars *TQ* and *AS* to *EM* and *EL* extended (in Figure 3a) or to *EF* and *EO* extended (in Figure 3b), Alhazen argues in both cases (a) and (b) that if *TQ*:*QO* ≥ *AS*:*SE*, then *AT* will look bigger than it would if only one refraction took place at the ether surface. That is, he argues that angle *LEM* (or *OEF*) will be greater than angle *GEK*, the latter being the angle through which *AT* would be seen across a homogeneous atmosphere.

The argument cannot, of course, claim to constitute an explanation of the moon illusion as a constant phenomenon, nor was it intended to be such an explanation. It was merely meant to show that under certain conditions the apparent size of a celestial magnitude may be increased by the interposition of a dense moisture – an idea which, as we shall see, Alhazen retained and incorporated in the *Optics*. Since the treatise *On the Appearance of the Stars* makes no allusion to the *Optics* or to the psychological explanation offered in it, it seems rea-

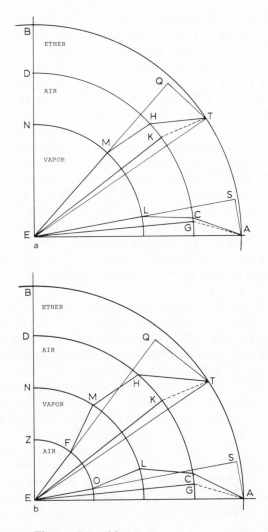

Figures 3a and b

sonable to assume that it was written some time before the composition of that relatively late work, but this cannot be certain.

> *Alhazen's treatise* "Solution of Difficulties in the
> *Almagest* . . ."

Apart from the *Optics*, Alhazen's most interesting remarks on the moon illusion are contained in a treatise which, in the Istanbul MS Fatih 3439, folios 142b–150b, bears the title "Solution of Difficulties in the *Almagest* Which a Certain Scholar Has Raised" (or: ". . . Which Some Scholars Have Raised"): *Ḥall shukūk fī al-Majisṭī yushakkiku fīhā baʿḍu ahl al-ʿilm.*[23] It consists of three distinct sections, the first of which (142b–145a) comprises five "difficulties" (*shukūk*) in Book I of the *Almagest*, and thus corresponds to title no. 38 in Ibn Abī Uṣaybiʿa's List III of Alhazen's writings.[24] Section II immediately follows under a new title, "Answers to Doubtful Questions (*masāʾil*) in the *Almagest.*" These are nine questions which are not confined to Book I of the *Almagest*, and which occupy pages 145a–148b. Question nine ends with the statement "This is the last of your [*sic*] questions."[25] Section III (148b–150b) consists of four additional difficulties (*shukūk*) without an introductory title, but the fourth (concerning lunar parallax) is described as "one of the difficulties raised by Abū al-Qāsim ibn Maʿdān," a scholar otherwise unknown to me. The three sections may thus have been originally separate compositions, a conjecture confirmed by the fact that the same problem is treated in all of them. But all appear to have been written after the *Optics*, which is explicitly mentioned in sections I and III.

Only some of the difficulties discussed in this treatise can be reported here, and even these will have to be briefly summarized. The first of the five difficulties that make up section I concerns a problem suggested by the *A*-passage, which Alhazen paraphrases (or quotes?) as follows: "The apparent increase in the size of the stars when they are on the horizon is not perceived in them because of their decreased distance [from us] while in that position, but because a moist vapor that surrounds the earth rises between them and the eye, thus causing them to be seen thus; just as what is thrown in water appears larger, and the lower it sinks the larger it becomes."[26] It is important to be clear as to what the suggested problem is. It is not the problem of why the horizon-star is larger than the zenith-star. Nor is it the problem arising from comparing the apparently enlarged star which is viewed from a *dense* medium to the enlarged object viewed from a *rare* medium. (Alhazen had offered a solution to the first problem in his *Optics*, and he refers to the second in section III of the present treatise.) Rather it is the problem of simply explaining why an object placed in water appears to increase in size when it sinks deeper. In Alhazen's view the problem existed because a true statement in the *A*-passage about the magnification of immersed objects was not accurately explained in Ptolemy's *Optica*: the fault, he said, lay, not with the *Almagest*, but with the *Optica*.

Thus, referring to Book V of Ptolemy's *Optica*, and basing himself, as he says, on Ptolemy's "adopted method," Alhazen sets out to show how an immersed object looks larger than it is because the refracted rays will contain a larger angle than that contained by rectilinear rays.

When, however, the object is lowered, the angle between the visual rays, which now have to be refracted less in order to reach the extremities of the object in its farther position, must become smaller.

Now if someone believes that the magnitude of an object is perceived according to the magnitude of the angle alone, then he will doubt Ptolemy's statement [in the *Almagest*] that the object increases in size as it sinks deeper. But the statement is subject to doubt only if the doubter relies exclusively on the angle. Such a reliance, however, would be a mistake; for Ptolemy has shown in the *Optics*, in the course of his discussion of size, that size is not perceived according to the angle alone, but according to the magnitude of the angle and of the [object's] distance and according to whether the object is inclined or frontally situated. Thus [Ptolemy] says in the second of his Propositions on size[27] that of two unequally distant objects and perceived through the same angle, the nearer never looks larger than the farther object, but is either seen as smaller [than the farther object] (when a sensible interval exists between them) or equal to it (when the difference between their distances [from the eye] is insensible). Thus if size were perceptible by means of the angle alone, then the two objects seen through the same angle, or through two equal angles, would always appear to be equal – which is not the case.

Moreover, we have shown in our book on *Optics* that size is perceptible by [comparing] the angle to the distance and also in accordance with the angle alone.

In the course of citing a number of observations supporting the view just explained, Alhazen refers further to Ptolemy's "*Optics*, V, Prop. 17,"[28] after which he concludes:

An object looks larger when it sinks deeper [in water] because its distance increases and because the distance of its image (*khayāl*) increases after it sinks and because the magnification entailed by the increase in distance is greater than the diminution entailed by the decrease in the angle [of vision]. And thus it evidently follows from the principles asserted by Ptolemy and from our own examples that an object must appear larger when it sinks deeper in water.

To represent the facts "more precisely" than in Ptolemy's *Optica*, Alhazen finally gives a proof of which the following is a simplified paraphrase.

Figure 4 is in the plane passing through the eye at *A* and the center of "the sphere of water" (i.e., center of the earth) at *I*, and intersecting the water-surface in arc *EZ*. *BG* and *WJ* are two positions of the same

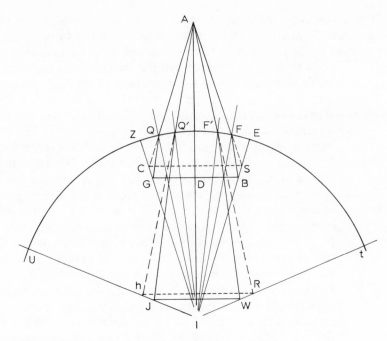

Figure 4

and similarly oriented object viewed from A – WJ being farther away from the eye than BG. The lines drawn from I are diameters of the sphere that cut the water-surface at right angles.

BG is first viewed by means of the refracted rays AFB and AQG. And since $\angle FAQ > \angle BAG$, BG will appear larger than it would if viewed by the rectilinear rays AB and AG.

At the lower position WJ, the object is viewed by means of refracted rays $AF'W$ and $AQ'J$, such that $\angle F'AQ' < \angle FAQ$. WJ should therefore look smaller than BG, if size were determined by the visual angle alone.

But, Alhazen maintains, the distance at which the object is seen is another factor which in fact outweighs the effect of the visual angle. To determine the locations of the images of BG and WJ he draws the normals to the refracting surface through the extremities of the two objects, and extends the incident rays to meet the normals at S, C and at R, h, respectively.

> Now Ptolemy has shown in the *Optics* that an object in water is seen at the location of the image; and if so, then it must be seen as having the magnitude of the image whether the angle of the image is greater or smaller, because the

[object's] extremities are seen at the image's extremities.
Therefore the magnitude *BG* will appear in water to be
larger than it is because its image is larger than it. This,
then, is the true explanation of the apparent magnification of
objects seen in water.
A similar argument applies to *WJ*.

Two queries related to our problem make up the first question in section
II.

Query 1: If refraction into a denser medium is a cause of the magnification of the horizon star, then where does this refraction take place
if the eye is placed *inside* the dense medium?   *Alhazen's answer:* It
has to be assumed that the moisture rising on the horizon does not
extend all the way to the observer's locality; so that a ray issuing from
the observer's eye will first travel through rare air before it strikes the
surface of the dense moisture. And it has to be assumed further that
this surface must be plane, though it need not be continuous. That is,
the refraction causing the magnification must be produced by parts of
the moisture that are distributed on a plane surface, in the same way
that the water droplets responsible for producing the rainbow are arranged in a regularly circular shape without this being the shape of the
misty air itself.

Query 2: Where should the star's apparent size be closer to its real
size – near the horizon or near the zenith?   *Answer:* At the latter
position, where the rays are refracted least.

There is no mention in section II of any psychological explanation,
from which it does not follow that this section antedated the *Optics*.
The answers given here are in fact offered as a *supplementary* explanation in the *Optics* (see the "Translation of Alhazen's text in *Optics*"
below).

The first of the four doubts raised by Abū al-Qāsim ibn Maʿdān, and
discussed in section III, assumes, like query 1 in section II, that the
dense vapor continuously surrounds the entire surface of the earth. In
support of his denial of this premiss Alhazen oberves that we are able
to endure contemplating the sun for a while at rising and setting, but
not when it is above our heads. This is "certain proof," he says, that
the dense moisture at the horizon does not extend into the atmospheric
areas directly above us. Alhazen thus maintains here that the magni-
fying effect of the moisture exists only for an observer located in the
inhabited region of the northern hemisphere, and that such a moisture
rises only from the water encircling that region. Concluding this section
he writes:

This being made clear, I say that what I have stated so far
are the answers to your questions.[29] However, other
difficulties exist throughout the *Almagest*, but I merely
answered the questions addressed to me. Among these
difficulties that concern the magnification of the stars at the
horizon is [the following]: the stars are in the heavens, and
the heavens are rarer than the air; and if a visible object lies
in the rarer medium while the eye lies in the denser
medium, then the object must be seen as smaller than it is;
and the greater the density of the medium close to the eye
the smaller the object will appear; therefore the stars should
appear smaller at the horizon than in the middle of the sky.
I have explained this matter in my book on *Optics*, in the
discussion on refraction, where I showed that the
magnification of the stars at the horizon has a universal
cause, other than the moisture, on account of which the
stars and their mutual distances appear to be larger at the
horizon than at the middle of the sky or at other altitudes
higher than the horizon. I have also shown [there] that the
moisture at the horizon adds to the increase in size that is
due to the universal cause. I have not gone into this here
because it is not one of the doubts you have raised.[30]

Alhazen's psychological explanation of the moon illusion

Alhazen offered his psychological explanation of the moon illusion at the end of the last chapter in Book VII of his *Optics* (*Kitāb
al-Manāẓir*), a chapter devoted to the "errors of sight" due to refraction. The explanation follows a discussion of the role of refraction in
viewing heavenly bodies. It is an original and sophisticated explanation
which deserves a full commentary. Such a commentary cannot be given
here, but, fortunately, Alhazen's text can stand on its own, being easily
understandable without extra help.

The following translation is made from the Istanbul MS Fatih 3216
(copied at Baṣra in A.H.476/A.D.1084), folios 131b–138b, corresponding
to page 280, line 54, to page 282, line 61, in F. Risner's edition of the
medieval Latin translation of the *Optics*.[31] Additions and corrections
made with the aid of this Latin text are enclosed in angle brackets.
Expressions in parentheses are in the original Arabic text; my own
additions are in square brackets. I have not noted omissions or departures in the Latin from the Arabic text.

I have used "magnitude" for *miqdār*, and "size" for both *ʿiẓam* and
miqdār, two words used indifferently in the Arabic text. "Distance"
and "interval" both stand for *buʿd/remotio*; "distance" always means

remoteness from the observer, unless otherwise specified. "To guess" or "conjecture" render *ḥadasa/perpendere, aestimare*. "Refraction angle" (*zāwiyat al-inʿiṭāf*) is the visual angle through which the object is viewed by refraction; "refracted angle" (*al-zāwiya al-munʿaṭifa*) is the angle through which the incident ray is refracted. I have translated *ʿilla* as "cause," "reason," and "explanation."

TRANSLATION OF ALHAZEN'S TEXT IN *OPTICS*,
BOOK VII

We say that sight will perceive any star at the zenith to be smaller than in any region of the sky through which the star travels; that the farther the star is from the zenith the larger its magnitude will appear than it does at the zenith; that the star looks largest at the horizon; and that the same is true of the intervals between the stars. Now this is found to be so in fact, namely, that the stars, and their mutual distances, appear to be smaller at the middle of the sky than when they are far from it, and that the star (or interval between two stars) appears largest at the horizon. It remains for us to show the reason why this is so.

We say: It has been shown in Book II of this work, in our discussion of size, that sight perceives size from the magnitudes of the angles subtended at the center of the eye and from the magnitudes of the distances of the visible objects and from comparing the magnitudes of the angles to those of the distances. We have also shown there that sight neither perceives nor ascertains the magnitudes of the objects' distances unless these distances extend along near and contiguous bodies; that distances that do not so extend are not ascertainable in magnitude by sight; and that when sight cannot ascertain the magnitudes of the objects' distances, then it fails to ascertain the objects' sizes. We showed there, too, that when sight fails to ascertain the distance of an object, then it makes a guess in regard to the distance's magnitude by likening it to the distances of familiar objects at which it can perceive objects similar to that object in form and figure, then perceives the size of that object from the magnitude of the angle subtended by it at the eye-center as compared to the distance it has conjectured. But the distances of the stars do not extend along near bodies. Sight does not, therefore, perceive or ascertain their magnitudes, but merely conjectures their magnitudes by assimilating the stars' distances to the

distances of very remote earthly objects which it can
perceive and whose magnitudes it conjectures.

Moreover, the body of the heavens is not seen as a sphere
whose concavity faces the eye; nor is sight aware of the
corporeality of the heavens, and only perceives of the
heavens a certain blue color. As to the heavens'
corporeality, extension in the three ⟨dimensions⟩, circularity
and concavity – these sight has no way of perceiving. And
when sight cannot identify something it likens it to one of
the familiar objects that resemble it. Thus it perceives the
sun and moon as flat, and perceives convex and concave
bodies from an excessively great distance as flat, and also
perceives arcs whose convexity or concavity faces the eye
as straight. For when sight does not perceive [in the case of
convex objects] the nearness of their middle points and the
remoteness of their extremities, or (in the case of concave
objects) the remoteness of their middles and the nearness of
their extremities, it likens the convex and concave surfaces
to flat surfaces and likens arcs to straight lines, because
most familiar objects have flat surfaces and straight edges.

Nor, when the form of a star reaches the eye, is sight
aware that it is a refracted form, or that it has been
refracted from a concave surface, or that the body in which
the star is is rarer than that in which the eye is. Sight rather
perceives the form of a star as it perceives the forms that
come to it in straight lines from objects located in the air.
The forms of visible objects, when they encounter a body
whose transparency differs from that of the body in which
they were, do not undergo refraction for the sake of the
[seeing] eye; nor will the eye be aware of their refraction or
of the surface of the differently transparent body; rather, the
refraction occurs in virtue of a natural property that is
peculiar to the forms of light and color. The refracted forms
of the stars thus reach the eye just as the forms of visible
objects in the air reach it, and sight perceives them just as it
does the forms of objects in the air.

Sight perceives the color of the sky and the extension of
that color in length and breadth without perceiving the shape
of the sky or identifying its figure by pure sensation. And
when sight perceives a color as extended in length and
breadth without perceiving its shape or identifying its figure,
then it perceives it as flat, because it assimilates it to
familiar flat surfaces that extend in length and breadth, such
as those of walls and the ground. Similarly, sight perceives

convex and concave surfaces from a large distance as flat, and also perceives the surface of the earth in wide areas as flat and is not aware of its convexity in the absence of hills and protrusions and depressions.

Sight, therefore, perceives the surface of the heavens as flat and perceives the stars in the same way as it perceives familiar objects scattered over wide areas of large dimensions, and assimilates the distances of the stars to those of familiar objects scattered over vast areas on the earth's surface the ends of which [areas] sight perceives as farther away than their middles and perceives those points that are close to the middle as less distant than those that are further removed from it. But if sight perceives under equal angles a number of objects scattered over a large area, while perceiving the magnitudes of the distances of those objects, then it will perceive the farther of those objects as larger. For it will perceive the size of the far object from comparing the angle subtended by that far object at the eye-center to a large distance, and will perceive the size of the nearer object from comparing the angle subtended by this near object (which is the same as the angle subtended by the far object) to a smaller distance.

Now this is found to be clearly the case, namely, that when two objects are viewed under the same angle (or under two equal angles), and their distances differ appreciably, then the farther object will look bigger. For let someone stand in front of a wide wall and raise his hand before one eye while closing the other, then look with one eye while holding his hand between it and the wall. His hand will screen a portion of the width of that wall. But since the hand and the [screened] width of the wall are seen at the same time, then they are seen through the same angle; and sight will at the same time perceive the [screened] width of the wall to be many times larger than the hand. And if the person moves his hand to one side so as to expose the hidden portion of the wall, and looks at the exposed portion and at his hand, he will perceive that portion to be many times larger than his hand. And he perceives his hand and the exposed portion by two equal angles. This experiment therefore shows that sight perceives size from comparing the angle to the magnitude of the distance.

Now sight perceives the surface of the heavens as flat and does not perceive its concavity, and it perceives the stars scattered over it. It therefore perceives separate and equal

stars as having unequal sizes, because it compares the angle
subtended at the eye-center by the extreme star near the
horizon to a large distance, while comparing the angle
subtended by the star at or near the middle of the sky to a
small distance. Thus it will perceive the star at or near the
horizon to be larger than the star at the middle of the sky or
near the zenith; and will perceive one and the same star (or
interval between two stars) at different points in the sky to
be of different magnitudes. It will thus perceive one and the
same star to be larger at or near the horizon than at or near
the middle of the sky, and will perceive the interval between
two given stars to be larger at the horizon than at or near
the middle of the sky, because it compares the angle
subtended at the eye-center by the horizon-star to be a large
distance and compares the angle subtended by the zenith-
star to a small distance. And there is no great discrepancy
between the two angles; rather, they are ⟨close⟩ though
different. And the case is similar with intervals between the
stars. But if the sense [of sight] compares two angles close
in magnitude to two distances of greatly different
magnitudes, then it will perceive the farther [object] to be
larger.

The proof of the truth of this explanation is that the angles
subtended at the eye-center by a given star from all regions
in the sky ⟨when these angles are contained by refracted
lines⟩ are equal ⟨to those through which the star is
perceived⟩ by means of straight, unrefracted lines. For the
eye being located at the center of the heavens, these angles
will not be greatly reduced by refractions of the stars'
forms. And if these reductions are not greatly different in
magnitude, then the difference between the refraction angles
through which the star (or interval between two stars) is
perceived in different locations will not be great. And if so,
then the size of the star (or interval between two stars) will
not seem to differ greatly at different points in the sky on
account of the difference between the angles. That the
refraction angles are not greatly smaller than the angles
contained by the straight lines has been shown in the
experiment described in the chapter on refraction, where it
was made clear that sight perceives the star by
refraction. . . .[32] From which it is manifest that the refracted
angles are too small to bring about a great difference
between the angles through which sight perceives the star at
different points in the sky. But there is a great difference in
size between the star (or interval between two stars) at the

horizon and at the middle of the sky. Therefore the difference between the refraction angles cannot be the cause of that difference in size at the various points in the sky. And it has been shown that sight perceives the sizes of visible objects by ⟨comparing⟩ the angles to the distances. Therefore, if the difference between the angles is small, and the difference between the distances is great, and a visible object appears larger from the greater distance – if all that is so, then the reason why the stars (and intervals between them) are seen to be larger at the horizons than at or near the middle of the sky is the conjecture made by the sense [of sight] regarding their greater distance at the horizons than when they are at the middle of the sky.

What sight perceives regarding the difference in the size of the stars at different positions in the sky is one of the errors of sight. It is one of the constant and permanent errors because its cause is constant and permanent. The explanation of this is [as follows]: Sight perceives the surface of the heavens that faces the eye as flat, and thus fails to perceive its concavity and the equality of the distances [of points on it] from the eye. And it has been established in the mind that flat surfaces that extend in all directions about the eye are unequally distant from it, ⟨and that those directly above are closer to the eye than those to the right and left of it⟩. Now sight perceives those parts of the sky near the horizon to be farther away than parts near the middle of the sky; and there is no great discrepancy between the angles subtended at the eye-center by a given star from any region in the sky; and sight perceives the size of an object by comparing the angle subtended by the object at the eye-center to the distance of that object from the eye; therefore, it perceives the size of the star (or interval between two stars) at or near the horizon from comparing its angle to a large distance, and perceives the size of that star (or interval) at or near the middle of the sky from comparing its angle (which is equal or close to the former angle) to a small distance; ⟨and it perceives a great difference between the distance of the middle and that of the horizon⟩. We have thus stated the cause on account of which sight errs in regard to the difference in the size of the stars (or their mutual intervals); and it is a constant, permanent and unchanging cause.

That is also the cause of sight's perception of the stars as small on account of their remoteness. For the star, when remote, subtends a small angle at the center of the eye; and,

failing to ascertain the magnitude of the star's distance, the
sentient [faculty] merely makes a conjecture in regard to this
magnitude, comparing the distances of the stars to the
distances of familiar but excessively far objects on the
surface of the ground. Thus [the sense faculty] compares the
angle produced by the star at the eye (which is a small
angle) to a distance like the earthly distances, and in
consequence of this comparison perceives the star as small.
If sight had a true perception of the magnitude of the star's
distance, it would perceive the star as large. Similarly, sight
perceives all excessively far objects on the surface of the
ground as small because it does not ascertain their
distances. We have thoroughly explained this in Book III of
this work.

And just as sight errs in regard to the magnitude of a
star's distance because it fails to ascertain it and because it
assimilates it to distances on the surface of the earth, so also
it errs in regarding the distances of the star at different
positions in the sky as unequal in magnitude, though these
distances are equal, because, again, it assimilates them to
certainly unequal distances to the right or left or in front of
it on the surface of the ground. And just as its error in
regard to the star's distance and size is permanent and
constant, so also its error in the difference between the
star's distances and size at various positions [in the sky] is
permanent and constant. For the form of these distances
does not vary in the eye from time to time but is always the
same, and it is sight that assimilates it to the distances of
familiar and excessively far objects on the surface of the
earth.

The enlargement of heavenly objects at the horizon may
frequently have another cause. This cause occurs when a
thick vapor stands between the eye and the star positioned
at or near the horizon, if the vapor is at or near the horizon
and does not continue to the middle of the sky but rather
forms a section of a sphere whose center is the center of the
world because it surrounds the earth. If such a section
terminates before [reaching] the middle of the sky, then the
surface of it that faces the eye will be plane. But if the
surface of the vapor facing the eye is plane, then the form[s]
of the stars (and intervals between them) will be seen behind
the vapor as larger than before the vapor occurred. Because
the form of the star will [first] occur at the place on the
heavens concavity from which it will be refracted to the eye.

Then, in the absence of the thick vapor, the form would extend from this place to the eye in straight lines. But, in the presence of the thick vapor at the horizon, this form will extend to the surface of the vapor that faces the [middle of] the sky, and thus occur in that surface. Sight will therefore perceive this form just as it would perceive objects placed in the vapor: that is, the form will extend through the thick vapor on straight lines then will be refracted at the surface of the vapor facing the eye, this refraction being away from the normal to the vapor's surface (which is a plane surface), since the air near the eye is rarer than the thick vapor. It follows from this that the form will appear to be larger than it would if it were seen rectilinearly. (This was shown in Proposition 1 of this chapter, namely, that when the eye is in the rarer medium and the visible object in the denser medium, and the surface of the denser medium is plane [then the object will iook bigger than it is].) Thus the form that occurs in the surface of the vapor facing the middle of the sky is the visible object, and the medium in which this form is is the thick vapor, and the eye is in the rarer medium of air.

The principal cause on account of which the stars (and their mutual intervals) are seen at the horizons to be larger than at or near the middle of the sky is the one stated earlier. It is the inseparable and permanent cause. When, however, a thick vapor happens to rise at the horizons, it increases their magnification. But this is an accidental cause which always occurs [only] in certain regions of the earth and occasionally in others but is not permanent.

Appendix: Edition of the Arabic text of the *PH*-passage

The Arabic text of the *PH*-passage presents the translator with certain textual difficulties that make it necessary to declare how one proposes to read it before attempting to put it into another language. (See "Notes on the translation of the *PH*-passage" above.) For the reading proposed here, I have relied on two manuscripts of the Arabic text of Ptolemy's *Planetary Hypotheses*: British Library MS Add. 7473 (pp. 92a–b) and Leiden University Library MS Or. 1155 (p. 25b), both of which can be consulted in the facsimile and variants published by Bernard Goldstein and referred to earlier. (In the notes following the Arabic text of the *PH*-passage, I refer to these two manuscripts as *B* and *L*, respectively.) I have also used a quotation of the same passage in Ibn al-Shāṭir's *Nihāyat al-suʾl* as preserved in six manuscripts: Bodleian Library, Marsh 139, pp. 36a–b; Hunt 547, 52b–53a; Marsh 290,

45b–46a; Marsh 501, 49a–b; Leiden University Library, Or. 194, 75a; Suleymaniye Library (Istanbul), 339, p. 138 (= 69b).*

{1} Wa ammā al-sababu alladhî min ajlihi ṣāra mā yaẓharu li-l-naẓari wa yutakhayyalu ilayhi min ʿiẓami jirmihā laysa ʿalā nisabi abʿādihā fa-yanbaghî lanā an naʿlama annhu al-ghalaṭu alladhî yadkhulu ʿalā al-baṣari min qibali ikhtilāfi al-manāẓiri. {2} Wa yatabayyanu dhālika fî jamîʿi mā yaẓharu wa yurā ʿalā buʿdin kathîrin. {3} Fa-kamā anna al-abʿāda anfusahā lā takūnu kammiyyatuhā maʿlūmatan mimmā yaẓharu li-l-ʿayni fa-lā al-tafāḍulu fî-mā bayna al-ashyāʾi al-mukhtalifati al-miqdāri minhā yuʿlamu ʿalā al-tanāsubi allatî hiya ʿalayhi, {4} li-jamʿi al-baṣari wa qabḍihi iyyāhu [wa] tanqîṣihi lahu ilā mā huwa lahu ashaddu ilfan. {5} Wa li-dhālika narā kulla wāḥidin mina al-kawākibi qarîban minnā akthara min ḥāli ḥaqîqatihi li-inḥiṭaṭi al-baṣari ilā al-abʿādi allatî qad iʿtādahā wa alifahā. {6} Ka-dhālika al-ḥālu fî al-ziyādāti wa al-nuqṣānāti allatî taʿriḍu li-l-ʿiẓami bi-ḥasabi ziyādati al-abʿādi wa nuqṣānihā, fa-innahā takūnu aqalla mina al-nisbati, ka-al-ḥāli fî al-abʿādi, li-ʿajzi al-baṣari, ka-ma qulnā, ʿan tamyîzi wa idrāki aqdāri kammiyyati tafāḍuli kulli nawʿin mimmā dhakarnā.

{1} al-manāẓiri] al-manẓari *L* {2} Wa yatabayyanu] wa nubayyinu *B*, *L – B adds* ikhtilāf {3} mimmā] fa-mā *B – fî-mā L / fa-lā] wa lā B, L /* yuʿlamu] *first letter undotted in B* {4} li-jamʿi] jamʿ *B / tanqîṣihi] all letters undotted in B – 3rd, 4th and 5th letters dotted as* q, y *and* ḍ *in L /* lahu (*second occurrence*)] *omitted in B* (4) narā] *no dots for the first letter in B / wa* alifahā] *B adds:* fî-mā (?)bayyannā (*as we have shown*), (?)*or:* fî-mā baynanā (*around us*) {6} aqalla] anqaṣa *L*

Notes

1 See Helen E. Ross and George M. Ross, "Did Ptolemy Understand the Moon Illusion?," *Perception* 5 (1976):377–85.

2 *Ptolemy's Almagest*, translated and annotated by G. J. Toomer, New York: Springer-Verlag, 1984, p. 39.

3 O. Neugebauer, *A History of Ancient Mathematical Astronomy*, New York: Springer-Verlag, 1975, part II, p. 896.

4 For Theon's text, see A. Rome, ed., *Commentaires de Pappus et de Théon d'Alexandrie sur l'Almageste*. Vol. II: Théon d'Alexandrie, *Commentaires sur les livres 1 et 2 de l'Almageste* (Cittá del Vaticano: Biblioteca Apostolica Vaticana, 1936), pp. 346–51. An analysis of this text, including a French translation of the passage examined here, is in A. Rome, "Notes sur les passages des *Catoptriques* d'Archimède conservés par Théon

---

* I am grateful to George Saliba for photographs of the relevant pages in the last five manuscripts.

d'Alexandrie," *Annales de la Société Scientifique de Bruxelles*, ser. A, 52 (1932):30–41.

5 Albert Lejeune, ed., *L'Optique de Claude Ptolémée dans la version latine d'après l'arabe de l'émir Eugène de Sicile* (Louvain: Publications Universitaires, 1956), bk. II, paras. 52–63, pp. 38–46, esp. para. 56, p. 40. (The text of the *Optica* in this edition will hereafter be referred to as "Ptol. *Opt.*" See also Lejeune, *Euclide et Ptolémée: Deux stades de l'optique géométrique grecque* (Louvain: Publications Universitaires, 1948), pp. 95–9.

6 Ptol. *Opt.*, V, para. 6, pp. 225–6; also paras. 70–71, pp. 262–3.

7 Bernard Goldstein, *The Arabic Version of Ptolemy's Planetary Hypotheses*, Transactions of the American Philosophical Society, n.s., vol. 57, pt. 4, Philadelphia, 1967. A third copy of the *Hypotheses* has since been reported to exist in the Osmania University Library at Hyderabad, Dn, no. 306; see S. Sezgin, *Geschichte des arabischen Schrifttums*, VI (Leiden: E. J. Brill, 1978), pp. 94–95. I have been unable to consult this copy.

8 For accounts of Ptolemy's procedure see, in addition to Goldstein's translation and commentary, Olaf Pederson, *A Survey of the Almagest* (Acta Historica Scientiarum Naturalium et Medicinalium, 30), Odense, Denmark: Odense University Press, 1974, pp. 393–6; O. Neugebauer, *A History of Ancient Mathematical Astronomy*, II, pp. 919–22.

9 Neugebauer, *A History of Ancient Mathematical Astronomy*, II, p. 896. Having noted that Ptolemy's incorrect explanation in the *A*-passage must have antedated the *Optica*, Neugebauer continues: "Indeed, in the 'Planetary Hypotheses' this explanation is no longer upheld and the said phenomenon is recognized as an optical illusion, caused by wrongly estimating size in relation to nearby terrestrial objects, a topic further studied in his 'Optics.'" Neugebauer refers to Goldstein's translation of the *A*-passage.

10 Bodleian Library MS March 139, p. 46a, lines 3–6.

11 Ibid., pp. 46a–b.

12 Ptol. *Opt.*, III, 59, pp. 115–16. (I have added the numerals in braces to facilitate references to the passage.) Lejeune supplies the following explanation for sentence {3}: "Parce que les distances verticales nous paraissent moindres que les distances horizontales et que nous interprétons les angles visuels en fonction de la distances supposée" (p. 116 n.53). But this piece of reasoning is not actually stated in the *O*-passage.

13 After raising the problem of interpreting "*minus sensit*" in sentence {1} (see first paragraph of "Commentary" on translation of *O*-passage), Ross and Ross go on to argue as follows: "But there is a more direct reason for doubting that Ptolemy intends any form of psychological explanation. It is clearly implied by the first sentence quoted that the images are already of different sizes by the time they are formed in the eye, since Ptolemy attributes the illusion to the visual ray itself, and not to the mind or brain in processing the sensory information it receives. So, if the ray is responsible, the change in size will already be present before any psychological factors can come into play" ("Did Ptolemy Understand the Moon Illusion?," p. 381). The argument loses force if the faculty of sight, and not what happens to the physical ray itself, is said to be responsible for the visual effect.

14 Ibid., p. 381.

15 Ibid.

16 See article "Ibn al-Haytham" in *Dictionary of Scientific Biography* VI (1972), pp. 189–210, esp. p. 205.

17 The Arabic text of the *Dubitationes* has been edited by A. I. Sabra and N.
    Shehaby as *al-Shukūk ʿlā Baṭlamyūs*, Cairo: Dār al-Kutub, 1971. An English
    translation of the section of this work that deals exclusively with Ptolemy's
    Optica is in A. I. Sabra, "Ibn al-Haytham's Criticisms of Ptolemy's
    Optics," *Journal of the History of Philosophy* 4 (1966):145–9. The reference
    to the *A*-passage occurs in the first section of the book, which is concerned
    with problems in the *Almagest*. Ibn al-Haytham's criticism here is the same
    as that quoted below from section III of his *Ḥall shukūk fī al-Majisṭī*.

18 This has survived in Istanbul MS Ahmet III 3329, copied in Jumādā II, 655
    (A.D. 1257), 123 fols.

19 Aḥmad's work is extant in two Istanbul MSS: Laleli 2759₂, and Ragip Paşa
    934, both of which are listed in Max Krause, "Stambuler Handschriften
    islamischer Mathematiker," *Quellen und Studien zur Geschichte der
    Mathematik, Astronomie und Physik*, Abt.B: Studien, Band 3 (1936):513–
    14. The relevant pages are 72a–74a in the first manuscript, and 54a–56a in
    the second.

20 The text of the whole argument in Ibn al-Haytham's *Commentary on the
    Almagest* is reproduced in the edition of the *Shukūk* cited in note 17 above,
    pp. 74–7.

21 The treatise *Fī Ruʾyat al-kawākib* is reproduced by Mullā Fatḥollāh
    Shirwānī (d. A.H. 891/A.D.1486) in an optical appendix included in his
    commentary on al-Ṭūsī's *Tadhkira*, as I have noted in a photograph of this
    appendix from MS 493 in the Central Library of Tehran University, kindly
    provided by Dr. Ḥusayn Maʿṣūmī. I have also consulted a transcription of
    the same treatise made by Anton Heinen from a manuscript in a private
    collection at Lahore; see Heinen's "On some hitherto unknown manuscripts
    of works by Ibn al-Haytham," to appear in the Proceedings of the Second
    International Symposium on the History of Arabic Science, held in Aleppo,
    1979.

22 Ibn al-Haytham was, of course, aware of Ptolemy's treatment of
    atmospheric refraction in Book V of the *Optica*. What he means here is that
    Ptolemy did not explain how refraction affects perception of the star's size.

23 I have not been able to consult two other manuscripts bearing the same
    title; see Sezgin, *Geschichte des arabischen Schrifttums* VI (Astronomie),
    p. 258 (no.13).

24 Cf. *Dictionary of Scientific Biography* VI, pp. 206–7.

25 "*wa huwa ākhiru mā dhakarahu min al-masāʾil.*" I take the third person
    singular form in *dhakarahu*, not literally but as the common polite form of
    address, thus assuming that sections II and III are pieces of correspondence
    with, possibly, the same person mentioned in section III (see first paragraph
    under "Alhazen's treatise . . . "; Abū al-Qāsim ibn Maʿdān).

26 Compare Ibn al-Haytham's quotation of the same passage in *al-Shukūk ʿalā
    baṭlamyūs*: "The sun is seen larger at the horizon than when it is in the
    middle of the sky only because a moist vapor that surrounds the earth
    occurs between it and the eye, thereby causing it to be seen thus – just as
    what is thrown in water appears larger, and the deeper it sinks the larger it
    becomes" (p. 5).

27 This is Proposition 7 in Book II in Lejeune's edition of Ptol. *Opt.*, p. 40:
    Veluti si fuerint due quantitates *ab*, *gd*, habentes eundem situm [i.e.,
    orientation] et subtendentes eundem angulum qui est *e*. Cum ergo
    distantia *ab* non sit equalis distantie *gd*, sed propinquior ea, *ab* utique
    numquam apparebit maior quam *gd* secundum quod decet propter

propinquitatem suam. Sed aut minor apparebit, quod fit cum distantia alterius ad altera habuerit sensibilem quantitatem; aut uidebitur equalis ei, quod fit cum quantitas diuersitatis distantie fuerit insensibilis.

Ibn al-Haytham's words are almost a verbatim reproduction of Ptolemy's text.

28 Corresponding to Proposition 98 in Lejeune's edition of Ptol. *Opt.*, V, 76, pp. 264–5.

29 *masā'ilihi* (his questions). For changing the form of this word from third to second person, see note 25, above.

30 *min shukūkihi* (of the doubts he has raised). For translating this in the second person, see note 25, above.

31 F. Risner, ed., *Opticae thesaurus. Alhazeni Arabis libri septem, etc.*, Basel, 1572.

32 The ellipsis stands for a sentence I cannot make out in Arabic or Latin. In the Latin translation, which is fairly literal, it reads as follows: "Et videt stellam fixam ex polo mundi, et remotio eius est ipso in una revolutione; nam haec diversitas invenitur parva; ex quo patet, etc." (Risner, *Opticae thesaurus*, p. 282, lines 5–6).

9

Roger Bacon and the origins of perspectiva in the West

DAVID C. LINDBERG

During the early Middle Ages optics did not exist as a discipline. By the end of the Middle Ages it had become a school subject, its content and methodology defined by a variety of standard texts and disseminated by a modest pedagogical tradition.[1] How did this transformation occur? How did *perspectiva* (as it was named) enter the Western educational system and establish itself as a discipline? What, moreover, were its methodology and content, and how did they come to be defined?

The answer to these questions might seem to lie simply in the translation of Greek and Arabic books in the twelfth century, which radically altered the intellectual life of the West. It is certainly true that Greek disciplines, with their Islamic extensions and refinements, descended suddenly on the West in the twelfth century; and optical matter was included. But the assimilation and appropriation of these new materials by Western scholars was no trivial accomplishment. The translations did not merely enlarge Western knowledge; they also complicated it. There were battles to be fought over the new learning, which did not fit snugly into its new social and intellectual environment. Moreover, the new learning was no simple unity, but a complex mélange of documents and philosophies that spoke with many different voices. On almost every subject, it presented Western scholars with competing methodologies and conflicting content. It was not possible simply to receive this heritage and assent to its doctrines unless one enjoyed self-contradiction. In order to assimilate, one had to accommodate.

In this chapter we examine the process of assimilation, and its accompanying struggles, in the formation of the discipline of optics or

This paper is based on research supported by the National Science Foundation and the Graduate School of the University of Wisconsin.

perspectiva. The focus of our attention will be Roger Bacon, the most influential actor in the proceedings. But in order to put his achievement in context, we must first sketch the historical background.

Optical knowledge in the early Middle Ages

Scholars of the early Middle Ages surely held opinions about the nature of light and the act of vision, but these did not constitute a school subject or a discipline to which a name was attached or a scholarly career might be devoted. If these opinions had any disciplinary affiliation, it would most generally have been with epistemology or metaphysics – with epistemology because of the preeminence of sight among the five external senses, with metaphysics because of the frequent use of light as analogue and metaphor by Neoplatonic metaphysicians. Light also figured in theological discussions, owing to its prominence in the creation account in Genesis, the biblical use of light metaphors (particularly in the Gospel of John) and the close connection between theology and metaphysics.[2] Beyond these disciplinary connections, light and sight were simply interesting natural phenomena, which might be objects of brief discussion in a handbook, an encyclopedia, or a book of marvels.

The level of understanding was low. Encyclopedias and handbooks commonly explored the mysteries of vision, reporting cases of the evil eye or double pupil, recounting Tiberius Caesar's ability to see in the dark or Strabo's ability to see 135,000 miles.[3] In the realm of theory, most authors were content with the claim that vision occurs by the extramission of a visual ray. Ocular anatomy also received occasional, superficial attention. And meterological phenomena involving light – lightning and the rainbow – were regularly discussed.[4]

The most sophisticated optical discourse of the early Middle Ages is found in the works of Augustine of Hippo (A.D. 354–430), who employed Neoplatonic emanationism, with its ubiquitous light metaphors, to explain the nature of the Trinity, the creation of the world, the relationship between body and soul, the process of sense perception, and the acquisition of knowledge. Moreover, in his *Literal Commentary on Genesis*, Augustine offers brief, but careful, discussions of the nature of visible light (maintaining its corporeality) and the act of sight (aligning himself with the extramissionists).[5]

Augustine's hegemony as an optical authority was first challenged in the twelfth century. The first half of Plato's *Timaeus* had been available in Latin since the fourth century but attracted little notice until the twelfth, when it became the focus of attention in the cathedral schools. In the translated portion of the *Timaeus*, Plato had discussed light and vision, arguing that visual fire emanates from an observer's eye and coalesces with daylight to form "a single homogeneous body" stretching from the eye to the visible object, able to transmit motions from the visible object to the soul, where they produce sensation. Twelfth-century scholars, such as William of Conches and Adelard of Bath, were quick to understand and elaborate on this Platonic account.[6] In so doing, they lent support to the theory of visual fire emanating from the eye and grappled with some of the perennial objections raised against extramission theories (how, for example, can corporeal substance emanating from the eye expand to fill all of the space up to the fixed stars?), but they did not fundamentally alter the character of optical discourse. Light and vision remained nothing more than interesting natural phenomena, which one might fruitfully discuss in a general work on nature.

The *Timaeus* also reinforced a particular view of the relationship between mathematics and nature. Augustine and Boethius, echoing Plato's own opinion, had argued that the reality underlying the material world is mathematical. Augustine, for example, commenting on Wisdom 11:12, had discussed God's creation of the visible world "according to measure, number, and weight";[7] and Boethius had pointed out in his *De arithmetica* that "all things constructed by primeval nature appear to be formed according to the pattern of numbers. For this was the principal exemplar in the mind of the Creator."[8] Now with a portion of Plato's *Timaeus* before them, twelfth-century scholars were inspired to a renewed appreciation of the power of mathematics to unlock nature's secrets.[9]

The translations

The translations of the twelfth century bestowed on Western scholars, for the first time, entire treatises devoted to optical phenomena. Among the earliest and, therefore, most influential were the *Optica* and *Catoptrica* of Euclid, translated perhaps about the middle, or early in the second half, of the twelfth century.[10] A far more substantial, but also far more difficult, treatise translated about the same time was the *Optica* of Ptolemy – difficult, in no small measure, because of defects in the Arabic version from which its translator, Emir Eugene, rendered his Latin version.[11] The translation of two short works – the *De aspectibus* of al-Kindi and the *De speculis* or *De aspectibus* of Tideus – by Gerard of Cremona in the second half of the twelfth century contributed to this growing optical corpus. Finally, late in the twelfth century or early in the thirteenth century an anonymous translator rendered the *De aspectibus* or *Perspectiva* of Alhazen (Ibn al-Haytham), which was to be one of the most powerful forces shaping Western optics.[12]

These treatises defined the subject matter of optics or *perspectiva*. The works of Euclid, Ptolemy, Tideus, and al-Kindi were concerned principally with vision and visual radiation, including the propagation of the latter. Alhazen's great tome gave equal prominence to light and also devoted modest attention to the anatomy and physiology of the eye.[13] Light, vision, and the eye: these became the central core of the science of optics. The rules of perspective were included under vision. Reflection and refraction were dealt with as accidents of either light or visual radiation, occurring as the result of encounter with an opaque or transparent surface. Psychological issues were inevitably raised in the quest for a theory of vision.[14] And meteorological phenomena, such as the rainbow, came to be viewed as important appendages of optics.[15]

While fixing the list of topics included within optics, this new literature also suggested a methodology (or perhaps several methodologies) for pursuing it. Euclid's *Optica*, as one might expect, bears a methodological resemblance to Euclid's *Elements*. The treatise begins with postulates, numbering between four and nine (depending on the particular version in hand), and proceeds to fifty-eight propositions that follow, more or less, from the postulates. Each proposition begins with an enunciation, which is followed by a proof or demonstration, more or less rigorous. The structure of Euclid's *Optica* is thus drawn from geometry. So, too, is the mode of analysis. Euclid reduces optics, insofar as possible, to the analysis of geometrical lines. The postulates affirm that rays proceed in straight lines from the eye, the collection of such rays constituting a visual cone, and that whatever intercepts a ray is seen. The rectilinearity of the rays, interrupted only by the accidents of reflection or refraction, makes it possible to represent the physical acts of radiation and vision by the lines of a geometrical diagram and thus to submit optical problems to geometrical analysis. It is true that certain of Euclid's postulates spill over into the physical realm, revealing, for example, that rays are truly physical agents and specifying the direction of their propagation, but it is generally acknowledged that Euclid held such nongeometrical content to a minimum.[16]

The physical content of optics was substantially enlarged by the works of Ptolemy and al-Kindi. Ptolemy's *Optica*, in its original form, apparently delved deeply into the physical aspects of vision and the nature of visual radiation; unfortunately, Book I was lost before the treatise reached Western hands, and what remained was principally a mathematical presentation of the rules of perspective and an analysis, largely mathematical, of binocular vision and the reflection and refraction of radiation.[17] Al-Kindi too was interested in the physics of vision, arguing vigorously for the extramission of rays from the observer's eye, while insisting that these rays form a continuous cone of radiation. His *De aspectibus* also discusses the physical nature of visual radiation, maintaining that it is a transformation of the transparent medium, induced by the visual power of the eye.[18] Nonetheless, *De aspectibus* remains visibly mathematical. Its pages are filled with geometrical diagrams, and the argument, besides being divided into propositions, is frequently about geometrical matters. Al-Kindi devotes considerable attention, for example, to demonstrating the rectilinear propagation of light. And in a proposition that was to prove highly influential in the West, he presents a geometrical argument to explain why sight is sharpest in the center of the field of vision. Indeed, despite his concern with the physics of vision, it was al-Kindi who taught his successors to analyze radiating surfaces into radiating points (from which lines may be drawn in all directions), and thus to render radiation from luminous bodies susceptible of geometrical treatment.[19]

Alhazen's *De aspectibus*, finally, was a comprehensive optical text, which integrated the physical claims of the natural philosopher, the anatomical and physiological concerns of the physician, and the geometrical analysis of the mathematician. Nonetheless, what stood out as new in *De aspectibus* was its fulfillment of the promise of geometrical optics. Alhazen certainly discussed the anatomy of the eye and the nature of visual radiation, but what was overwhelmingly apparent to his readers was the successful application of geometry to the whole of optics, including the path of external radiation within the observer's eye (for Alhazen was an intromissionist) – the latter constituting a minor mathematical triumph. Alhazen also achieved impressive success in devising geometrical solutions to a series of very difficult problems of image-formation in mirrors and refracting media.[20]

A supporting literature, offering a somewhat different approach to optical phenomena, became available about the same time. Medical works, such as Hunain ibn Ishaq's *De oculis* (which was based very closely on Galen) and Avicenna's *Liber canonis*, enlarged the understanding of ocular anatomy and physiology and strengthened the notion that they were topics worthy of attention. And the writings of Aristotle and his commentators (especially the Muslims Avicenna and Averroës) provided powerful reinforcement of physical and psychological concerns. Indeed, these works left the definite impression that the important issues were not mathematical at all.[21] In Aristotle's opinion, mathematics and physics were autonomous branches of theoretical knowledge, each with its own subject matter and appropriate explanatory principles; and it was wrong to allow mathematics to intrude on physics (into which the study of sensation clearly fell).[22] Moreover, in his *De anima* and *De sensu*, Aristotle demonstrated by example how to approach optical phenomena through physical, rather than mathematical, analysis.[23]

Unfortunately, the lesson was muddled by other parts of the Aristotelian corpus, where a different methodology was specified or employed. In his *Meterologica*, Aristotle submitted the rainbow, in some measure, to geometrical analysis.[24] And in his *Physics*, he made influential remarks, of a methodological sort, that placed optics on the boundary between physics and mathematics – identifying optics (along with harmonics and astronomy) as a "middle science," that is, as one of the more physical branches of mathematics, and noting that it studies mathematical lines "*qua* physical, not *qua* mathematical."[25]

The debate over methodology in the thirteenth century

Western scholars, then, were called on to deal witih a complex heritage that embodied opposing tendencies: While agreeing, more or less, on the topics to be included within the new discipline, the newly translated literature offered conflicting notions of the method or methods by which it was to be pursued. The first Western scholar to struggle seriously with the methodological issues thus raised was Robert Grosseteste (ca. 1168–1253), distinguished theologian, commentator on philosophical and theological works, translator, teacher, and bishop. Grosseteste was certainly familiar with Euclid's *De speculis* and probably with Euclid's *De visu* and al-Kindi's *De aspectibus*, although almost certainly not with Ptolemy's *Optica* or Alhazen's *De aspectibus*.[26] We must also keep in mind that Grosseteste's education in the twelfth-century schools introduced him to Platonic philosophy and, in particular, to Plato's view of the relationship between mathematics and nature. And finally, his own scholarly research brought him into contact with Aristotle's *Posterior Analytics* and physical treatises, where the intricate relationship between physics and mathematics was spelled out.[27]

What Grosseteste learned from this diverse literature was that optical questions could be treated mathematically. This is perhaps clearest in his *De lineis, angulis, et figuris*, not itself an optical text, but a treatise on the radiation of force in general. Here he follows al-Kindi in reducing natural action to the radiation of force and submits the radiation of force, in turn, to geometrical analysis.[28] Both light and visual radiation are examples of this radiating force, and it is clear that the geometrical rules applicable to radiating force in general are equally applicable to its optical manifestations. Grosseteste begins this treatise with a ringing proclamation of the utility of geometrical analysis:

> The usefulness of considering lines, angles, and figures is
> very great, since it is impossible to understand natural
> philosophy without them. They are useful in relation to the
> universe as a whole and its individual parts . . . Indeed, they
> are useful in relation to activity and receptivity, whether of
> matter or sense; and if the latter, whether of the sense of
> vision, where activity and receptivity are apparent, or of the

other senses, in the operation of which something must be added to those things that produce vision. . . .

Now, all causes of natural effects must be expressed by means of lines, angles, and figures, for otherwise it is impossible to grasp their explanation. This is evident as follows. A natural agent multiplies its power from itself to the recipient, whether it acts on sense or on matter. This power is sometimes called species, sometimes a likeness, and it is the same thing whatever it may be called.[29]

Grosseteste proceeds to point out that power may come from an agent to a recipient along either straight or broken lines, the former giving rise to stronger action than the latter. Broken lines may be either reflected or refracted. Reflection occurs at equal angles. Refraction is toward or away from the perpendicular depending on the relative densities of the two media involved and, of course, the direction of propagation. Grosseteste also discusses secondary radiation, apparently also propagated in straight lines. Finally, he describes the actions that result from various pyramids of radiation.[30]

In another treatise, *De iride*, Grosseteste undertakes to define the science of *perspectiva*. After noting that Aristotle, in his *Meteorologica*, recounted the facts of the rainbow, which properly concern the physicist, but not the explanation, "which concerns the student of perspective," Grosseteste proceeds: "We state that perspective is the science based on visual figures and that this subordinates to itself the science based on figures containing radiant lines and surfaces, whether that radiation is emitted by the sun, the stars, or some other radiant body."[31] The geometrical arguments of the perspectivist explain the facts supplied by the physicist.

Grosseteste addressed this methodological issue again, more abstractly in his commentary on Aristotle's *Posterior Analytics*. There we find a theoretical discussion of *quia* and *propter quid* knowledge and the subordination of the one to the other:

> *Propter quid* [or causal] knowledge and *quia* [or factual] knowledge differ in another way, namely, that the one is acquired through one science and the other through another science. And such sciences, the one yielding *propter quid* knowledge and the other *quia* knowledge regarding the same thing, are those that have a relationship to each other such that the one is subordinating and the other subordinated. For instance, the science devoted to radiating lines and figures is subordinate to geometry, which is devoted simply to lines and figures.[32]

One should not be misled by such remarks to suppose that Grosseteste called for a total reduction of optics to geometry. On the contrary,

Grosseteste was well aware of the physical analysis of light and vision in Aristotle's *De anima* and *De sensu*, and more especially of the implications of Aristotle's remarks in the *Physics* and *Posterior Analytics* about the relationship between physics and mathematics. As Grosseteste understood the matter, causal or *propter quid* demonstration, which leads to the highest kind of knowledge, is available in physics only or mainly through a mathematical cause. However, there is a lower form of demonstration (*quia* demonstration) that employs no mathematical cause but remains entirely within the realm of physics. In short, Grosseteste recognized that light has a physical nature, worthy of analysis, and he did not believe that this physical nature would yield to geometrical analysis.[33]

Despite his appeal for the mathematization of optics, Grosseteste did little himself to meet the need. In *De lineis* he discusses the propagation of power along direct, reflected, and refracted lines; he states the law of equal angles for reflection; and he briefly describes the refraction of radiation as it passes from one medium to another. In *De iride* he develops a "half-angle" law of refraction, according to which the angle between the ray and the perpendicular is half as great in the denser medium as in the rarer, and argues that the rainbow is produced by refraction.[34] If this is geometrization, it is of a most elementary kind; lines of propagation are defined and their orientation with respect to the reflecting or refracting surface specified, but there are no serious geometrical arguments.[35]

The sources that inspired Grosseteste's appeal for the mathematization of optics were also available to Albert the Great (ca. 1200–1280). Albert understood the Platonic position and (before his career was over) had read the works of Euclid and al-Kindi and even acquired a passing acquaintance with Alhazen's *De aspectibus*, but refused to convert to the mathematical religion.[36] Against the Platonic subordination of physics to mathematics, Albert maintained, with Aristotle, that the two disciplines are autonomous and that each must proceed according to its own principles. In an often quoted passage from his paraphrase of Aristotle's *Metaphysics*, Albert wrote: "We must here beware of the error of Plato, who said that natural things are based on mathematical things and mathematical things on divine things, just as the third cause is based on the second and the second on the first; and therefore he said that the principles of natural things are mathematical, which is altogether false."[37] The causes of physical phenomena must be found within physics.

Albert thus rules out the subordination of physics to mathematics; he does not, however, rule out all cooperative effort. Indeed, he notes in his *De sensu* that certain aspects of vision must be investigated through the science of "*perspectiva*, which science cannot be completed without a consideration of things that pertain to geometry."[38]

Moreover, Albert incorporates pieces of "perspectivist" doctrine in his own analysis. In defending the Aristotelian intromission theory of vision against its opponents, for example, he employs the visual cone or pyramid of the Euclidean tradition:

> It should be noted that every act of sight takes place under a pyramidal figure, the base of which is on the visible object and the vertex in the center of the crystalline humor. Thus since the eye is spherical, all lines extended from its base to its vertex are perpendicular to [the surface of] the eye and [also] the crystalline humor, from the circumference of which they are drawn to its center like the radii of a sphere. Now these pyramids are as numerous as the visible objects, since light proceeds from a visible object to the visual power with the visible form; and when the eye is turned here and there, the pyramid moves according to the different visible things.[39]

Albert also uses the concept of the visual pyramid to explain clarity of vision, noting that if a visible object is close to the eye, it forms a wide visual pyramid, the edges of which fall obliquely on the eye (that is, deviate substantially from the axis), whereas the same object farther away forms a narrower visual pyramid, the parts of which remain closer to the central axis.[40]

However, Albert does not allow mathematics to interfere with attempts to resolve the truly important issues. The main thrust of his voluminous writings on vision is to disprove the absurd notion of the extramissionists (Plato, Empedoceles, Euclid, and al-Kindi) that visual rays issue from the eye.[41] Even when discussing visual pyramids, Albert quickly abandons geometry and turns to the physical nature of the pyramid, arguing that it is not a body of light (*lux* or *lumen*), as the ancients supposed, but "a rectilinear alteration" of the transparent medium through which sight occurs.[42]

We must be careful not to exaggerate the differences between Grosseteste and Albert. We must not suppose that Grosseteste embraced the mathematization of optics without reservation, while Albert totally repudiated it. Grosseteste acknowledged that mathematization has its limits, and Albert used mathematics when it seemed beneficial to do so. Indeed, one will find approximately as much geometrical optics in Albert's writings as in Grosseteste's. The difference is in the nature of their commitment to it. Grosseteste was a promoter of the mathematical program; Albert, sensitive to what seemed excessive enthusiasm on the part of Grosseteste and the Platonists, found himself pulling in the other direction.

Bacon's science of *perspectiva*

Albert dealt with optical matters in books written in the 1240s and 1250s.[43] By the early 1260s, he was already being denounced as an ignorant fool by Roger Bacon (ca. 1220–ca. 1292). That the "un-named master" against whom Bacon directed his diatribes was Albert has been convincingly demonstrated by Jeremiah Hackett.[44] Aside from the fact that Albert and his ilk became theologians without ever receiving a proper education in philosophy, Bacon's principal griev-ance was Albert's (alleged) ignorance of mathematics and *perspectiva*. In his *Opus tertium*, Bacon wrote that "he who multiplied volumes" ignores *perspectiva* and the multiplication (that is, propagation) of spe-cies or virtues; "and not only is *he* ignorant [of these matters], but so too are all of the philosophical rabble, who err through him. For if you write to him, [inquiring] what he treats concerning these roots, you will find him incompetent regarding them."[45] In his *Opus minus*, Bacon continued the attack: "Again, since he is ignorant of *perspectiva* – for truly he . . . knows nothing about it – he has no possibility of knowing anything of value in philosophy."[46]

Bacon's own theoretical position was an extension, or development, of Grosseteste's, enriched by a much more thorough immersion in the new Greek and Arabic sources. Bacon was surely influenced by Gros-seteste's writings, as well as the sources available to Grosseteste (Plato, Euclid, Tideus, al-Kindi, Aristotle, and Avicenna), but he also had at his disposal the works of Ptolemy and Alhazen, where the promise of geometrical optics had been much more completely fulfilled. These works inspired in Bacon a great vision of the wonders of which mathe-matical science is capable. His writings are laced with the claim that mathematics is the gateway to knowledge, the key to a mastery of all the sciences. In the *Communia mathematica* we read: "No science can be grasped without this science [mathematics], and . . . nobody can perceive his ignorance in other sciences unless he is excellently informed in this one. Nor can things of this world be known, nor can man grasp the uses of body and things, unless he is imbued with the mighty works of this science."[47] In the *Opus maius* he proclaims:

> There are four great sciences, without which other sciences
> cannot be grasped or knowledge of things obtained;
> however, these sciences being known, anybody can progress
> gloriously and without difficulty or labor in the power of
> wisdom, not only in human, but also in divine sciences. . . .
> And the gate and key of these sciences is mathematics,
> which holy men discovered at the beginning of the world
> . . . and which has always been used by holy and wise men
> more than any other science. Its neglect for the past thirty
> or forty years has ruined the whole educational system of
> the Latins, since he who is ignorant of it cannot know the
> other sciences or the things of this world. . . . And, on the
> contrary, knowledge of this science prepares the mind and

elevates it to sure knowledge of all things, so that if one grasps the basics of wisdom concerning this science and applies them correctly to knowledge of other sciences and things, he will be able to know all things that follow, without error or doubt, easily and powerfully.[48]

Finally, in the *Opus tertium* he points out that mathematics "is the first of the sciences, without which the others cannot be known," that "the causes of natural things cannot be given except by means of geometry," and indeed that the Devil himself brought about the condemnation and neglect of mathematics because without its service theology and philosophy are useless.[49]

Nonetheless, Bacon's enthusiasm for mathematics was tempered by an occasional acknowledgment of its limitations. For example, in the *Opus maius* he admits, following Grosseteste, that mathematics is required for causal or *propter quid* demonstrations, while *quia* demonstrations (demonstrations through the effect) employ physical principles. "There are two modes of argumentation in relation to natural things," he writes,

> one by demonstration proceeding from causes, the other by demonstration proceeding to an effect. . . . But only cause leads to true knowledge, or at least it does so far better than effect. . . . Therefore, since in natural things demonstration by cause is obtained by means of mathematics, and demonstration by effect is obtained through natural philosophy, the mathematician is better able to obtain true knowledge of natural things than is the natural philosopher.[50]

Bacon clearly prefers *propter quid* demonstrations, which employ a mathematical cause and lead to true knowledge, but he does not deny all utility to the lesser *quia* form of demonstration.[51]

Interesting though Bacon's methodological prescriptions undoubtedly are, it was primarily his practice that influenced the subsequent course of the discipline. How closely did practice conform to precept? That Bacon found ways of applying geometry to optical questions is apparent from even a superficial examination of his works. Here, for the first time in the West, we find original optical treatises containing geometrical diagrams in profusion. Bacon's *De multiplicatione specierum* (which, strictly speaking, deals with all forms of radiation, whether visible or not) contains thirty-nine diagrams; the *Perspectiva* has forty-six and *De speculis comburentibus* twenty-eight.[52] Nor are these diagrams merely decorative; many support serious and sustained geometrical argumentation. Bacon uses geometry to elucidate the theory of direct vision, tracing the path of radiation through the humors of the eye; he develops a theory of binocular vision along geometrical

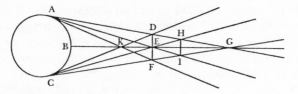

Figure 1

lines, and rules of perspective; and he exhaustively analyzes the phenomena of reflection and refraction, dealing with surfaces of various shapes, situated variously in relation to the observer and observed object.

One of the best examples of effective use of mathematical analysis is Bacon's sustained attack on the problem of radiation through small apertures, undertaken in his *De speculis comburentibus*. The problem was the classic one of explaining how radiation from a spherical body such as the sun, passing through a small triangular or rectangular aperture, can produce a circular image. This problem had evoked a certain amount of commentary beginning with the *Problemata* of Pseudo-Aristotle; al-Kindi and the author of the Pseudo-Euclidean *De speculis* had looked briefly into its geometry, but no source available to Bacon had attempted a thorough geometrical analysis.[53] Bacon undertook an exhaustive investigation of the various cones and pyramids of radiation passing through the aperture. There is no need to follow the argument in detail, but Figure 1 illustrates its basic structure.[54] If *ABC* is the sun and *DEF* a triangular aperture, then pyramid *ACG*, which conforms to the shape of the aperture, could explain why the image close to the aperture bears the shape of the aperture. And cone *ACE*, which comes to a vertex in the middle of the aperture, will give rise to a vertically opposite cone *EHI*, which bears the round image of the luminous source and could explain why the image far from the aperture is round. Unfortunately, the outermost radiation is that belonging to pyramid *ACK*, the vertically opposite counterpart of which, *KDF*, conforms in shape to the aperture. There seems no geometrical explanation of an image of the same shape as the source, and Bacon is forced to look for reasons why radiation between *CD* and *CH* does not contribute to the image. In the long run the analysis did not succeed, as Kepler was to make clear; but it was thorough, intelligent, influential, and, above all, geometrical. It taught Bacon's successors that the solution to the problem was to be sought in a geometrical analysis of the modes of radiation.

But largely geometrical though it may have been, the analysis was not *purely* geometrical. Bacon's attempt to explain why radiation be-

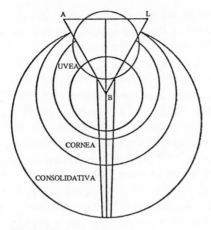

Figure 2

tween *CD* and *CH* is invisible rests on arguments about the strength of radiation and assumptions about the sensitivity of the visual faculty. Ray *CD*, he argues (still proceeding geometrically), is a "solitary" ray – that is, once beyond the aperture, it is intersected by no other ray issuing from the luminous source. This assures us, he argues (now abandoning mathematics for the psychology or physiology of sight),[55] that the ray will not be visible because solitary rays are too weak to stimulate vision. What must be clear is that any question about optical phenomena calls ultimately for an understanding of the eye and vision; and that inevitably introduces nonmathematical considerations.

Bacon's theory of vision is itself a nice illustration of the struggle to extend mathematical analysis to its limit – and, at the same time, of the recognition that there is indeed a limit, beyond which one must employ other modes of analysis. Bacon represents the path of radiating light by means of straight lines and shows how the vision-producing rays form a cone of light emanating from the visible object as base toward an apex in the observer's eye. In order to account geometrically for erect visual images, Bacon must show that rays proceed through the eye without intersection; this he accomplishes by means of a single refraction at the posterior surface of the crystalline humor or lens, which rectifies the incoming radiation without altering its internal configuration and directs it toward the opening of the optic nerve. Except for a few details, this is a highly mathematized scheme. Bacon even imposes a geometrical model on the structure of the eye, treating each of the tunics of the eye as a segment of a spherical surface and arranging their centers along a central axis. Figure 2, from Bacon's *Opus maius*, illustrates both the spherical anatomy of the eye and the geometry of radiation.[56] That this scheme is taken over from Alhazen in no way alters its value as an illustration of Bacon's mode of operation and the mathematical program that he came to represent.

But mathematics does not explain everything. Why are the vision-producing rays those that fall perpendicularly on the eye? The answer is to be found in the nature of light, which is weakened when refracted, and in the nature of the visual faculty, which perceives only unweakened radiation.[57] The laws of refraction themselves derive ultimately from what must be regarded as metaphysical, rather than mathematical, principles. There was the further problem of the mixing of species in the medium between object and observer, which resisted a purely mathematical solution.[58] In the course of developing his theory of vision, Bacon also had to face the delicate issue of ignoble, corporeal objects apparently acting on the noble senses (instruments of the soul); his proposed solution, which had nothing at all to do with mathematics, was that vision sends forth its own radiation, which "alters and ennobles" the corporeal medium and "excites" the species of the visible object, thus making the incoming radiation "commensurate with the nobility of the animated [eye]."[59]

Bacon's analysis of the refraction of light, already touched upon, will serve as a final example. Having imbibed the geometrical rules of refraction from his Greco-Arabic sources, Bacon proceeded to apply them to a variety of specific cases involving plane and spherical refracting surfaces. In the *Opus maius* he supplies ten drawings, with accompanying text, that cover most of the variations: The eye is situated first in the rarer medium, then in the denser; in the case of a spherical refracting surface, the eye is placed first on the concave, then on the convex, side of the interface; and the eye or the visible object (whichever is located on the concave side) is situated first inside and then outside the center of curvature.[60]

All of this is entirely geometrical. But it is, of course, only the application of geometrical rules that are themselves rooted in physical natures and metaphysical principles. Bacon begins his explanation of these rules by reference to the physics of impact and of motion through resisting media: Light passing from one resisting medium to another is deflected according to the resistances of the two media and the angle of incidence on the interface. This behavior is further elucidated by various mechanical analogies: A direct blow, for example, is more powerful than a deflected or oblique one. But why is obliquely incident radiation bent toward the perpendicular when it falls on a denser medium, away from the perpendicular when it falls on a less dense medium? Here physical principles no longer suffice, and Bacon has recourse to the metaphysical principle of uniformity. Light selects that direction of passage in the second medium which will most nearly maintain uniformity of strength; thus, if the second medium is denser than the first, the radiation compensates for the weakening effect of the denser medium by assuming an orientation within that medium closer to the path of greatest strength, the perpendicular.[61]

It should be noted how far short this argument falls of Bacon's stated aim of discovering a mathematical cause for facts provided by the physicist or natural philosopher, in order to achieve a *propter quid* demonstration. The data of refraction look remarkably like geometrical rules, while the causes seem to be physical and metaphysical. What we observe in Bacon's optical work, here and elsewhere, is not reduction of the discipline to mathematical demonstrations, however desirable that goal (considered in the abstract) may have seemed, but a thorough integration of mathematical, physical, metaphysical, psychological, and sometimes even anatomical, concerns. Bacon's theory of vision, for example, is set firmly into a physical and psychological context. His *Perspectiva*, where visual theory is most fully treated, begins with an analysis of the soul, sensation, and cognition, and these concerns furnish the setting for the entire work. *De multiplicatione specierum*, although containing its share of geometry and geometrical analysis, aims primarily to set forth the nature of species and the physics of their propagation through the medium.

It is, in fact, quite difficult to identify demonstrations that conform strictly to either the *propter quid* or the *quia* model. Typically, Bacon offers a geometrical analysis of a perceptual experience, but then proceeds to ground this geometrical analysis in physical natures. If any kind of causation seems to prevail, it is physical. Even *De speculis comburentibus*, the work in which Bacon comes closest to finding mathematical causes for physical phenomena, appeals in the end to the nature of light and the visual apparatus. For all his praise of mathematics, Bacon's practice of optics suggests that the ultimate realities, or at least the ultimate explanatory terms, are physical and metaphysical.

Bacon's failure to live up to his stated methodological ideals should not surprise or disappoint us. He was neither the first nor the last methodologist to meet such a fate. As methodologist, Bacon followed Grosseteste and the Platonic tradition in shaping a theory of science that enlarged the role of mathematics in demonstration over that assigned to it by Aristotle. When Bacon turned from methodology to optics, he no doubt hoped to display the possibilities of this mathematical program; but that goal inevitably competed with others. To understand Bacon's achievement in optics I believe that we must see it *primarily* as an attempt to sort through the principles of optics as received from his Greek and Islamic predecessors – weeding out error, eliminating conflict and contradiction, reconciling optical doctrine with philosophical and theological presupposition, imposing order, and introducing clarity and precision.[62] Bacon wished also to demonstrate that optics *of itself* had special prerogatives and offered entrance to some of nature's innermost secrets. And finally, lest this great achievement be lost to humanity, it was necessary to demonstrate its utility for Christendom. The breadth of Bacon's vision excluded any single-minded attempt at mathematization.

Nonetheless, Bacon did achieve considerable success in geometrizing optics. No Westerner before him went as far. Faced with a variety of opinions on the applicability of mathematics to optics, he followed Alhazen and Grosseteste in giving mathematics maximum play – even while recognizing that there were problems beyond its reach. His optical works are heavy with geometry, much of it borrowed from Greek and Islamic models, of course, but selected and disseminated by Bacon. It was Bacon who showed Western scholars how to do optics the mathematical way.[63]

Notes

1 On optics in the late medieval curriculum, see David C. Lindberg, *John Pecham and the Science of Optics* (Madison: University of Wisconsin Press, 1970), 30–31; Lindberg, "Introduction" to the reprint edition of *Opticae thesaurus Alhazeni Arabis libri septem . . .*, ed. Friedrich Risner, Basel, 1572 (New York: Johnson Reprint, 1972), xxiii.

2 On the use of optical knowledge in epistemology, metaphysics, and theology, see Clemens Baeumker, *Witelo, ein Philosoph und Naturforscher des XIII. Jahrhunderts (Beiträge zur Geschichte der Philosophie des Mittelalters*, III.2) (Münster: Aschendorff, 1908), 284–523; Klaus Hedwig, *Sphaera Lucis: Studien zur Intelligibilität des Seienden im Kontext der mittelalterlichen Lichtspekulation (Beiträge zur Geschichte der Philosophie und Theologie des Mittelalters*, n.s., XVII) (Münster: Aschendorff, 1980); and David C. Lindberg, "The Genesis of Kepler's Theory of Light: Light Metaphysics from Plotinus to Kepler," *Osiris*, n.s. 2 (1986):5–42.

3 Pliny *Natural History* VII.21, VIII.32, XI.54. Gaius Julius Solinus, *The Excellent and Pleasant Worke Collectanea Rerum Memorabilium*, trans. Arthur Golding (London, 1587), fols. Ei[v], Fiiii[v], Nii[r].

4 Pliny *Natural History* II.60, XI.54–55. Isidore of Seville *Etymologies* XI.1.36–37, XIII.8.2, XIII.10.1.

5 Augustine *De trinitate* IV.20, VIII.3; *De Genesi ad litteram* I.16.31, IV.34.54, VII.13.20, VII.15.21, XII.16.32; *De quantitate animae* 23; *Contra Academicos* III.11.16.

6 *Timaeus* 45b–46c; David C. Lindberg, *Theories of Vision from al-Kindi to Kepler* (Chicago: University of Chicago Press, 1976), 3–6, 91–94.

7 *De Genesi ad litteram* IV.3.7, trans. John Hammond Taylor, 2 vols. (Ancient Christian Writers, 41–42) (New York: Newman Press, 1982), I, 108.

8 I.2, in *Patrologia cursus completus, series Latina* ed. J.-P.Migne, 221 vols. (Paris: Migne, 1844–64), LXIII, col. 1083b.

9 Few of them went as far as Plato had gone, for most were also influenced

by Aristotelian reservations about the efficacy of mathematical analysis; see David C. Lindberg, "On the Applicability of Mathematics to Nature: Roger Bacon and His Predecessors," *British Journal for the History of Science* 15 (1982):7–10.

10 The former appeared in at least three independent translations, one from the Greek and two from the Arabic; and both treatises enjoyed wide circulation. David C. Lindberg, *A Catalogue of Medieval and Renaissance Optical Manuscripts* (Toronto: Pontifical Institute of Mediaeval Studies, 1975), 46–55; Wilfred R. Theisen, "*Liber de visu*: The Greco-Latin Translation of Euclid's *Optics*," *Mediaeval Studies* 41 (1979):51–60; Lindberg, *Theories of Vision*, 210–11.

11 Albert Lejeune, ed., *L'Optique de Claude Ptolémée, dans la version latine d'après l'arabe de l'émir Eugène de Sicile* (Louvain: Publications Universitaires, 1956), 9*–13*. Ptolemy's authorship of the *Optica* is called into question by Wilbur Knorr in a forthcoming paper, "Archimedes and the pseudo-Euclidean *Catoptrics*: Early Stages in the Ancient Geometric Theory of Mirrors," *Archives internationales d'histoire des sciences*. For our purposes, the identity of the author is unimportant.

12 Lindberg, *Theories of Vision*, 209–11.

13 Ibid., 11–32, 61–86.

14 A. Mark Smith, "Getting the Big Picture in Perspectivist Optics," *Isis* 72 (1981):568–89. Smith has performed an important service in calling attention to the cognitive side of the optical tradition; I am not entirely persuaded, however, of his claim that the raison d'être of *perspectiva* was to provide "a comprehensive model of visual induction" (p. 588). It appears to me that *perspectiva* was a discipline diverse in purpose and content, addressed to a number of pressing issues, of which cognition was (for some of its practitioners) the most important one, but hardly the only one.

15 The content of medieval *perspectiva* can be conveniently examined in John Pecham's *Perspectiva communis*, one of the standard texts; see Lindberg, *Pecham and Optics*.

16 For an analysis of Euclid's *Optics*, see Albert Lejeune, *Euclide et Ptolémée: Deux stades de l'optique géométrique grecque* (Louvain: Bibliothèque de l'Université, 1948); Lindberg, *Theories of Vision*, 11–14. See also A. Mark Smith, "Saving the Appearances of the Appearances: The Foundations of Classical Geometrical Optics," *Archive for History of Exact Sciences* 24 (1981):73–100.

17 Lejeune, *Euclide et Ptolémée*; Lejeune, *Recherches sur la catoptrique grecque* (Brussels: Palais des Académies, 1957).

18 Al-Kindi, *De aspectibus*, in Axel Anthon Björnbo and Sebastian Vogl, "Alkindi, Tideus und Pseudo-Euklid. Drei optische Werke," *Abhandlungen zur Geschichte der mathematischen Wissenschaften* 26, pt. 3 (1912):3–41. Lindberg, *Theories of Vision*, chap. 2.

19 The alternative is a unified or "wholistic" process of radiation, to which geometry could be applied only with the greatest difficulty; for further discussion, see Lindberg, *Theories of Vision*, 26–30, 58–60; Smith, "Getting the Big Picture," 578–80.

20 Lindberg, *Theories of Vision*, chap. 4; A. I. Sabra, "The Physical and the Mathematical in Ibn al-Haytham's Theory of Light and Vision," in *The Commemoration Volume of Biruni International Congress in Tehran* (Tehran: High Council of Culture and Art, 1976), 439–78.

21 Lindberg, *Theories of Vision*, chap. 3; Lindberg, "The Intromission–

Extramission Controversy in Islamic Visual Theory: Alkindi versus Avicenna," in *Studies in Perception: Interrelations in the History of Philosophy and Science*, ed. Peter K. Machamer and Robert G. Turnbull (Columbus: Ohio State University Press, 1978), 137–59.

22 Friedrich Solmsen, *Aristotle's System of the Physical World* (Ithaca, N.Y.: Cornell University Press, 1960), 259–64; Lindberg, "Applicability of Mathematics to Nature," 5–7.

23 *De anima* II.7; *De sensu* II.

24 *Meteorologica* III.5.

25 *Physics* II.2. 194^{a}10–11, trans. R. P. Hardie and R. K. Gaye, in *The Works of Aristotle Translated into English*, ed. David Ross and J. A. Smith, 12 vols. (Oxford: Clarendon Press, 1908–52), II.

26 Lindberg, *Theories of Vision*, 116–17; A. C. Crombie, *Robert Grosseteste and the Origins of Experimental Science 1100–1700* (Oxford: Clarendon Press, 1953), 116–17; Ludwig Baur, *Die Philosophie des Robert Grosseteste* (*Beiträge zur Geschichte der Philosophie des Mittelalters*, XVIII, 4–6) (Münster: Aschendorff, 1917), 110–11.

27 Lindberg, "Applicability of Mathematics to Nature," 10–14.

28 David C. Lindberg, *Roger Bacon's Philosophy of Nature* (Oxford: Clarendon Press, 1983), xlix–liii.

29 *A Source Book in Medieval Science*, ed. Edward Grant (Cambridge, Mass.: Harvard University Press, 1974), 385.

30 Ibid., 386–88.

31 Ibid., 389, with revisions. I am grateful to W. R. Laird for calling my attention to a mistake in the published translation of this text. In denying that Aristotle offered a geometrical explanation of the rainbow, Grosseteste presumably means only that Aristotle did not go far enough or reach a correct conclusion; it was surely clear to Grosseteste that Aristotle had *attempted* at least a partial geometrical analysis.

32 *In Aristotelis posteriorum analyticorum libros* (Venice: Gregorius de Gregoriis, 1514), I.12, fol. 14r, Cf. Crombie, *Grosseteste*, 91–94.

33 William A. Wallace, *Causality and Scientific Explanation*, 2 vols. (Ann Arbor: University of Michigan Press, 1972–74), I, 28–47; James A. Weisheipl, O.P., "Classification of the Sciences in Medieval Thought," *Mediaeval Studies* 37 (1965):73–74; Lindberg, "Applicability of Mathematics to Nature," 10–14.

34 Bruce S. Eastwood, "Grosseteste's 'Quantitative' Law of Refraction: A Chapter in the History of Non-Experimental Science," *Journal of the History of Ideas* 38 (1967):403–14; Eastwood, "Mediaeval Empiricism: The Case of Grosseteste's Optics," *Speculum* 43 (1968):306–21; Crombie, *Grosseteste and Experimental Science*, 117–27; Lindberg, *Theories of Vision*, 100–102. For translations of these two treatises, see *Source Book*, ed. Grant, 385–91.

35 And there are no diagrams, at least in the extant manuscripts. However, a reference in *De lineis* to refraction "to the right" suggests that Grosseteste may have had one in mind; see *Source Book*, ed., Grant, 387.

36 On Albert's sources, see Lindberg, *Theories of Vision*, 106–7. On Albert's career, see James A. Weisheipl, O.P., "The Life and Works of St. Albert the Great," in *Albertus Magnus and the Sciences: Commemorative Essays 1980*, ed. James A. Weisheipl (Toronto: Pontifical Institute of Mediaeval Studies, 1980), 13–51.

37 *Opera omnia*, ed. Bernhard Geyer, 40 vols. in progress (Münster:

Aschendorff, 1951–), XVI, part 1, p. 2, lines 31–35. For an excellent discussion of Albert's position on mathematics, see A. G. Molland, "Mathematics in the Thought of Albertus Magnus," in *Albertus Magnus and the Sciences*, ed. Weisheipl, 463–78; also Benedict M. Ashley, O.P., "St. Albert and the Nature of Natural Science," in *Albertus Magnus*, ed. Weisheipl, 94–102.

38 Chap. 14, *Opera omnia*, ed. A. Borgnet, 38 vols. (Paris: Vivès, 1890–99), IX, 35. For a corrected text, see Cemil Akdogan, "Optics in Albert the Great's *De sensu et sensato*: An Edition, English Translation, and Analysis," Ph.D. dissertation, University of Wisconsin (1978), 101–2.

39 *De sensu et sensato*, chap. 14, in *Opera omnia*, ed. Borgnet, IX, 35; for a better, but still imperfect, text, see Akdogan, "Optics in Albert the Great's *De sensu*," 101.

40 *Opera omnia*, ed. Borgnet, IX, 35; Akdogan, "Optics in Albert the Great's *De sensu*," 102.

41 *Summa de creaturis* II, quest. 22, in *Opera omnia*, ed. Borgnet, XXXV, 210–28; Lindberg, *Theories of Vision*, 104–7.

42 *De sensu et sensato*, chap. 14, *Opera omnia*, ed. Borgnet, IX, 35.

43 Weisheipl, "Life and Works of St. Albert," 22, 35–36.

44 Jeremiah M. G. Hackett, "The Attitude of Roger Bacon to the *Scientia* of Albertus Magnus," in *Albertus Magnus and the Sciences*, ed. Weisheipl, 53–72.

45 *Fr. Rogeri Bacon Opera quaedam hactenus inedita*, ed. J. S. Brewer (London: Longman, Green, Longman, & Roberts, 1859), 38.

46 Ibid., 327. Note that although Grosseteste and Albert were about equally accomplished in *perspectiva*, Bacon was an enthusiastic admirer of the one and an intemperate critic of the other.

47 *Opera hactenus inedita*, ed. Robert Steele and Ferdinand M. Delorme, 16 fascicles (Oxford: Clarendon Press, 1905–40), XVI, 7.

48 IV.1.1, in *The "Opus Majus" of Roger Bacon*, ed. John Henry Bridges, 3 vols. (London: Williams & Norgate, 1900), I, 97–98.

49 *Opera*, ed. Brewer, 105, 111, 269.

50 *Opus maius*, IV.4.15, ed. Bridges, I, 168–69. For amplification, see Lindberg, "Applicability of Mathematics to Nature," 17–18.

51 Indeed, he admits in *Opus maius* IV.1.3, ed. Bridges, I, 106, that *propter quid* demonstrations are not available in metaphysics, moral philosophy, logic, or grammar.

52 The *Perspectiva* is part V of the *Opus maius*, edited by Bridges. Editions and translations of *De multiplicatione specierum* and *De speculis comburentibus* appear in Lindberg, *Bacon's Philosophy of Nature*.

53 On the history of this problem before Bacon, see David C. Lindberg, "The Theory of Pinhole Images from Antiquity to the Thirteenth Century," *Archive for History of Exact Sciences* 5 (1968):154–62.

54 For a full analysis, see David C. Lindberg, "A Reconsideration of Roger Bacon's Theory of Pinhole Images," *Archive for History of Exact Sciences* 6 (1970):214–23; Lindberg, "Laying the Foundations of Geometrical Optics: Maurolico, Kepler, and the Medieval Tradition," in David C. Lindberg and Geoffrey Cantor, *The Discourse of Light from the Middle Ages to the Enlightenment* (Los Angeles: William Andrews Clark Memorial Library, 1985), 13–25.

55 It is not clear which. Bacon frequently fails to draw a clear line of demarcation between psychology and physiology.

56 *Opus maius*, V.1, dist. 3, chap. 3, ed. Bridges, II, 24. Bacon's theory of vision is developed in ibid., V.1, distinctions 2–7, II, 12–53. For analysis, see Lindberg, *Theories of Vision*, 107–16.

57 *Opus maius*, V.1, dist. 6, chap. 2, ed. Bridges, II, 37–39; Lindberg, *Theories of Vision*, 109.

58 *Opus maius*, V.1, dist, 6, chap. 3, ed. Bridges, II, 39–42.

59 Ibid., V.1, dist. 7, chap. 4, ed. Bridges, II, 52–53; Lindberg, *Theories of Vision*, 114–16.

60 *Opus maius*, V.3, dist. 2, chaps. 2–3, ed. Bridges, II, 148–53. Bridges' figures are erroneous or incomplete in seven out of ten cases; for corrected drawings and a translation of the associated text, see *Source Book*, ed. Grant, 426–30. For additional figures, see Bacon's *De multiplicatione specierum*, in Lindberg, *Bacon's Philosophy of Nature*, 109–11, 241–43.

61 *De multiplicatione specierum* II.3, in Lindberg, *Bacon's Philosophy of Nature*, 111–17. For analysis, see Lindberg, "The Cause of Refraction in Medieval Optics," *British Journal for the History of Science* 4 (1968):30–34.

62 On several occasions Bacon explicitly announces his goal to be the extraction of opinions from the *auctores*; see the prologue to the revised version of his *De multiplicatione specierum*, in Lindberg, *Bacon's Philosophy of Nature*, 347; *Part of the Opus Tertium of Roger Bacon*, ed. A. G. Little (Aberdeen: Aberdeen University Press, 1912), 20. In both of these passages, he proceeds to list his principal sources.

63 For his immediate influence, see Lindberg, *Theories of Vision*, 116–20. For the long-term influence of the Baconian tradition on Maurolico and Kepler, see Lindberg, "Laying the Foundations of Geometrical Optics," esp. 49–53.

10

Mathematics and experiment in Witelo's Perspectiva

SABETAI UNGURU

That the "perspectiva magistri Witelonis de Viconia"[1] represents the most advanced, influential, and comprehensive optical treatise written by a Westerner until Kepler's *Ad Vitellionem paralipomena* (1604) and *Dioptrice* (1611) is by now generally accepted by historians of optical thought. That it is also largely unoriginal is similarly the consensus of researchers.[2] Conceptually displaying all the main earmarks of its great model, Alhazen's *Perspectiva* (*De aspectibus*), it stands out, however, in its systematic approach, encyclopedic character, and its inclusion of a first mathematical book, lacking in Alhazen, setting out the mathematical presuppositions needed for the remaining nine books. It also does not shirk using, whenever appropriate, an experiential, experimental approach.

In this chapter I wish to look into Witelo's uses of mathematics and experiment, and to draw the necessary conclusions about such usage in medieval optics.

I

Systematic compiler that he was, Witelo assembled, in Book I of his *Perspectiva,* the mathematical propositions needed in the development of the remaining nine optical books. In doing so he displayed an uncommon knowledge and mastery of geometry for his time, the 1270s, gathering together in his book, from myriad sources, all the mathematical information needed, beyond Euclid's *Elements,* in matters optical. His sources are, in addition to Euclid's *Elements* and Apollonius' *Conic Sections* (both of whom he mentioned by name, Witelo being one of the first Western scholars to cite Apollonius and to display some familiarity with the chief work of the "Great Geometer"[3]), first and foremost, Alhazen's *Perspectiva,* from which Witelo literally transcribed many propositions, typically increasing their word content and

269

diluting their conceptual sharpness; Campanus of Novara's edition of the *Elements;* Theon's additions in his recension of the *Elements;*[4] Eutocius' Commentary on Archimedes' *On the Sphere and the Cylinder;* Pappus' *Mathematical Collection;*[5] Jordanus' *Geometria;* Theon's recension of Euclid's *Optics;* Euclid's *Optica* and *Catroptica;* Theodosius' *Spherics;* and Serenus' *De selectione cylindri.* By all accounts, this is an impressive lot.

In addition to revealing a more systematic strategy in the elucidation of optical topics, the presence of a mathematical book as a preliminary ingress to what is basically an optical treatise betrays, on the one hand, a sharp awareness of the essential difference between mathematics and physics and, on the other hand, the conviction of the applicability of mathematics to the natural world. These conclusions are borne out by the not so rare instances when Witelo takes an optical proposition from Alhazen, reformulates it in geometrical terms, then proves it as a purely geometrical theorem.[6] Furthermore, sometimes one and the same proposition will appear in geometrical garb in Book I and later as an optical proposition in some of the remaining books.[7] Witelo's justification for such a procedure, which necessarily results in tedious repetitions, clearly implies both his conception of the first book as a mathematical treatise and his clear-cut distinction between mathematical and physical science.[8] Indeed, in the introduction to the *Perspectiva,* Witelo states unambiguously:

> And so we betook ourselves to divide the present undertaking into ten partial books. Indeed, desiring to infer by mathematical demonstration [the behavior of] any visible thing as pertains to the phenomenon of its visibility, and in this manner to go about [our job] with greater certainty, to the extent of our possibility, we produced this independently standing book, the only exceptions being the [inferences] depending on [propositions drawn] from the *Elements* of Euclid and a few [coming] from the *Elements of Conics* of Apollonius of Perga, which are the only two we have made use of in this discipline, as will become clear shortly in [the course of our] progression. Consequently we sent forward, in the first book of this science, axioms which are necessary to this branch of knowledge beyond the *Elements* of Euclid; and in this [book] we [also] make clear those two [conclusions] which were demonstrated by Apollonius. For some of those [things] which we have sent forward in this book are also contained in that [other] book which we call *De elementatis conclusionibus,* in which we drew up all [the items] that were seen by us and that reached us from men

posterior to Euclid for the need of particular disciplines universally conceived.[9]

As befits a mathematical opus, Book I displays a Euclidean structure with definitions, postulates, and propositions. The sixteen concepts defined by Witelo are: *pole, convex,* and *concave* line or *surface, perpendicular line* to a convex or concave surface, *intersecting circles, great* and *small circles of a sphere, equal spheres, parallel spheres* or *circles, tangent spheres, intersecting spheres, tangent plane surface* to a sphere, *denomination of a ratio,* and *compound ratio.* The five "postulates" begin with a theorem ("That [two] equal angles constructed about the same point have the same distance from equal [segments taken on their uncommon] sides [to their common side]"),[10] continue with the "expanded" equivalent of Euclid's first postulate (which, in addition to requiring that a line be determined by any two points on it, posits the determination of a surface by "any" [!] two lines),[11] then require that any two tangent plane surfaces coincide, that two plane surfaces do not determine a body, and, finally, that equal ratios be compounded and divided in similar ratios and have the same denominations.

There seems to be no definite criterion – or reasonably limited set of criteria – for the ordering of all the 137 propositions following the postulates. They represent a compilation of widely different geometrical theorems having to do with the mathematical knowledge that Witelo deemed necessary for dealing efficiently with the optical problems put forward in the remaining nine books of the *Perspectiva.* They contain mainly theorems of plane geometry but also a moderate number of theorems of solid geometry. They deal with properties of parallel and perpendicular lines, ratios and proportions, triangles, parallelograms, and circles; they prove some elementary attributes of planes and lines, intersection of two planes, perpendicular and oblique lines and planes, construction and division of angles, angles at the circumference and at the center, arcs and chords of circles, angles between tangent and chord, tangents to a circle, inscribed hexagon, and intersecting circles. They also discuss the sphere, great and small circles in the sphere, intersection of a sphere and a plane, tangent to a sphere, tangent plane to a sphere, concentric spheres, tangent and intersecting spheres, and so forth. Other topics discussed are cones and cylinders, elementary properties of conic sections, and some rather involved theorems – taken over from Alhazen – having to do with intersections of conic sections, circles, and lines; also a number of theorems discuss properties of what *we* would call harmonic pencils. The final propositions deal with complicated problems of construction involving lines, circles, and conic sections.

Witelo's mathematics is in many instances quite primitive, suffering from a chronic lack of rigor,[12] even by some thirteenth-century standards, namely, those of the two great mathematical prodigies of the century, Leonardo Fibonnaci and Jordanus Nemorarius. Nevertheless, with the exception of these two outstanding figures, the mathematics of Witelo's time was dominated by a group of, generally speaking, minor "mathematicians"; the likes of Sacrobosco, Alexander de Villa Dei, Vincent of Beauvais, and Roger Bacon. Compared with most of them, Witelo does not fare badly. He clearly was a child of his times[13] – perhaps even better than his times. He was no great mathematical genius; as a matter of fact, he was no genius at all. As a mathematician he was not even great. He wrote a mathematical book in which he assumes the knowledge of the *Elements*. Yet from time to time he tries to prove what is basically a proposition of the *Elements*, and this almost always leads him to circularity.[14] His list of definitions is not complete. He included in it those definitions which, for one reason or another, seemed to him most useful in the proofs to follow. As he relied heavily on Euclid in his demonstrations, the knowledge of whom he took for granted at least methodologically, he could appeal whenever necessary to a Euclidean definition not appearing in his list, exactly as he appealed systematically to Euclid's propositions in carrying out his proofs. Although this may be rightly taken by a modern mind to represent an unpardonable logical flaw, one must not forget that Witelo did not set out to write a book on Foundations of Mathematics. Indeed, the very idea of such a possibility in the thirteenth century is anachronistic.

As a natural philosopher who wrote an optical treatise destined to become one of the standard sources of optical thought in the centuries to come, Witelo merely prefaced his treatise with a mathematical book meant to give the reader the basic mathematics needed to comprehend the balance of his treatise. Although this procedure betrays a more systematic approach than that of Alhazen, his model, one should not expect it to compete with Euclid for the title of "the prototype of modern mathematical form."[15] Furthermore, Witelo's postulates do not fulfill the requirements of completeness and independence. As with his definitions, Witelo simply put down a number of postulates which seemed to him to be required by the propositions he was planning to prove. When in the course of his demonstrations the need arose to use something not granted by his postulates, this in itself could not have been a great reason for worry. As knowledge of Euclid was presupposed by Witelo, he could always evoke the required Euclidean proposition, thus bypassing the need to achieve completeness in his own postulates. As a matter of fact, strictly speaking, both his definitions and his postulates are quite superfluous. I do not doubt, however, that Witelo was aware of the incompleteness of his list of "basic assumptions." He could not avoid realizing this during his proofs. But, being systematically minded and wishing to structure his mathematical book on the pattern of the *Elements,* he decided to begin it anyhow with "some" definitions and postulates. Anything lacking in those, his ancient model would readily supply. Finally, it should be stated that many of Witelo's proofs in Book I are too loose to be taken seriously,[16] those which are rigorous being generally copied from the various sources that stood at his disposal.

What, then, is the significance of the mathematical book of the *Perspectiva?* Foremost, is its very existence as a preliminary background to optics. Next is that in the barrenness surrounding him, Witelo displayed in his book an uncommon knowledge and understanding of geometry. It is correct to say specifically about the first book what Montucla stated of the whole *Perspectiva:* "It reveals . . . in its author a knowledge of geometry unusual for the times when he lived."[17] Although essentially a compilation from widely different sources – some of which Witelo was among the first in the West to use, along with Eutocius, Apollonius, and quite plausibly Pappus – the *arrangement* of those sources and their *understanding* are clearly Witelo's achievements. His erudition, his drive to put to use all the information he could gather, coupled with his talents of exposition (which went together, however, with an undeniable bent toward loquacity), are also merits that explain the great success of his bulky treatise. I do not doubt that in order to write the first book (and the remaining nine as well) Witelo had to undertake extensive studies of mathematical (and physical) works over a long time and that these studies represent the foundation on which the many compilations constituting the *Perspectiva* have been intelligently assimilated in an impressive work. It is also possible that, in spite of the mainly compilatory character of Witelo's work, some of the theorems proved in the first book (the minor ones, probably) may be original with Witelo and exhibit a certain independent mathematical ability. Finally, another aspect of the significance of the (entire) *Perspectiva* is the great and lingering influence it exerted during the coming centuries on writers on geometrical optics and on others who cited it profusely and, eventually, included it in the university curriculum.[18]

II

A direct intimation about the nature of Witelo's mathematical knowledge and competence, which would corroborate essentially the above description, can be gained from firsthand exposure to one of his optical propositions and its moorings in the mathematical propositions of the first book. As such a proposition, II.9 has been selected because of its relative brevity, its reliance on more than one proposition of Book

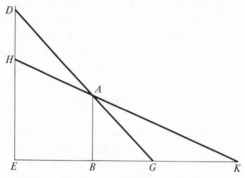

Figure 1. (*Persp.*, Book II, Fig. 5)

I, and its reasonable typicality. The enunciation of II.9 is: "The intersecting lines produced from the ends of the parallel altitudes of a higher luminous body and a lower umbrageous body are [respectively] proportional to their altitudes; from which it is obvious that the same altitude of an umbrageous body throws a larger shadow from a lower light than from a higher light."[19] And here is the proof:

Let line *AB* be the altitude of a certain umbrageous body ["Figure 1"] and let there be another altitude of a luminous body, namely *DE*, parallel to it. And let line *DE* be longer than line *AB*. And let lines *EB* and *DA* be drawn, which, [when] extended, will intersect on some side [of the parallel lines], in point *G*, by the sixteenth [prop.] of the first [book] of this [treatise]. I say that the ratio of line *GB* to line *GE* and [that] of line *GA* to line *GD* will be like [that] of line *AB* to line *DE*.

Indeed since line *BA* is parallel to line *DE* by hypothesis, it is therefore plain, by I.29 [*Elements*], that angle *GBA* is equal to angle *GED* and angle *GAB* is equal to angle *GDE*. Moreover angle *BGA* is common to both triangles *DGE* and *AGB*. Therefore, by VI.4 [*Elements*], the ratio of line *GB* to line *GE* is like [that] of line *BA* to line *ED*. Consequently, by the 5th [prop.] of the first [book] of this [treatise], inversely, the ratio of line *GE* to line *BG* will be like [that] of line *ED* to line *AB*. The proposed [thing] is therefore plain, since it can be proved in the same manner about lines *GA* and *GD*. And from this it is clear that the same altitude of the umbrageous body throws a longer shadow from a lower light than from a higher light.

In fact let it be that some luminous body is situated in point *H*. And let ray *HA* fall in a point of line *EG*, which is

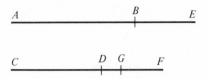

Figure 2. (*Persp.*, Book I, Fig. 5)

K. And, as previously, the ratio of *EK* to *BK* will be like [that of] *HE* to *AB*. But, by V.8 [*Elements*], the ratio of *HE* to *AB* is smaller than [that] of *DE* to *AB*. Therefore, by V.I1 [*Elements*], the ratio of *EK* to *BK* is smaller than [that of] *EG* to *BG*. Hence shadow *BK* has increased considerably with respect to shadow *BG*, as is clear by V.10 [*Elements*] and by the fourth [prop.] of the first [book] of this [treatise]. And on this account it happens that the lunar shadows are always longer than the solar shadows; and such is [the case] concerning any other higher and lower luminous bodies. The proposed [thing] is therefore clear.[20]

The proposition is straightforward and requires no comment. It uses three propositions from Book I, the mathematical book, namely 4, 5, and 16, which we should now consider.

[Proposition] 4. An addition of equal lines to two unequal lines of known ratio having been made, the ratio of the longer to the shorter is diminished [thereby].

Let there be two unequal lines, *AB* and *CD* ["Figure 2"], of known ratio. And let line *AB* be longer than line *CD*. Also let line *BE* be added to the same *AB* and line *DF* to the same *CD*, and let lines *BE* and *DF* be equal. I say that the ratio of line *AE* to line *CF* is smaller than [that] of line *AB* to line *CD*. For since three lines are given, namely *AB* and *CD* and *BE*, let there be constructed, according to the preceding [proposition], a line proportional to line *BE*, according to the ratio of the lines *AB* and *CD*, and let [this] line be *DG*. Since line *AB* is longer than line *CD*, it is evident that line *BE* is longer than line *DG*. Thus line *DF* is [also] longer than line *DG*. Therefore let a line equal to the same *DG* be cut off from line *DF*, by I,3 [Euclid]. Since the ratio of line *AB* to line *CD* is equal to [that] of line *BE* to line *DG*, by V,12 [Euclid], [it follows that] the ratio of the whole line *AE* to the whole line *CG* will be equal to [that] of line *AB* to line *CD*. But, by V,8 [Euclid], the ratio of line *AE* to the greater line *CF* is smaller than [that] to the smaller line *CG*. Hence the ratio of line *AB* to line *CD* is

Figure 3. (*Persp.*, Book I, Fig. 6)

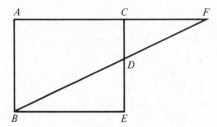

Figure 4. (*Persp.*, Book I, Fig. 17).

greater than [the ratio] of line *AE* to line *CF*. And this is what was proposed.[21]

[Proposition] 5. If the ratio of the first [magnitude] to the second is equal to that of the third to the fourth, [then] by inversion the ratio of the second to the first will be equal to [that] of the fourth to the third.

For let *A* be the first, *B* the second, *C* the third and *D* the fourth ["Figure 3"], and let the ratio of *A* to *B* be equal to [that of] *C* to *D*. I say that by inversion the ratio of *B* to *A* will be equal to [that] of *D* to *C*. Since the ratio of *A* to *B* is clearly equal to [that] of *C* to *D*, by V,16 [Euclid], alternately, the ratio of *A* to *C* will be equal to [that] of *B* to *D*. Hence the ratio of *B* to *D* is equal to [that] of *A* to *C*. Hence, by the same V,16 [Euclid], alternately, the ratio of *B* to *A* will be equal to [that] of *D* to *C*, that is [the ratio] of the second to the first is equal to [that] of the fourth to the third, which is what was proposed."[22]

[Proposition] 16. If straight lines are drawn through the ends of two parallel and unequal lines, it is necessary that they meet on the side of the smaller line.

Let there be ["Figure 4"] two lines *AB* and *CD*, parallel and unequal, and let line *CD* be less than line *AB;* and let lines *AC* and *BD* be drawn through the ends of the same. I say that those lines *AC* and *BD* will meet beyond the line

CD [i.e., on the other side of *CD* with respect to *AB*]. For let line *CD* be extended beyond point *D*, up to point *E*, and let line *CE*, by I,3 [Euclid], be equal to line *AB*, and let line *BE* be drawn. Now, this line *BE*, by I,33 [Euclid], is parallel to line *AC*. Hence, by the second [prop.] of this [book], as line *BD* meets line *BE* in point *B*, it is clear that the same meets line *AC*, which is parallel to line *BE*. But it is also necessary that it meets [*AC*] on the [other] side of line *CD* [with respect to *AB*], which *CD* is less than line *AB*, [all this] by the fourteenth [prop.] of this [book], or by VI,2 [Euclid].[23] What was proposed is therefore clear; indeed its point of intersection, which is *F*, will be beyond line *CD*.[24]

Once more, no comments are needed beyond what has already been said.

III

We should now examine the uses of experiment in the *Perspectiva* and the connections, if any, between experiment and mathematics in Witelo's methodology. The examples will be drawn mostly, though not exclusively, from Books II and III, which are truly representative of Witelo's attitude toward experiment. The conclusions I shall be reaching can be summed up as follows:

a. There was experiment in medieval optics.
b. There was awareness on Witelo's part of the difference between doing things experimentally and otherwise – mathematically, for instance.
c. "Proving" things experimentally is not a rarity in optics.
d. Witelo's theory of vision is medieval in spite of his experimental approach (where used).
e. Witelo's "experimental-inductive" methodology fits snugly in his theory of vision and does not lead to any breakthrough.
f. To have a breakthrough you need a new theory (Kepler), not experiments.
g. As D. C. Lindberg has shown, Kepler's achievement in optics cannot be understood without a thorough understanding of the medieval background from which he grew.
h. The background – including the experiments, for whatever they are worth – is necessary, but obviously not sufficient.

It is necessary now to say a few words concerning the use of the term "experiment." Even though *experimentum* and *experientia* were used almost interchangeably throughout the Middle Ages all the way to the Renaissance, a growing sharpening of conceptual distinctions took place gradually, and various scholars distinguished between (1) mere experience, observation, and (2) artificially set-up empirical situations, not occurring normally in nature, for example, searched-for, sought-after experience. The latter is *experiment*. Indeed, Francis Bacon, that much maligned "naive inductivist," put it this time too, as in so many other cases, pregnantly and succinctly: "It remains but mere experience, which, when it occurs, is called a chance happening; if it is sought after, [however], it is called experiment."[25] The following identity is, then, established:

Experientia quaesita ≡ experimentum

Various gropings toward this recognition include Peter Maricourt's "industria manuum" (it was Peter Peregrinus whom Roger Bacon referred to in the *Opus tertium* as "dominus experimentorum"), Roger Bacon's own "scientia experimentalis," Thomas Aquinas' "scientia experientie" or its identical twin "scientia experimentalis," Nicolas of Cusa's "experimenti statici," etc., etc. It is common knowledge that qualitative experiments (and not only qualitative) were already known in ancient times, in medicine and alchemy and mechanics, among other departments of knowledge, while truly quantitative experiments make increasingly their appearance in the eleventh, twelfth, and thirteenth centuries in Islamic science and in its Western counterpart, especially through the use of the balance (Abu-r-Baihan and his determination of specific weights) and other measuring devices (Roger Bacon, Witelo, and Nicolas of Cusa).

Illustrating the variety and fluidity of thirteenth-century attitudes to experiment are the views of Saint Thomas and Roger Bacon. The former, while failing to discriminate between "scientia experientie sive scientia adquisita" and "scientia experimentalis," remained largely within the Aristotelian framework when it came to his understanding of experiment (experience) as arising out of repeated memories:

> Again, Isidore says that devils know many things by experience. But experimental [experiential] knowledge is discursive; for out of many memories comes one experience and from many experiences [experiments] there results one universal, as is said [by Aristotle] at the end of the *Posterior Analytics* and at the beginning of the *Metaphysics*.
> Therefore the knowledge of angels is discursive.[26]

And: "Again, Isidore says that devils know many things by experience. For experience consists of many memories, as is said in the *Metaphysics*. Therefore there is in them [i.e., angels] also a potentiality for memory."[27] This, needless to state, is a far cry from *experiment* and *experimental science* as harbingers of modern science. Such examples can be multiplied: "experience originates from the senses;"[28] "in actions [in matters of conduct] experience causes not only knowledge [science] but also a certain habit, as a result of usage [custom] which facilitates action [conduct]."[29] "It must be said that to tempt [i.e., to put to a test] is properly speaking to make a trial of something. Now we make a trial of something in order to know something about it; hence the immediate goal of every tempter is knowledge."[30] Finally, for the use of *experimentum* in the sense of "proof" we have: "Si ergo ratio humana sufficienter experimentum praebet."[31] All these examples are drawn from the *Summa Theologiae*. They are only confirmed and strengthened by Saint Thomas's other works.

Roger Bacon, on the other hand, waxes grandiloquent when speaking of experiment and experimental science. He also seems to reach Nirvana when discussing mathematics. The trouble is that it is not easy to reconcile Bacon's sophisticated theory with his extremely simplistic practice in these matters. According to him, experimental science is the most powerful of the sciences in its own right, but also a tool for use in the other sciences. As a tool, it has a double "dignity": the verification of the conclusions of the other sciences and the attainment of new truths which, though lying within the domains of those sciences, are outside the competence of their methods of investigation. Additionally, in its own right, it discovers marvels in nature and harnesses them to human needs, especially ecclesiastical, through the practical arts.

Following in Aristotle's footsteps,[32] Roger Bacon sees the origin of science in man's inborn curiosity. Furthermore, knowledge (science) is attainable in three ways, by authority, reason, and experience:

> But authority has no savor, unless the reason for it is given,
> and it does not give understanding but belief. For we believe
> on the strength of authority, but we do not understand
> through it, nor can reason distinguish between sophism and
> demonstration, unless we know how to test the conclusion
> by works, as I will . . . show in the experimental sciences.[33]

Through experience we face literally truth. Our questions are answered directly by experience. Moreover, the fundamental sense in experimental science is vision. Besides, experience is needed in all the sciences as a precondition for achieving understanding with absolute certainty. This is so even in the "abstract" science of mathematics in which drawing of figures and counting are appealing to the senses and providing the test for the truthfulness achieved by demonstration. Error-free is not the same as doubt-free. Bacon interprets Aristotle's statement that "proof is reasoning that causes us to know" as necessarily entailing "provided the proof is accompanied by appropriate experience."[34] This is the key for Bacon's belief in the value of experience to mathematics. Mathematics contains certifying experience as an *intrinsic* ingredient of its basic operations (drawing of figures and counting). Experience and mathematics are mutually necessary to one another. They are essentially complementary in Bacon's scheme: Mathematics supplies errorless truth while experience furnishes doubtless certainty. Concerning the sequence "belief – experience (experiment) – reason", Bacon says:

> Hence in the first place there should be readiness to believe, until in the second place experience follows, so that in the third reasoning may function. For if a man is without experience that a magnet attracts iron, and has not heard from others that it attracts, he will never discover this fact before an experiment. Therefore in the beginning he must believe those who have made the experiment or who have reliable information from experimenters [experts].[35]

Also, in mathematics itself, "without mathematical [i.e., astronomical] instruments nothing can be known, and these instruments are not made by the Latins and could not indeed be made for 200 pounds, or even 300."[36] As a coordinator of knowledge, mathematics specifies the quantifiable elements in nature. It grants the scholar the needed certainty, since it can confirm his observational perceptions. Only through mathematics can he determine what properties and causes in nature are susceptible of scientific investigation. Moreover, mathematics is the basis of all logical categories. It follows, then, that logic, too, is subordinated to mathematics.

Experience, of its own nature, is confined to the concrete, the particular. Mathematics, of its own nature, is confined to the universal. According to Bacon, a knowledge of both the particular and the universal is essential for a complete understanding of science and it was therefore perfectly possible to say both that experience is essential for all science and that mathematics is the "key to the sciences."

In practice, Bacon follows his own theoretical precepts very little. Most of the "experiments" he cites are little more than commonsense observations. Besides, he attached minimal importance to experiment as a tool in induction, thinking of it almost exclusively as a test of preconceived – or prereceived – ideas which he considered essential first steps to scientific understanding.[37]

IV

There are numerous instances of appeal to and use of experiments and experience in Witelo's *Perspectiva*. The very first proposition of Book II is a case in point. It reads: "All luminous rays, as well as the multiplication of forms, stretch forth in straight lines."[38] The proof is exceedingly long, occupying almost two full pages in the Risner edition of 1572. It starts as follows: "What is proposed here can be made known not by a demonstration but rather instrumentally; indeed, the diversity [of the attempts] of the ancients to prove this made use of a variety of instruments, while we are using that which we describe below, which we believe to harmonize better and more legitimately with what is proposed here."[39]

Witelo continues with a very detailed description of the construction of his (Alhazen's) instrument, which in Alhazen is called *"organum refractionis"* and which Witelo uses both for the determination of the rectilinearity of propagation of light and colors and for his measurements of angles of refraction between various media, an outline of which is given by Crombie.[40] What interests us here, however, are simply those passages which show Witelo's "hands-on," pragmatic, matter-of-fact approach. Some instances should suffice:

> Let there be taken, therefore, a round bronze vessel, sufficiently thick (like the mother of an astrolabe), the width of whose bottom is one cubit . . . and let the height of its edge be equal to the width of two inches . . . and let the vessel be placed . . . in a lathe, and let it be shaped by turning until its periphery be truly round both extrinsically and intrinsically, and let its plane surfaces be levelled, and let the column-like body, which is the middle of the back, be [also] made round. . . . And then let the vessel be brought back to the lathe and let three parallel circles be marked in it. . . . And so let the middle one of these circles be divided into 360 parts and, should it be possible, into minutes. Then . . . let a round opening be drilled. . . . Then let a plane, somewhat thick, bronze plate be taken. . . . And let it be smoothed down . . . and let the plate be bored through with a round opening. . . . Then let the little plate be fastened to the bottom of the vessel. . . . Then let a bronze quadrangular rule be taken . . . and let its surfaces be equalized till they are equal to [the surface] of a rectangle. . . . Then . . . let there be made a round opening, whose size is fit for the body which is in the back of the instrument . . . and let it be such that it revolve in the same instrument. . . . And let two little strips be made . . . which should be consolidated over the extremities of the ruler . . . and let a pin be inserted holding together the ruler with the instrument.[41]

This is more than enough. We can, then, clearly say with Witelo: "Et hoc est propositum."[42]

This is also, perhaps, the place to point out that Witelo uses this very *"organum refractionis"* in Book X to prove *experimentally,* as he puts it, his various conclusions about refraction. Thus, in proposition X.4, where he deals with the relationship between the angles of incidence and refraction, he says: "What is proposed here can be proved instrumentally, so that [there be] a demonstration [that] is expressed sensibly by means of an instrument."[43] Furthermore, the enunciations of propositions X.5,6, and 7 read respectively:

> "To determine experimentally the sizes of the angles of refraction from air to water."[44]
>
> "To determine experimentally the sizes of the angles of refraction from air or water to plane or convex glass and vice versa."[45]
>
> "To find out experimentally the sizes of the angles of refraction from air or water to concave glass and vice versa."[46] And concerning propositions X.10 and 11 he states respectively: "This is clear by experience" and "What is proposed here is obvious by reason and experience."[47]

Finally, in X.42 one finds the statement "What is proposed here is sufficiently clear from the preceding; but it pleases me to show the same experimentally and [to unravel] the universal cause by means of a particular example;"[48] in X.48 the proof ends with the words "But in experiencing these [things] there is also great latitude, which we, [however], leave to such of curious mind;"[49] and in his study of the rainbow,[50] there is repeated appeal to experience and experiment in proving his conclusions.

I should not omit to point out here that some of the claims made on behalf of Witelo's empiricism, especially by Crombie, are often quite exaggerated, if not outright wrong. A case in point is represented by proposition X.8, in which Witelo reproduces in tabular form results he allegedly obtained experimentally in measuring angles of incidence and refraction between different media – air, water, and glass (all six possible arrangements). It is clear in this particular case that Witelo lied and Crombie was taken in. As shown by Albert Lejeune and David Lindberg,[51] Witelo's table is copied in its entirety from Ptolemy's *Optics* and additionally garnished with absurd data when purportedly measuring angles of refraction from the denser to the rarer medium, which could *not* have been obtained by actual measurements. Witelo's "empirical" numbers, then, are due to (1) his misunderstanding of the "reciprocal law" of refraction (which he nevertheless duly states after Alhazen) and (2) his failure to grasp the phenomenon of total internal reflection. Clearly, empiricism has its dangers to which neither perspectivists nor historians of optics are immune.

Pursuing now our survey of Witelo's appeal to experiment, I would like to illustrate it with proposition II.5, which, because of its shortness, lends itself to being reproduced in full:

> Lights and colors do not blend in transparent bodies but penetrate [them] separately.

The reason for this thing is to be shown experimentally. Let many locally distinct candles be placed in a certain place and let them all be opposite to one aperture leading to a darkened place and let a certain non-transparent body be placed opposite the aperture in the darkened place. And so the lights of the candles appear over that body separately according to the number of the candles, and any of those [lights] appears opposite to one candle, in accordance with the straight line passing through the aperture and through the middle of the light of the candle. And if one candle is covered completely only one light, [which is] opposite to that candle, will be destroyed and should the candle be uncovered, the light returns. And so it is obvious that in the middle of the aperture, where all or many [lights] penetrate one another in one point, the lights do not blend in the same point but are separate according to their essences; and on account of this, when extended later, they are distinguishable locally according to their diversity, by the places in which they fall. And since light traversing colored things gets colored by those colors, as was supposed, it is obvious that if light penetrates separately, the colors too, which are carried with the light, will penetrate separately. What is proposed is therefore clear.[52]

Since the empirical setup is obvious and the proposition speaks for itself, no further comment is really necessary.

The next use of experiment in Book II is in proposition 24, which reads: "Every luminous body illuminates a smaller place from which it does not escape more strongly than a greater space."[53] The reader can be spared the "proof" this time, but not without first being made aware that, in this case, Witelo calls his procedure *per exemplum*.[54] Our next instance is II.42. In it Witelo proves that a perpendicular ray penetrates a body denser than the one it previously traversed without being refracted. The proof is long and does not really concern us. It is performed "instrumentally" according to the following statement: "The proof of this proposition rests more on an instrumental endeavor than on other demonstrations."[55] Then follows a long and detailed description of the experimental arrangement and procedure involved in establishing the truth of the proposition by means of Alhazen's *obiectum refractionis,* in which the practitioner is referred to as the *experimentator*[56] and which ends with the following words: "From which it is clear that the passage of light through the body of water is by straight lines, by XI.1 [Euclid], and this is what we intended to show experimentally about the intended proposition."[57]

In proposition II.43, Witelo discusses the behavior of oblique rays issuing from a rarer into a denser medium, namely their refraction toward the perpendicular. Thus he says: "This proposed theorem can also be shown experimentally [to be true],"[58] which he indeed does, relying on the procedural setup of the previous proposition.

The following proposition, II.44, occupying almost two full pages in Risner's edition, deals with the lack of refraction of perpendicular rays coming from a denser medium (glass) to a rarer one (air or water). The exceedingly lengthy and loquacious proof, a gem of Witelo's phatic style, involving, this time, too, an experimental arrangement, proceeds according to the initial statement: "The proposed theorem can be similarly shown [to be true] by instrumental experience."[59]

The next two propositions of Book II (45 and 46) are both tackled and disposed of experimentally. Thus, II.45 discusses the refraction of oblique rays from a denser to a rarer medium according to the methodological statement "What is now proposed here is to be shown by instrumental experience in conformity with the preceding [propositions]."[60] II.46 reads: "It is necessary that every incident and refracted ray be situated in the same plane surface."[61] The proof, relatively succinct for a change, starts with the declaratory statement "But even that which is now proposed can be shown experimentally"[62] and proceeds directly by reliance on the experimental set up of II.43, ending effectively with the words "this experiment [test] is sufficiently evident each time."[63] Finally, II.47 represents an attempt at summing up and taking stock of the results of propositions II.42–II.45. In it Witelo makes it abundantly clear that he is fully aware of the peculiar experimental approach he adopted in the preceding propositions, which he desires, nevertheless, to supplement here by means of more general, physical, and philosophical considerations, coupled with common-sense inferences applied to the nature of light and motion. Thus he says:

> That which was proved so far instrumentally by means of
> specific trials, we intend to enhance by a natural
> demonstration. Indeed all natural motions which are made
> according to perpendicular lines are stronger because they
> are intensified by the universal celestial virtue flowing into
> every body lying beneath [the heavens] along the shortest
> straight line.[64]

And in the course of his discussion, which relies, among other things, on reasoning by *reductio ad absurdum,* one also finds the following remarks: "And so it is plain, according to the previous reasoning concerning the strength of the perpendiculars and by [various] instrumental trials, [specifically] by the 42nd and 44th [propositions] of this [book], that the ray [which is] incident . . . perpendicularly penetrates the whole body,"[65] and "Natura autem frustra nihil agit."[66] It is certain, therefore, that Witelo engages in experiments fully aware of their *sui generis* nature in his work, the overwhelming character of which is theoretical and deductive.

Compared with Book II, Book III has little, if any, experiments, but plenty of observations and experience. To illustrate, appeal to the latter meets the eye in proposition 3, where it is proved that the eye is spherical and in which it is said: "And since this is false and against a notion which is clear to the [visual] sense, because it is possible for a thing greater than the eye itself to be seen, it is plain that it is not possible for the surface of the visual organ to be plane;"[67] in proposition 4, reading "The eye is the spherical organ of visual power constituted of three humors and four tunics issuing from the substance of the brain [and] arranged spherically,"[68] in which one finds the statements "In what manner the eye may be the organ of visual power, we leave to the labor of another part of [natural] philosophy. That it is indeed spherical is necessary by the preceding proposition and also [follows] from [the fact] that it is of a watery nature, the property of which is always to become rounded. . . . Moreover the assiduous concern of the anatomists has taught [us] thoroughly that the eye is constituted of three humors and four tunics,"[69] and "Hence it is thus clear that the humors and tunics of the eye are fixed in spherical channels, and the statement of the proposed definition of the eye is clear, according to the experience of all those who have written thus far about the anatomy of the same";[70] in proposition 6, in which it is established that the visible forms *act* on the eye and in which it is said, "Indeed the eye suffers from strong light as, [for example], at the sight of the solar body or of another strong light, say the light reflected to the eye from a polished body or from another exceedingly white body";[71] and, finally, in proposition 16, reading, "Vision does not take place without pain and suffering endured by the substance of the eye. From which it is clear that the eye ought to be of an adequate disposition in its health in order to prosecute completely the [process of] vision"[72], in which appeal is repeatedly made to observation of the eye's reaction to strong light.

These are not all the instances in Book III, or in the rest of the *Perspectiva,* in which experience and/or experiments make their pivotal appearance. But for obvious reasons, the foregoing examples should suffice.

One other concern that is discernible in this book and that certainly deserves to be pointed out is Witelo's effort to show, primarily in proposition III.60 but with reverberations in the following few propositions, that, in a sense, the naked eye is blind:

> In fact the distinction between those two whitenesses does
> not itself pertain to the sensation of whiteness, because the

sensation of whiteness stems from the whitening of the eye's surface, which is accomplished by any whiteness, while the distinction of those [two] whites is achieved on account of the difference of action of those two whites in the same eye. Hence that distinction does not [stem] from the sense alone, but it is [also a result stemming] from another power of the soul, which we call the discerning [power]. *And it is the same with the comparison and discrimination of other sensible forms; indeed nothing of those [forms] is perceived by sight alone, but by the collaboration of reasoning and the discerning virtue. For sight by itself has no discerning power, but [only] the discerning power of the soul distinguishes all those [things] by means of sight.* The proposed [thing] is, therefore, clear.[73]

All this shows, I submit, among other things, Witelo's methodological sophistication. It is clear that he is no blind, naive empiricist.

V

On the whole, Crombie's assessment of Witelo's methods seems to me accurate even if one can discern now and then a tendency to exaggerate Witelo's "inductivism" and to lend too much credence to his rhetorical "manual and technical skill" or to his "quantitative experiments."[74] It is thus quite all right to speak of "Witelo's effective combination . . . of experiment with geometry, of manual skill with rational analysis and synthesis,"[75] as it is to point out that "The methods by which the various modes of operation of these [visual] forms were to be investigated were, according to Witelo, by observation and experiment and by mathematics,"[76] even though one might quarrel with this particular enumerative order.

Furthermore, Crombie's thorough study of Witelo's theory of the rainbow led the former to state convincingly:

The precise manner in which the different colours were produced by the incorporation of darkness in the rays Witelo then investigated by means of experiments with refraction through crystals and spherical glass vessels filled with water.[77]

He [Witelo] went on to describe further experiments on refraction [by producing artificial rainbows through light passing through hexagonal crystals and spherical glass flasks filled with water]. The subject was unexplored and experiment was the guide: "For the colour or visible form is carried to vision only by the nature of light which it contains; and to what has been said the careful inquirer will be able by experiment to add many things."[78]

However, needless to emphasize, Witelo remained a medieval perspectivist in spite of his thoroughness, erudition, *and* alleged "experimental-inductivist" bent. His theory of vision is medieval, not modern. In the overwhelming majority of instances, his approach is not that of the inductivist, the experimenter, but, on the contrary, of the deductivist, the theoretician. His use of experiments, which is certainly present, is not, however, undertaken in order to originate previously unknown conclusions, but rather in order to establish, support, or test the accuracy and appropriateness of convictions held beforehand. What Crombie said about Witelo's discussions of lenses applies throughout Books II and III and – I venture the not totally uneducated guess – throughout most of the *Perspectiva:* Witelo's conclusions stem from theoretical considerations alone, while in many cases the requisite experimentation is left to the curious reader (*ingenio perquirentis*).[79] And so it is not at all surprising that even though there were "careful inquirers" after Witelo who were able, perhaps also by "experiment to add many things,"[80] what they did not, and could not, add by experiment, as long as they remained thoroughly committed to all the elements of the medieval theory of vision, was its modern modification, the Keplerian theory.

This theory, to state the obvious, is *not* an outgrowth of an experimental, inductivist methodology. As A. Koyré has put it, "*too much methodology is dangerous and . . . more often than not . . . results in sterility. . . .* No science has ever started with a *tractatus de methodo* and progressed in the application of such an abstractly devised method."[81] Kepler's breakthrough and the beginning of modern optics are no exceptions to Koyré's assessment. On the contrary. David S. Lindberg has shown[82] that Kepler's contribution to optical thought can best be understood by seeing it as the continuation and significant modification of the medieval theory, of which it is the culmination. No role to speak of is played in it either by induction or experimentation as such.

Thus, and I am relying heavily on Lindberg's book, Kepler denied that the impressions allegedly left by light on the lens, which was seen as connected directly to the retina and the optic nerve by means of the aranea, are transmitted to the brain through the intermediary of the retina and the optic nerve. This he did by pointing out, first, that the capsule of the aranea surrounding the lens was not connected to the retina, as was shown by Felix Platter (1536–1614), the lens being in touch only with the uvea, and, second, by negating that sight was a form of touch, light and color traversing the eye instantaneously.[83] Moreover, the shape of the lens differs in its posterior surface from the shape assigned to it by Witelo and the perspectivists, in order to avoid intersection of rays at the center of the eye and, consequently, image inversion. The posterior surface of the crystalline humor is more curved, gibbous than the anterior surface (i.e., its radius of curvature is smaller than that of the anterior surface) and the image thrown on the retina is not only inverted but also reversed with respect to the object. Also, clearly, the eye sees more than the hemisphere of rays perpendicular to it, as can be verified by the images of objects at the very periphery of the visual field that enter the eye obliquely. So very oblique rays do leave an impression in the eye. What, then, about not so oblique rays, but such that while oblique, are very close to the perpendicular? Shouldn't they leave a quite powerful impression, because only slightly weakened? How, then, avoid total confusion on the very basis of the perspectivist theory of the visual image itself?

As shown by Lindberg, in anatomical ocular matters Kepler was, like Alhazen earlier, parasitic on the results of contemporary anatomists. He used primarily the works of Felix Platter (who made the retina and the optic nerve the *principal* seat of sensitivity of the eye) and Johannes Jessen (his Prague friend who, though writing later than Platter, stuck to the old views of crystalline photosensitivity). Concerning the geometry of the eye and its influence on light rays, Kepler bases his conclusions on an analysis of the functioning of spherical transparent lenses, assuming the anterior part of the eye (aqueous humor and the anterior surface of the crystalline humor) to be almost spherical in combination, so that rays entering it almost perpendicularly would undergo practically no refraction after they had already been refracted once at the cornea. Thus, to each point in the visual field, serving as the apex of a cone, there corresponds the base of that cone on the anterior surface of the lens; all the rays forming the base are again refracted when they emerge from the posterior surface of the lens (which is hyperbolic, not spherical) into the vitreous humor, away from the perpendicular, reaching together a new vertex at a point on the retina that serves as the new apex of another cone having the same base as the first. There is thus a one-to-one correspondence between points of the object and points of the *retinal* image by means of two cones having a common base and two vertices, such that the image-vertices are inverted and reversed with respect to the object-vertices. After that the image formed on the retina will be interpreted in the brain by means of the visual spirit carried by the optic nerve, this being a physicopsychological problem and not an optical one, thus not lending itself to geometrical analysis.

Kepler's contribution, then, according to Lindberg, is the result of many factors, among which one must mention seeing the eye as a *camera obscura,* a powerful and skillful talent at ray geometry and, crucially, a complete grasp and penetration of the perspectivist tradition, coupled with the readiness to answer the outstanding optical problems of the seventeenth century within the overall framework set by perspectivist optics. Having rejected, for good reasons, the perspectivist cone of perpendicular rays, Kepler still had to come up with a one-to-one correspondence between object-points and image-points. This he did by the "two-cone" mathematical analysis of ray geometry and the theory of retinal image, in which he cleaved to the main ingredients of perspectivist optics, which were, in keeping with Lindberg, "the principle of punctiform analysis, the laws of propagation of light, the requirement that one point in the visual field must stimulate one and only one point within the eye, basic conceptions about the nature of seeing, and the commitment to a methodology that incorporates mathematical, physical, and physiological reasoning."[84]

Let me conclude this brief excursus on Kepler with the following quotation from Lindberg's book:

> Kepler was the culminating figure in the perspectivist tradition. . . . That his theory of vision had revolutionary implications, which would be unfolded in the course of the seventeenth century, must not be allowed to obscure the fact that Kepler himself remained firmly within the medieval framework. The theory of the retinal image constituted an alteration in the superstructure of visual theory; at bottom it remained solidly upon a medieval foundation. Kepler attacked the problem of vision with greater skill than had theretofore been applied to it, but he did so without departing from the basic aims and criteria of visual theory established by Alhazen in the eleventh century . . . his theory of vision was not anticipated by medieval scholars; nor did he formulate his theory out of reaction to, or as a repudiation of, the medieval achievement. Rather, Kepler presented a new solution (but not a new kind of solution) to a medieval problem, defined six hundred years earlier by Alhazen. By taking the medieval tradition seriously, by accepting its most basic assumptions but insisting upon more rigor and consistency than the medieval perspectivists themselves had been able to achieve, he was able to perfect it.[85]

VI

It seems to me, then, that in optics (perspective), as in other branches of theoretical knowledge, experiment plays only a minor, derivative role.[86] Moreover, I tend to agree with Koyré that until the development of scientific technology, theoretical and practical achievements were largely independent in the domain of optical thought. The progress of optics in the thirteenth and fourteenth centuries was not determined by methodological considerations but rather, and crucially, by the availability of Alhazen's *De aspectibus* (*Perspectiva*), with its new departures in optical thinking, and by the sharpening and convergence of issues growing out of its sophisticated analysis.

Witelo's use of experiment, like that of Alhazen, his model, stands neither at the origins of a methodological revolution nor issues from such a revolution. As my examples have abundantly shown – and they represent only a lean chrestomathy – Witelo employs experiment in a plain, straightforward, nondramatic way, an almost pedestrian way. It is clear to the careful reader of his *Perspectiva* that experiment enjoys no special status with him. On the contrary. It seems to crop up when no satisfactory alternative is in sight. I would say that Witelo *prefers* the theoretical, deductive approach and that experiment is his second-best choice. There is no fanfare or particular pride in its use, which is quite natural and normal under the given circumstances. It results in no departure from Witelo's theoretical commitments, which it is rather meant to buttress and authenticate. Witelo seems to me to appeal to experiment *faute de mieux;* he would rather give a mathematical proof if he could and if it were at all suitable. And so I find myself, again, in basic agreement with A. Koyré:

> As for myself, I don't believe in the explanation of the birth
> and development of modern science by the human mind
> turning away from theory to *praxis*. I have always felt that it
> did not fit the real development of scientific thought, even in
> the seventeenth century; it seems for me to fit even less that
> of the thirteenth and fourteenth. I don't deny, of course,
> that in spite of their alleged – and often real –
> 'otherworldliness,' the Middle Ages, or to be more exact, a
> certain, and even a rather large number of people during the
> Middle Ages, *were* interested in techniques; nor that they
> gave to mankind a certain number of highly important
> inventions. . . . Yet, as a matter of fact, the invention of the
> plough, of the horse harness, of the crank, and of the stern
> rudder had nothing to do with scientific development; even
> such technical marvels as the Gothic arch, stained glass, the
> foliot or the fusee of late medieval clocks and watches did
> not depend on, nor result in, any progress in corresponding
> scientific theories. Strange as it may seem, even such a
> revolutionary discovery as that of firearms has had no more
> scientific effect than it had scientific bases. Bullets and
> cannon balls brought down feudalism and medieval castles,
> but medieval dynamics resisted the impact. Indeed, if
> practical interest were the necessary and sufficient
> precondition of experimental science – in our sense of the

word – this science would have been created a thousand years, at least, before Robert Grosseteste, by the engineers of the Roman Empire, if not by those of the Roman Republic.[87]

The implications of what I have presented here strike me as both obvious and nonseminal. They should not be controversial. My excuse, then, for this evidential presentation is simple: I thought it worth saying. My methodological guide was G. K. Chesterton. In Kingsley Amis' words in his introduction to a selection of Chesterton's stories,[88] it is said that

Chesterton's stance on most matters could be summed up very roughly as follows. What is simple, generally agreed, old and obvious is not only more likely to be true than what is complex, original, new and subtle, but much more interesting as well: a prescription calculated to alienate almost any type of progressive thinker.

As to Witelo's use of an experimental methodology in his *Perspectiva,* my reaction can best be stated in the inimitable words of Ambrose Bierce, the author of *The Devil's Dictionary:* "Example is better than following it" and "Where there is a will there is a won't."

Notes

1 Explicit of Bern MS 61, fol. 318v., Bürgerbibliothek Bern. In the most recent biography of Witelo, published by Jerzy Burchardt (*Witelo Filosofo della Natura des XIII Sec: Una Biografia* (Wroclaw: Ossolineum, 1984), appearing in the series *Conferenze Pubblicate a Cura dell'Accademia Polacca delle Scienze, Bibliotheca e Centro di Studi a Roma,* No. 87), the author adopts afresh the old suggestion that Witelo ended his life as a Premonstratensian monk at the Abbey of Vicogne, but not before, Burchardt argues, completing some fourteen years of diplomatic employment, mainly in the service of the Silesian prince Henryk Probus IV. Burchardt's biography develops further ideas propounded in his article "Zwiazki Witelona z Wroclawiem," *Sobotka* 4 (1974): 445–456, and supplements this with a rendition of four documents bearing on his hero's life. The biography is exceedingly learned, full of details having any conceivable bearing on Witelo's life and career, involving an incredible amount of research effort in its preparation, but leaving, largely, the main signposts of that life and career untouched.
2 For some of the latest scholarship on this issue, see D. C. Lindberg's "Introduction" to a facsimile reprint of the Risner, 1572, edition of the *Perspectivae* of Alhazen and Witelo: *Opticae Thesaurus Alhazeni Arabis Libri Septem, Nuncprimum Editi, Eiusdem Liber de Crepusculis et Nubium Ascensionibus. Item Vitellonis Thuringopoloni Libri X* (New York: Johnson Reprint, 1972), pp. V–XXXIV. (Hereafter, references to the *Perspectivae* of Alhazen and Witelo in this edition will be given, respectively, as *Opt. Thes. Alh.* and *Opt. Thes. Wit.*); Sabetai Unguru, *Witelonis Perspectivae Liber Primus.* Book I of Witelo's *Perspectiva.* An English Translation with

Introduction and Commentary and Latin Edition of the mathematical book of Witelo's *Perspectiva* (Wroclaw: Polish Academy of Sciences Press, *Studia Copernicana* XV, 1977); A. Mark Smith, *Witelonis Perspectivae Liber Quintus*. Book V of Witelo's *Perspectiva*. An English Translation with Introduction and Commentary and Latin Edition of the First Catoptrical Book of Witelo's *Perspectiva* (Wroclaw: Polish Academy of Sciences Press, *Studia Copernicana* XXIII, 1983).

3 See Sabetai Unguru, "A Very Early Acquaintance with Apollonius of Perga's Treatise on Conic Sections in the Latin West," *Centaurus* 20, no. 2 (1976):112–128.

4 Cf. my edition of Book I, cited above in note 2.

5 Cf. my "Pappus in the Thirteenth Century in the Latin West," *Archive for History of Exact Sciences* 13, no. 4 (1974):307–324.

6 Compare for instance, proposition I.21 in Witelo with V.5 in Alhazen (*Opt. Thes. Wit.*, pp. 10–11, and *Opt. Thes. Alh.*, p. 128).

7 Props. I.113 and I.114, for instance, reappear respectively as props. VII.45 and VII.44 (*Opt. Thes. Wit.*, pp. 43–45, 294–297).

8 This is how Witelo justifies his *modus operandi* in the case of prop. VII.44: "Hoc quod hic proponitur demonstrandum, patet per 114 th. 1 huius: ut autem huic nostro proposito conclusio mathematica sensibiliter applicetur, eandem demonstrationem duximus iterandam" (*Opt. Thes. Wit.*, p. 294).

9 *Opt. Thes. Wit.*, p. 2:
   > Praesens itaque negotium decem libri partialibus duximus distinguendum. Volentes enim omne ens visibile, ut suae visibilitati passio accidit, mathematica demonstratione concludere, et hac via eatenus (ut nobis est possibile) certius ambulare: librum hunc per se stantem effecimus, exceptis his, quae ex elementis Euclidis, et paucis, quae ex conicis elementis Apollonij Pergaei dependent, quae sunt solum duo, quibus in hac scientia summus usi, ut in processu postmodum patebit. In primo itaque huius scientiae libro axiomata praemittimus, quae praeter elementa Euclidis huic scientiae sunt necessaria: et in hoc ea duo, quae demonstrata sunt ab Apollonio, declaramus. Plurima tamen et horum, quae in hoc libro praemittimus, continentur in eo libro, quem de elementatis conclusionibus nominamus, in quo universaliter omnia conscripsimus, quae nobis visa sunt, et quae ad nos pervenerunt a viris posterioribus Euclide, pro particularium necessitate scientiarum universaliter conclusa.

10 "Aequales angulos super idem punctum constitutos aequalem continere distantiam aequalium linearum" (*Opt. Thes. Wit.*, p. 4).

11 "Item inter quaelibet duo puncta lineam, et inter quaslibet duas lineas superficiem posse extendi" (ibid).

12 One should keep in mind, however, the necessary historicity of the concept of rigor itself and its close dependence on, and embodiment in, the reigning mathematical culture. The growth of mathematics may be viewed legitimately, among other ways, also as the successive replacement of standards of rigor in a strictly monotonic increasing sequence.

13 "Le treizième siècle n'est pas été un siècle de génie . . . " (J. E. Montucla, *Histoire des mathematiques* (Paris 1799), vol. I, p. 506).

14 See, for instance, props. 14 and 29 (*Opt. Thes. Wit.*, pp. 8, 13).

15 Howard Eves, *An Introduction to the History of Mathematics*, rev. ed. (New York: Holt, Rinehart and Winston, 1964), p. 121.

16 Cf., for instance, props. 1, 14, 19, 20, 23, 24, 29, 32, etc. (*Opt. Thes. Wit.*, pp. 5, 8, 10, 11–13, 14).

17 Montucla, *Histoire des mathematiques*, p. 508: "Il indique . . . dans son auteur, une connaissance de géométrie rare pour le temps où il vivait."

18 See Lindberg's "Introduction" cited above, note 2.

19 Cf. my forthcoming edition and translation of Books II and III of the *Perspectiva* in the series *Studia Copernicana;* also, *Opt. Thes. Wit.*, p. 65: "A terminis aequidistantium altitudinum corporis luminosi altioris, et corporis umbrosi bassioris productae lineae concurrentes, sunt suis altitudinibus proportionales. Ex quo patet, quod eadem altitudo corporis umbrosi ex lumine bassiori longiorem proijcit umbram quam ex lumine altiori."

20 Cf. my forthcoming edition [note 19]; also *Opt. Thes. Wit.*, p. 65:
Sit altitudo corporis umbrosi cuiuscunque linea *AB*: et sit altitudo alia illi aequidistans ipsius corporis luminosi, quae sit *DE*: sitque linea *DE* maior quam linea *AB*: producaturque linea *EB* et *DA*, quae protractae concurrent ad aliquam partem in puncto *G* per 16 t 1 huius. Dico, quod erit proportio lineae *GB* ad lineam *GE*, et lineae *GA* ad lineam *GD* sicut lineae *AB* ad lineam *DE*. Quia enim linea *BA* aequidistat lineae *DE* ex hypothesi: palam ergo per 29 p 1, quoniam angulus *GBA* est aequalis angulo *GED*, et angulus *GAB* aequalis angulo *GDE*: angulus quoque *BGA* communis est ambobus trigonis *DGE* et *AGB*: ergo per 4 p 6 est proportio lineae *GB* ad lineam *GE*, sicut lineae *BA* ad lineam *ED*: ergo per 5 t 1 huius, erit econtrario proportio lineae *GE* ad lineam *BG*, sicut lineae *ED* ad lineam *AB*. Palam ergo est propositum: quoniam eodem modo demonstrari potest de lineis *GA* et *GD*. Et ex hoc patet, quoniam eadem altitudo corporis umbrosi ex lumine bassiori longiorem proijcit umbram quam ex lumine altiori.

Esto enim quod aliquod corpus luminosum sit in puncto *H*: Cadatque radius *HA* in punctum lineae *EG*, quod sit *K*: eritque per premissum modum proportio *EK* ad *BK*, sicut *HE* ad *AB*: sed per 8 p 5 proportio *HE* ad *AB* est minor quam *DE* ad *AB*: sed proportio *DE* ad *AB* est, sicut proportio *EG* ad *BG*, ut patuit: ergo per 11 p 5 proportio *EK* ad *BK* est minor quam *EG* ad *BG*. Multum ergo excrevit umbra *BK* respectu umbrae *BG*, ut patet per 10 p 5 et per 4 t 1 huius. Et ex hoc accidit, quod umbrae lunares semper sunt longiores quam umbrae solares: et ita est de aliis corporibus luminosis altioribus et bassioribus quibuscunque. Patet ergo propositum.

21 Unguru, *Witelonis Perspectivae Liber Primus*, p. 50; for the Latin text see ibid., p. 217.

22 Ibid.; for the Latin text see ibid., p. 218.

23 Actually, prop. VI.2 does not warrant the conclusion, whereas the fifth postulate, or Witelo's equivalent, I.14, does.

24 Unguru, *Witelonis Perspectivae Liber Primus*, p. 56; for the Latin text see ibid., p. 223.

25 "Restat experientia mera, quae, si occurrat, casus; si quaesita sit, experimentum nominatur" (*Novum organum* I LXXXII).

26 "Praeterea, Isidorus dicit quod daemones per experientiam multa cognoscunt. Sed experimentalis cognitio est discursiva; ex multis enim memoriis fit unum experimentum, et ex multis experimentis fit unum universale, ut dicitur in fine *Posteriorem* et in principio *Meta*. Ergo cognitio angelorum est discursiva" (*Summa theologiae* I.58,3, quoted after the Blackfriars Edition, vol. 9 [1968], p. 150).

27 "Praeterea, Isidorus dicit quod angeli multa noverunt per experientiam.

Experientia autem fit ex multis memoriis, ut dicitur in *Meta*. Ergo in eis est etiam memorativa potentia" (ibid., I.54,5, p. 86).

28 Ibid., I.64,1, p. 282: "experientia a sensu oritur."

29 "experientia in operabilibus non solum causat scientiam, sed etiam causat quemdam habitum, propter consuetudinem, qui facit operationem faciliorem" (ibid., I.II.40,5, vol. 21 [1965], p. 16).

30 "Dicendum quod tentare est proprie experimentum sumere de aliquo. Experimentum autem sumitur de aliquo, ut sciatur aliquid circa ipsum; Et ideo proximus finis cujuslibet tentantis est scientia" (ibid., I.114.2, vol. 15 [1970], p. 76).

31 Ibid., II.II.2.10. What Thomas says here is that "if human reason provides sufficient proof," then "the merit of faith is altogether taken away."

32 "Πάντες ἄνθρωποι τοῦ εἰδέναι ὀρέγονται φύσει." (*Met*. A.980a22).

33 "Tamen auctoritas non sapit nisi detur eius ratio, nec dat intellectum sed credulitatem; credimus enim auctoritati, sed non propter eam intelligimus. Nec ratio potest scire an sophisma vel demonstratio, nisi conclusionem sciamus experiri per opera, ut . . . in scientiis experimentalibus demonstrabo" (*Compendium studii philosophiae*, in J. S. Brewer, ed., *Opera quaedam hactenus inedita* (London, 1859), p. 397.

34 Cf. N. W. Fisher and Sabetai Unguru, "Experimental Science and Mathematics in Roger Bacon's Thought," *Traditio* XXVII (1971):353–378, passim.

35 "Unde oportet primo credulitatem fieri, donec secundo sequitur experientia, ut tertio ratio comitetur. Si enim inexpertus magnetem trahere ferum nec audiens ab aliis quod trahat in principio debet credere his qui experti sunt, vel qui ab expertis fideliter habuerunt" (J. H. Bridges, ed., *The "Opus Majus" of Roger Bacon* [London, 1900], II, p. 202).

36 "sine instrumentis mathematicis nihil potes sciri, et instrumenta haec non sunt facta apud Latinos, et non fierent pro ducentis libris nec trecentis" (*Opus tertium*, in J. S. Brewer, ed., *Opera quaedam hactenus inedita*, p. 35).

37 Fisher and Unguru, "Experimental Science," pp. 371–372, 366, 376–77.

38 "Radij quorumcumque luminum et multiplicationes formarum secundum rectas lineas protenduntur" (*Opt. Thes. Wit.*, p. 61).

39 "Hoc quod hic proponitur non demonstratione sed instrumentaliter potest declarari; diversitas tamen antiquorum ad hoc probandum pluribus usa est diversis instrumentis, nos vero utimur isto quod hic subscribimus, quod regularius huic proposito credimus convenire" (ibid.).

40 A. C. Crombie, *Robert Grosseteste and the Origins of Experimental Science 1100 – 1700* (Oxford, 1953), pp. 220–23.

41 *Thes. Opt. Wit.*, pp. 61–63:

Assumatur itaque vas aeneum rotundum convenienter spissum, ad modum matris astrolabij, cuius fundi latitudo sit unius cubiti . . . et altitudo hore eius sit aequalis latitudini duorum digitorum . . . et ponatur hoc vas . . . in tornatorio, et tornetur quousque periferia eius sit extrinsecus et intrinsecus vere rotunditatis, et adaequentur plane superficies ipsius, et corpus columnare, quod est in medio dorsi, fiat rotundum. . . . Et deinde reducatur vas ad tornatorium, et signentur in ipso tres circuli aequidistantes. . . . Dividatur itaque medius istorum circulorum in 360 partes, et si possibile fuerit, per minuta: deinde. . . . perforetur foramen rotundum. . . . Deinde accipiatur lamina aenea plana aliquantulum spissa. . . . planeturque adeo . . . et perforetur lamina

foramine rotundo . . . deinde consolidetur parva lamina fundo vasis. . . .
Deinde accipiatur regula aenea quadrangula, . . . et adaequentur
superficies eius, donec fiant aequales rectangulae. . . . Deinde . . . fiat
foramen rotundum, cuius amplitudo sit capax corporis, quod est in dorso
instrumenti . . . fiatque taliter, quod revolvatur in ipso instrumentum
. . . fiantque duae pinnulae . . . quae consolidentur super extremitates
regulae . . . et immittatur cuspis continens regulam cum instrumento.

42 Ibid., p. 63.

43 "Quod hic proponitur potest instrumentaliter demonstrari, ita ut
demonstratio auxilio instrumenti sensibiliter exprimatur" (ibid., p. 407).

44 "Quantitates angulorum refractionis ex aere ad aquam experimentaliter
declarare" (ibid., p. 408).

45 "Quantitates angulorum refractionis ex aere vel aqua ad vitrum planum vel
convexum, et econverso experimentaliter declarare" (ibid., p. 410).

46 "Quantitates angulorum refractiones ex aere vel aqua ad vitrum concavum,
vel econverso experimentaliter invenire" (ibid., p. 411).

47 "Hoc patet per experientiam" and "Quod hic proponitur, patet ratione et
experientia" (ibid., p. 414).

48 "Quod hic proponitur, patet satis ex praemissis: sed et idem placuit
experimentaliter declarare, et universalem caussam particulariter
exemplare" (ibid., p. 440).

49 "Sed et in horum experimentatione est maxima latitudo quam relinquimus
ad talia curiosis" (ibid., p. 444).

50 For which, see Crombie, *Robert Grosseteste*, D. C. Lindberg, *Theories of
Vision from al-Kindi to Kepler* (Chicago, 1976); W. A. Wallace, O. P., *The
Scientific Methodology of Theodoric of Freiberg* (Fribourg, 1959); and Carl
B. Boyer, *The Rainbow: From Myth to Mathematics* (New York, 1959).

51 A. Lejeune, *Recherches sur la catoptrique grecque* (Brussels, 1957), pp.
153–155, and D. C. Lindberg, "Introduction" to *Opt. Thes.* (see note 2,
above), pp. XX–XXI.

52 *Opt. Thes. Wit.*, p. 64:
Luces et colores in corporibus diaphanis non admiscentur adinvicem,
sed penetrant distincti . . . Huius rei experimentaliter declarandae
caussa, ponantur in loco aliquo candelae multae localiter distinctae, et
sint omnes oppositae uni foramini pertranseunti ad locum obscurum, et
opponatur foramini in loco obscuro aliquod corpus non diaphanum.
Luces itaque candelarum apparent super illud corpus distincte secundum
numerum candelarum, et quaelibet illarum apparet opposita uni candelae
secundum lineam rectam transeuntem per foramen et per medium
luminis candelae: et si cooperiatur una candela, destruetur unum lumen
oppositum illi candelae tantum, et discooperta candela, revertitur lumen.
Palam itaque quod luces in medio foraminis, ubi se intersecant omnes
vel plures in puncto uno, non admiscentur in eodem puncto, sed sunt
distinctae per sui ipsarum essentias: et ob hoc cum ulterius
protenduntur, tunc secundum locorum, quibus incidunt, diversitatem
localiter distinguuntur. Et quoniam lux res coloratas pertransiens,
illarum coloribus coloratur, ut suppositum est: palam, si lumen penetrat
distinctum, et colores, qui feruntur cum lumine, penetrabunt distincti.
Patet ergo propositum.

53 "Omne corpus luminosum minus spatium, a quo non egreditur, fortius
illuminat quam spatium maius illo" (ibid., p. 70).

54 Ibid.

55 "Huius propositionis probatio plus experientiae instrumentorum innititur, quam alteri demonstrationum" (ibid., p. 76).

56 Ibid.

57 "Ex quo patet, quod transitus lucis per corpus aquae est secundum lineas rectas per 1 p 11. Et hoc est, quod circa propositam propositionem experimentaliter intendimus declarare" (ibid., p. 77).

58 "Experimentaliter etiam et hoc propositum theorema potest declarari" (ibid.).

59 "Instrumentali similiter experientia propositum theorema potest declarari" (ibid., p. 78).

60 "Hoc quod nunc hic proponitur, est conformiter prioribus per instrumentalem experientiam declarandum" (ibid., p. 80).

61 "Omnem radium incidentem et refractum in eadem plana superficie consistere est necesse" (ibid., p. 81).

62 "Sed et id, quod nunc proponitur, potest experimentaliter declarari" (ibid.).

63 "Satis evidens est haec experimentatio omni tempore" (ibid.).

64 "Illud, quod particularibus experientijs hactenus instrumentaliter probatum est, naturali demonstratione intendemus adiuvare. Omnes enim motus naturales, qui fiunt secundum lineas perpendiculares, sunt fortiores, quoniam coadiuvantur virtute universali coelesti secundum lineam rectam brevissimam, omni subiecto corpore influente" (ibid.).

65 "Palam itaque secundum rationem praemissam fortitudinis perpendicularium et per experientias instrumentales per 42 et 44 huius, quoniam radius incidens . . . perpendiculariter, penetrat totum corpus" (ibid., p. 82).

66 Ibid.

67 "Et quoniam hoc est falsum et contra suppositionem, quae patet sensui, quoniam possibile est rem maiorem ipso oculo videri: palam, quia non est possibile, ut superficies organi visivi sit plana" (ibid., p. 85).

68 "Oculus est organum virtutis visivae sphaericum, ex tribus humoribus et quatuor tunicis a substantia cerebri prodeuntibus sphaerice se intersecantibus compositum" (ibid.).

69 "Quomodo sit oculus virtutis visivae organum, negotio alterius partis philosophiae relinquimus: quod autem sit sphaericus, necessarium est per praecedentem propositionem: et etiam ex eo, quod est naturae aqueae, cuius proprietas est semper rotundari. . . . Quod autem sit oculus ex tribus humoribus et quatuor tunicis compositus, diligens anatomizantium cura edocuit" (ibid.).

70 "Sic ergo patet, quod humores et tunicae oculi sphaerice se intersecant: et patet declaratio definitionis propositae oculi secundum omnium eorum experientiam qui de ipsius anatomia hactenus scripserunt" (ibid., p. 87).

71 "Laeditur enim visus ex forti luce, ut in aspectu corporis solaris vel alterius lucis fortis, ut lucis reflexae ad oculum a corpore polito, vel ab alio corpore valde albo" (ibid., pp. 87–88).

72 "Visio non fit sine dolore et passione a substantia oculi abijciente. Ex quo patet, visum oportere convenientis dispositiones in sanitate esse ad hoc, ut complete exerceat visionem" (ibid., p. 91).

73 Ibid., p. 112, my emphasis:
> Distinctio vero inter illas duas albedines non est ipse sensus albedinis: quoniam sensus albedinis est ex dealbatione superficiei visus, quae fit ab utraque albedine: distinctio autem illarum albedinum fit propter diversitatem actionis illarum duarum albedinum in ipsum visum. Non est ergo illa distinctio a solo sensu, sed est ab alia virtute animae, quam

dicimus distinctivam. Et similiter est de comparatione et distinctione aliarum sensibilium formarum: nihil enim istorum accipitur solo visu, sed ratione et virtute distinctiva coadiuvantibus: visus enim per se non habet virtutem distinguendi, sed virtus distinctiva animae distinguit omnia illa mediante visu. Patet ergo propositum.

74 Crombie, *Robert Grosseteste* (see note 40, above), p. 218.
75 Ibid., p. 214.
76 Ibid., p. 216.
77 Ibid., p. 230.
78 Ibid., p. 232, quoting Witelo X.83, *Opt. Thes. Wit.,* p. 474.
79 *Opt. Thes. Wit.,* p. 439. This appears in prop. X.40.
80 Cf. The quotation appearing in text to note 78, above.
81 A. Koyré, "The Origins of Modern Science: A New Interpretation," *Diogenes* 16 (1956):1–22, at 14–15.
82 *Theories of Vision* (see above, note 50).
83 Ibid., p. 188.
84 Ibid., p. 281 n. 122.
85 Ibid., pp. 207–208.
86 I am struck by this statement as almost tautological.
87 Koyré, "Origins of Modern Science" (note 81, above), p. 12.
88 *G. K. Chesterton Selected Stories* (London, 1972), p. 12.

PART IV. MEDICINE

11

The two faces of a medical career: *Jordanus de Turre of Montpellier*

MICHAEL R. McVAUGH

I

How important did the medieval natural philosopher think it to confront his philosophy with experience of the natural world? This is a question that preoccupied fourteenth-century academics less than it has twentieth-century historians, who see in such a self-conscious comparison of fact and theory the germ of the scientific revolution to come. Granted that the *Posterior Analytics* explained how a science of nature depended on sense-data, did that explanation lead some natural philosophers actively to collect these data, or did it remain for all an epistemological abstraction? Oddly, in considering this question historians have not yet paid much attention to the one sector of the scholastic community that we know did routinely address and manipulate such data: the academic physicians. Did they apply medical theory in their practice? Did they alter (or abandon) theory as a result of practice? Such problems remain virtually unexplored.

This neglect may in part be due to a widespread unwillingness to conceive of the scholastic physician as being a practitioner as well as a lecturer. A long tradition of historical writing has viewed scholastic medicine as divorced from, and even opposed to, the empirical orientation of the ordinary healer. The implicit argument has been that scholastic medical theory was false and sterile; if academic physicians had really been actively practicing medicine, they would have recognized this sterility and developed new ideas; they did not, ergo etc. However, more recent scholarship has made it clear that in France and Italy the students and professors in university medical faculties were routinely and deeply involved in medical practice.[1] Furthermore, there is good evidence that even the lowliest practitioners, lacking any academic training, still chose to interpret their empirical findings in the light of whatever scraps of scholastic terminology and theory they could pick up. A newly expanding theoretical framework was a dynamic factor in medicine by the late thirteenth century or early fourteenth century, and to some extent would have shaped the understanding of virtually every healer, however trained.[2]

Another powerful reason for historians' failure to investigate the relation of theory to practice in medieval medicine is that detailed accounts of physicians' activities do not exist. Nothing like the sixteenth- and seventeenth-century case notes of Turquet de Mayerne or Richard Napier can be found in the thirteenth and fourteenth centuries. In fact, usually we cannot even say which physicians were reputed to be the ablest of their day, let alone talk about the intricacies of their practice. The academically trained physician may potentially be a particularly significant figure for our understanding of medieval attitudes toward experience; unfortunately, that potential can only be realized in instances where we know enough about an individual physician to evaluate him both as practitioner and as theoretician, and such instances are rare indeed.

One physician who comes near to meeting these qualifications is Jordanus de Turre, a master in the medical faculty of Montpellier during the first third of the fourteenth century. To be sure, we have relatively little concrete knowledge about the details of his career.[3] He can have been born no later than about 1280, but he first appears to us in the summer of 1313 among ten or so named medical masters regulating admission to the faculty.[4] In 1318 he assembled a collection of his most successful treatments, *recepte*, for his son Johannes, "eunti ad practicam";[5] in the late fall of the same year he was summoned urgently to Barcelona to treat the grave illness of Jaume II of Aragon.[6] Two years later Jordanus was accused of disturbing the peace of the university, of declaring that he valued its chancellor (Guillem de Beziers) less than "quantum pro uno ansere," and was temporarily expelled.[7] But by 1326 he had returned to the faculty as *decanus* and had completed an ambitious study of pharmaceutical theory, *De adinventione graduum*, which developed themes that had been prominent in debates at Montpellier in the previous generation. Jordanus was still at Montpellier nine years later when he was recalled to Barcelona, this time to consult on the condition of the dying Alfons IV. Thereafter nothing is known of him.[8]

Even so summary an account of a life can indicate something of its character. For Jordanus to have been twice singled out from the dozen or so masters of Montpellier to treat a monarch two hundred miles away, near the beginning as well as toward the end of his career, strongly suggests that he had early acquired a considerable reputation for clinical skills. His decision to collect and publish his *Recepte* –

obviously not meant simply as a private communication to his son – indicates his own pride and confidence in those skills. Most of the other surviving texts attributed to Jordanus have the same general character, and consider at greater length the treatment of a specific illness – "Ad curam hydropisis," "Regimen ad curandum epilepticum," and so forth.[9] He seems to have enjoyed a particular reputation for understanding leprosy: a generation after Jordanus's disappearance from the record, Guy de Chauliac cited "magistrum Jordanum in monte pessullano" as his authority on the symptoms of this disease, and in fact two different texts *de lepra* have been attributed to him.[10]

Another clear indication of Jordanus's contemporary reputation is the incorporation into recipe collections of a number of cures ascribed to him.[11] These collections, which appear to derive from early-four-teenth-century Montpellier, bring together recipes attributed to a dozen or so masters at the school, among which Jordanus's contributions bulk largest. Some of these latter can be found more or less in the same form in his *Recepte* of 1318, and may have been copied out of that work, but by no means all of these recipes or *experimenta* are of this sort. The suggestion is strong that some at least were passed down independently, particularly the more circumstantial ones – like the treatment prescribed "pro quodam Lambardo qui patiebatur febre ter-tiana."[12] The preservation of such individualized treatments speaks to the value placed on Jordanus's practice by his contemporaries. So, too, does a mid-fourteenth-century copy of a regimen (for a patient *dispositus ad paralisim et colericam passionem*) drawn up by four Montpellier physicians, of whom Jordanus is named first.[13] Such scraps of evidence, inconclusive though they are, suggest that Jordanus was regarded as one of the best practitioners of his day in France and Spain, and that his cases were exemplars of treatment.

What may be called the other face of Jordanus's medical career, his academic-philosophical life, is known to us solely through his one surviving work of scholastic medicine, *De adinventione graduum in medicinis simplicibus et compositis*. This work has so far been found in only one (fifteenth-century) manuscript, MS Vat. lat. 2225, which mixes medical texts with writings by mathematicians and natural philosophers like Campanus de Novara and Giovanni Marliani.[14] The *explicit* in this copy reads, in part, "tractatus iste . . . factus et editus ad utilitate studentium medicine maxime in studio montis pessulani et ad informacionem Johannis filii nostri supradicti finitus fuit . . . die mercurii pre quadragesima per magistrum Jordanem [*sic*] de Turre magistrum in artibus et medicina et decani studii medicorum montis pessulani anno domini M.CCC⁰.XXV⁰."[15] The text as provided by this lone manuscript is quite poor, with innumerable copyist's errors and omissions, and can only be construed with patient thought. Inaccessible as *De adinventione* thus is on several counts, a summary of its plan and contents may therefore be interesting.

In order to make such a summary meaningful, however, it is necessary first to give a quick account of the problem to which *De adinventione* was addressed, the problem of determining the degree of a compound medicine. Since classical times it had been generally understood that an individual drug – a simple medicine – had a qualitative nature of complexion characterized by the dominance of one active quality, hotness or coldness, and one passive quality, dryness or wetness; the strength of this qualitative dominance was measured on a scale running from temperateness (neutrality) to the fourth degree, where the different degrees were established by an increasingly strong effect on the sense of touch.[16] In the 1290s, Arnald of Villanova (then a member of the medical faculty at Montpellier) attempted to go further and to determine a technique by which a physician could fix the degree of a compound medicine – could calculate, that is, the sum of the effects of the individual simple medicines making up the compound. He had two Arabic models to choose between, the inventions of al-Kindi and Averroës, who had both interpreted the degree of a medicine as derived from a relation between the number of qualitative "parts" of hot and cold that it contained. They had agreed that the ratio in the temperate medicine is 1:1, and that it is 2:1 in a medicine of the first degree; for al-Kindi, however, the ratio thereafter increased geometrically for subsequent degrees ($2° = 4:1, 3° = 8:1, 4° = 16:1$), while for Averroës it increased arithmetically ($2° = 3:1, 3° = 4:1, 4° = 5:1$).

Arnald adopted al-Kindi's geometric relationship, but for purposes of actual calculation combined this with a further feature of Averroës' system. Al-Kindi had supposed that equal weights of all medicines were comparably effective, so that, for example, one dram of a medicine hot in the third degree would precisely neutralize one dram of a medicine cold in the same degree. From Averroës' work Arnald developed a quite different thesis, namely, that all medicines of the same degree share the same "dose," a smallest quantity that will produce the characteristic effect of that degree, and that medicines can properly be compared only in those doses or in multiples thereof. As an example, in order to reduce a dose (two drams) of sandalwood, which is cold in the second degree, to temperate, by adding honey, which is hot in the second degree, we must use the dose proper to honey, two ounces (Arnald had proposed as a subsidiary refinement that the dose of foodstuffs was eight times that of medicines of the same degree). Arnald's system allowed the physician to calculate the strength of any compound medicine from its constituents, measuring it in degrees and subdivisions of a degree – he posited three equal subdivisions or *mete* within each degree. Although Arnald left Montpellier in 1300, returning only briefly in the years 1305–8 (he died in 1311), it is clear that even in his absence from the school his ideas on a mathematical pharmacy excited much discussion and indeed controversy among the other masters there. It is in this tradition that Jordanus is taking part, some thirty years after Arnald first proposed his theory.[17]

Jordanus introduces *De adinventione* by explaining that he has been asked to clarify this *difficilis et pulchra* subject by his son John and his friends in the *studium* of Montpellier, and will treat it under five (as it eventually turns out, six) *distinctiones* and their chapters. To begin with, he explains (I–1) why it is necessary to assign degrees to simple medicines (i.e., because it is sometimes desirable to alter a patient's unhealthy complexion by giving him a medicine opposite to his complexion in quality but equal in strength), and he then goes on (I–2) to argue on similar grounds for the necessity of degrees for compounds. Those who claim that simples can do anything that compounds can do are mistaken, he insists, for compound medicines can be designed to respond to a number of different problems simultaneously.

With a need for medicinal degrees established, Jordanus proceeds to set out (II–1) three current definitions of "gradus," or degree: as a quality's difference from an abstract norm; as a specific, perceptible change worked upon the human body by a primary quality; or as a specific series of ratios (whether arithmetic, geometric, or unspecified) between opposing qualities. He develops (II–2) this last definition by explaining more fully what geometric and arithmetic proportionality entail in the systems of al-Kindi and Averroës, and concludes that the latter is the *verior calculatio*, because it corresponds better both to sensation and to patterns of natural increase in, for example, music and geometric figure. He repeats (II–3) the different ratios used in the two systems to characterize the four degrees, and describes the four bodily physiological changes that correspond to them: an imperceptible, cumulative *mutatio*; a perceptible *impressio*; a *lesio*; and a *corruptio*.[18] Finally, he treats (II–4) the subdivisions or *mete* of the degree, proving against Arnald of Villanova that we must admit four *mete* rather than three into a degree (though later he concedes a way of admitting three instead), and making the point that they should be related by the same proportionality as the degrees themselves, whether arithmetic or geometric.

Now Jordanus takes up what has evidently become a particularly difficult problem for Montpellier to resolve: How does the concept of quantity apply to degrees? He first distinguishes (III–1) between quantity as applied to extended matter (whether magnitude or multitude, in Aristotle's terms) and what he calls "quantitas virtualis," which measures the intensity of something like whiteness or hotness that can undergo intension and remission, a quantity that is "de gremio forme sive agentis." Next (III–2) he addresses the subject of dose, the quantity of a medicine that has been determined to produce that medicine's characteristic effect, and he constructs his discussion in the form of a refutation of Arnald of Villanova's teaching in the *Speculum medicine* of 1308. Arnald, he argues, was wrong on a number of specific points, all stemming from his failure to understand that it is impossible to assign the same fixed dose to all medicines (or foods) of a given degree: Each medicine has its own proper dose that must be determined individually. But will increasing dosage increase degree too, as Averroës seems to say? Certainly not per se, replies Jordanus, since extensive quantity cannot be a *principium agendi*; it is only *quantitas virtualis* that increases intensity per se, while the greater medicinal effect of larger doses arises *per accidens*.

Having dealt with the general issues raised by compounding medicines, Jordanus can turn to the actual calculations necessary to determine the degree of a compound. Curiously, despite his announced belief that the arithmetic is the *verior calculatio*, he describes the geometric method first (IV–1), and in much greater detail, working out as an example the degree in al-Kindian terms of a medicine compounded from thirteen simples, and showing how to compute it whether assuming three or four *mete* to a degree. His explanation of the arithmetic approach (IV–2) uses no examples and treats only a four–*meta* degree.

Jordanus concludes *De adinventione* with two *distinctiones* that attempt to resolve specific questions about "degree" in general (V) and about particular degrees (VI), questions that let us understand a little more fully what theoretical problems were preoccupying the Montpellier faculty in his day. Two of them (V–1, V–3) have to do with the human complexion and its relation to sensation and degree, but the one to which Jordanus devotes most attention – and about which, as he says, "fuit hactenus magna controversia in hoc studio" – is again the question of whether quantity can increase or decrease degree (V–5). He argues once more that while the amount of a medicine or the duration of its application may alter its degree *per accidens*, only a change in "quantitas virtualis" can alter degree per se.

This narrow and technical work is particularly revealing of Jordanus as academic master. If the implications of his introduction are to be trusted, he had a reputation among his colleagues and students not merely as an able practitioner but as a particularly sensible interpreter of theoretical issues. Certainly no one lacking intellectual self-confidence would have been likely to assail "magnus Arnaldus" so repeatedly and directly on such issues, for Arnald de Villanova had had a predominant influence in shaping Montpellier's recent intellectual life. The medical sources Jordanus cites – Averroës' *Colliget*; Avicenna's *Canon*; Galen's *Tegni, De simplici medicina, Megategni, De*

morbo et accidenti, and *De sectis* – tell us something about his formation in that subject. Most of these titles had entered the Montpellier
curriculum in 1309, and as a master Jordanus would have had to comment upon them routinely, thus developing a command of the new
Galenic literature as well as of the older *articella* in which he was
probably first trained.[19] Contemporaries are quoted as well, not only
Arnald but figures from other medical traditions, Peter of Spain and
Francis of Piedmont (d. 1319); his range of sources makes it clear that
Montpellier's thought – or at least Jordanus's – did not develop in
isolation from Italy or the north.

But the references that stem from Jordanus's prior education in the
liberal arts are still more revealing. The use of the *Categories, Metaphysics* (with Averroës' commentary), *Ethics* (with Eustratius' commentary), *Physics, Meteorology*, and *De anima* – not to mention Avicenna's *Sufficientia* – makes plain a broad and thorough formation in
Aristotelian philosophy. Jordanus makes great use, too, of the ingredients of the quadrivium: of arithmetic, geometry, and music, though
not astronomy. They are particularly prominent when he explains the
nature of geometric and arithmetic proportionality, and when he defines the ratios variously characteristic of the four degrees in the systems of al-Kindi and Averroës as examples of different species of inequality: *multiplex, superparticularis, superpartiens*, and so forth. To
be sure, this much of the language of arithmetic theory had already
been applied to pharmacy in al-Kindi's work. However, Jordanus goes
further, defending the arithmetic sequence over the al-Kindian geometric one by insisting that arithmetic increase is what occurs naturally;
he illustrates this by citing the increasing sequence triangle-quadrangle-
pentagon in geometry, and the sequence melodia-diapason-diatesseron
in music. This language must surely have been drawn from the elementary quadrivial literature, presumably from Boethius (whom Jordanus cites, along with Euclid), and probably his *De institutione arithmetica*, where all these examples can be found. Jordanus later
suggests that the infinite divisibility of a line implies the existence of
an infinite number of medicinal degrees, and, indeed, of the stages in
any quality capable of intension and remission. His comparison between geometric extension and qualitative intensity is particularly interesting because it foreshadows later developments in natural philosophy,[20] but its starting point, the assumed infinite divisibility of a line,
again shows Jordanus's acquaintance with at least certain basic teachings of an arts curriculum.[21] Indeed, we can perhaps conclude that
Jordanus's pointed use of quadrivial doctrine was meant to emphasize
his own competence in this regard to medical readers who might not
have had training in the liberal arts. In Jordanus's day, the requirements
for a medical degree at Montpellier included five years of study for
those who had the M.A., and only one more – six – for those who did

not.[22] Significantly, Jordanus was careful to identify himself as "magistrum in artibus et medicina" in the *explicit* to *De adinventione*, as though to insist that both types of learning were essential for a thorough development of medical theory.[23]

II

Anyone who pursued a medical curriculum in a medieval university was bound to be sensitized to the peculiar epistemological character of his subject – particularly if, like Jordanus, he had already studied the liberal arts. Medicine was unique in that it was both *scientia*, as Avicenna had explained at the beginning of the *Canon*, and *ars*, as Hippocrates had labeled it in the opening phrase of the *Aphorisms*. For medicine to be *scientia* meant, to an Aristotelian familiar with the *Metaphysics, Ethics,* and *Posterior Analytics*, that it incorporated universal truths derived necessarily by logical demonstration from accepted premises. *Ars*, on the other hand, referred to a capacity to act reasoningly on the outside world so as to make or do something that need not otherwise exist or happen.[24] For medicine to be an *ars* implied therefore that it dealt not with the necessary but merely with the desirable and possible. The need to relate these two quite different interpretations of the character of medical knowledge was only one of the issues that had made the *scientia/ars* distinction a widely discussed subject in faculties of medicine by 1300. Is it really possible, for example, for medicine or, indeed, any study of natural things to yield demonstrable and universal truths, since such studies must depend upon a knowledge of particulars? In the *Metaphysics*, Aristotle seemed to distinguish medical practice from medicine-as-*ars* on just these grounds: The physician who possesses the art will recognize the universal even if perhaps not the individual, and hence will be wiser than the mere practitioner, though he may often be unable to cure.[25] Again, medical knowledge that we would call broadly physiological was usually understood as *scientia*; could therapeutic wisdom be adjudged "scientific" as well? We today must resist the temptation to see the two terms as corresponding to our "theory" and "practice": *ars*, like *scientia*, was *doctrina*, a knowledge that can be taught, not an activity. But in contrasting them the medical student was nevertheless brought to reflect on the contrast between learning from books and learning from patients.

Another contrasted pair of words that the medical student encountered early in the curriculum – *theorica* and *practica* – suggests a closer fit to the modern terminology, but in fact (at least within the schools) both these terms still technically referred to *doctrina*. The assertion that medicine could be divided into these two realms was a commonplace of medical education by virtue of its prominence in a wide variety of staple texts. It was featured in Johannitius' *Isagoge* and Haly Abbas's *Pantegni* for twelfth-century readers, and by the end of the thirteenth century was being learned from Avicenna's *Canon* as well. The language of the *Pantegni* was perhaps a little equivocal, and did not make it perfectly explicit that *practica* is, like *theorica*, an aspect of textbook learning: "All medicine is either *theorica* or *practica*. *Theorica* is a complete knowledge of things that must be mastered by the intellect alone, bearing in mind the needs of practice (*rerum operandarum*); *practica* demonstrates the subjects of theory in the realm of the senses and of manual operation, in accordance with the understanding gained in previous theory."[26] But "demonstrates" (*demonstrare* in some manuscripts of the work, *monstrare* in others) seems to imply the exposition of *practica* in the schoolroom; it is *operari* and *operatio* that more obviously refer to modern "practice."

This point is made much more clearly at the very outset of the *Canon*, where Avicenna begins by announcing that medicine is a science:

> Someone may complain, "Medicine is divided into *theorica* and *practica*, but you are putting it all into *theorica* when you say that it is a *scientia*." [I answer:] When we say that part of medicine is *theorica* and part is *practica*, you must not suppose that we are thereby dividing it into theory (*scire*) and practice (*operari*), as many studying this phrase appear to think. Rather . . . , both these divisions are *scientia*: one [*theorica*] is *ad sciendum principia* and the other [*practica*] is *ad sciendum operandi qualitatem*. [Here Avicenna gives examples of theoretical and practical *scientia*.] Hence when you understand (*sciveris*) the character of these two divisions, the *scientiam scientialem* and the *scientiam operativam*, you will have obtained mastery *even if you have never actually practiced* (*adeptus eris etsi nunquam operatus fueris*) [author's emphasis].[27]

Avicenna's intended usage is plain: *scire* and *operari* are the two faces of medicine, while *theorica* and *practica* are subdivisions of the former. To be sure, the *Canon*'s distinction between *practica* and *operatio* was not always rigidly maintained by early-fourteenth-century writers, who can be found using the former word to refer loosely to what the physician does when he deals with patients. But when they try to explain themselves carefully, they invariably return to the Avicennan usage and set *practica* off from *operatio*.[28] For example, Jordanus addressed his *Recepte*, colloquially, to a son entering "ad practicam." But in the *De adinventione* he was careful to distinguish between *theorica* and *practica* in the technical, Avicennan sense: The *medicus theoricus*, he says, studies how to describe medicinal intensity on a scale of numerical degrees, while the *medicus practicus* knows how to do so on a scale of sensory impressions (fol. 65ra) – a distinction corresponding obviously to the *theorica* and *practica scientia* of the *Canon*.

Avicenna's analysis of medicine's aspects brought home to the medical student the distinction between classroom and sickroom learning more directly than did comparisons of medicine as *scientia* and *ars*. However, the student was not thereby led to think about how *scire* and *operari* might be related, for it was acknowledged that there is no demonstrable, teachable, relation between the two. Innumerable particulars condition a person's state of health, and there is no way to ensure that *scientia practica* can be fitted successfully to every case, since each case to which a physician is summoned is to some extent individual and unique. All that can be taught is the contents of *practica* – the norms, rules, and canons that guide and direct *opus*, embodied in such things as *consilia, experimenta*, and lists of drugs – and the way in which they depend upon *theorica*.[29]

In Jordanus's day, the most reflective discussion of how the two facets of a physician's activity were joined was Arnald of Villanova's *De intentione medicorum*, written at Montpellier at the beginning of the 1290s, when Jordanus was perhaps launching upon his medical training.[30] The physician's task, explained Arnald, is to preserve or restore health. "Corresponding to *(iuxta)* these two *labores operandi* there must be two subdivisions within that *doctrina* or part of medicine that is called operative or *practica*. . . . There is another kind of *doctrina* necessary to the art of medicine, in which we learn how to understand *(cognosci)* health and illness with its causes, and this part is called cognitive or *theorica*."[31] Arnald has added to the Avicennan analysis the point that there is a "correspondence" between *scientia* or *doctrina* and *opus*. What more definite can be said about this correspondence? "In every *doctrina cognitiva* . . . [the physician] uses those words with which he can most quickly and with fewest obstacles be directed *ad recte operandum*. . . . Since the *opus* concerns particulars, an awareness *(notitia)* of the particular and individual is more valuable to *opus* than a universal and general *speculatio*, and consequently in his *doctrina* [the physician] descends to the individual as far as possible."[32]

These are Arnald's only observations about the nature of the correspondence, and it should be remarked that they run *in one way*, from *doctrina* to *opus*. For it is *opus* that justifies the physician, not *doctrina*; it is simply not relevant to Arnald's purpose to reflect on whether *recte operandum* may have consequences for *doctrina*.

To summarize, then: An academically trained physician of 1300 would have acknowledged two aspects to his work, *scire* and *operari* – though with some variability in the terminology. The first maxim of

his profession – Galen's "Intentio medicorum sola operatio est"[33] –
together with its tendency to see medicine-as-*scientia* as demonstrable
truth, ensured an intellectual movement from *scientia* to *opus*, from
theory to practice: As Arnald argued, "[doctrina] cognitiva, que prima
est, non est nisi propter operativam et ad illam ordinatur." Hence we
may occasionally find scholastic physicians suggesting how theory may
broadly guide (though not logically determine) the particulars of practice; but the nature of their training was such as to discourage them
from thinking that the reverse was also true. The study of experimental
medicine was not, I think, something that could easily have been introduced into a medieval medical school. In any case, of course, all
this is something of an abstraction. The texts may tell us how the
scholastic physician understood theory and practice to be linked in
medical activity, but they do not tell us whether they really were linked
in this way.[34] It is precisely at this point that it becomes important to
examine the activity of particular physicians whom we know as both
theoreticians and practitioners – like Jordanus de Turre.

III

How are we to explore the interaction of the two halves of
Jordanus's professional life? What sort of evidence would reveal to us
the shaping of the practitioner by the theoretician, or vice versa? Unfortunately, we cannot investigate in the way that would at first seem
most natural, by looking to see whether he applied or tested the doctrines of *De adinventione* in his practice. Although Jordanus's *Recepte*
of 1318 list several hundred compound medicines, these cannot be assumed to embody his views of 1326 on the mathematization of pharmacy, and, more seriously, they are all the wrong sort of medicines.
Jordanus makes it clear in *De adinventione* (fol. 60vb) that the mathematical procedures he is describing are for defining the degree and dose
of alterative medicines, those medicines the physician employs when
he wants merely to alter the patient's *complexio* or qualitative balance,
"since the dosage of other [kinds of] medicines – laxatives, theriacs,
those that cure some particular or universal illness by *virtus specifica*
– can be learned *ab auctoribus*" (of whom Avicenna is Jordanus's
favorite). The *Recepte* preserve just such nonalterative remedies, compounds found by experience to produce a specific effect. There is no
need to record individual alterative medicines, precisely because they
can be compounded in accordance with general rules.[35]

Yet our inability to study how Jordanus may have applied these views
in actual practice need not be a serious bar to understanding his scientific personality. The confrontation of a particular doctrine or theory
with experience was bound to be less decisive for the medieval physician, it seems to me, than for the medieval architect or navigator –

to mention two cases used by Guy Beaujouan to exemplify a potentially fruitful interaction.[36] The aim of the physician's activity, we must remember, was his patient's health, a state that is peculiarly difficult to define in any objective manner. The regularly successful outcome of the physician's treatment of most illnesses (which as we now recognize are self-limiting) would have convinced him that his beliefs were sound. When Jordanus de Turre was summoned to Jaume II of Aragon late in the autumn of 1318, the monarch was in the grip of a quartan fever (presumably malarial) that had first attacked him in September; his condition had worsened to the point that in November he was preparing resignedly for his death.[37] Of his doctors' treatment of the illness we know only that toward the middle of November they prescribed an electuary containing powdered gold and pearls, one conceivably recommended by Jordanus himself.[38] Almost immediately the king's health stabilized and then began slowly to improve, although only in May 1319 did he announce himself fully recovered. We may imagine that Jordanus and his colleagues took this episode to confirm the medicinal claims made for gold and pearls by such authorities as the *Antidotarium Nicolai*.[39] But is this application of theory to practice really comparable in its success to the use of geometry in architecture of which Beaujouan is speaking? I think we must acknowledge that much of medieval medical theory lay outside the realm of testability, and that its applicability to the pursuit of health could not easily have been disproved by experience.

In any case, Jordanus displays a concern in *De adinventione* for the actual process of compounding that suggests he did indeed expect such elaborate alterative medicines to be prepared. While his account of thirteen precautions to be observed in selecting simples for compounding – collect them at the right time and place, use them while they are fresh, and so on – is a traditional one, this scarcely means it was not intended to be taken seriously. More convincing, however, is the discussion following his argument that, contrary to Arnald of Villanova's teaching, it is impossible to lay down laws establishing one common dose for all medicines of a given degree and that, on the contrary, each must be determined individually:

> When reason proves inadequate for this, and prior
> experience is lacking, what should be done (*quid operandum
> erit*)? . . . The physician should select from as temperate a
> locality in his region as he can find the most temperate
> individual he can find in that area; then select a medicine
> hot in the first degree, the best in that degree in all respects,
> and administer it to his subject – who should be healthy,
> neither full nor hungry, not *subdolens*, neither chilled nor
> heated, on a temperate day and hour – administer to him, I

say, some medicine in the first degree, in quantity let us say half a dram. Note what effect this quantity has, considering urine, pulse, the color and warmth of the body, and other pertinent signs of hotness. Determine whether he perceives hotness or not; if he does, perceptibly, the quantity is already too great, since [a first-degree medicine] ought not to do this on its first administration. But if he does not, let it be given to him again under the same conditions and see what happens. If it is given four or five times and has no effect upon his senses, leave him to himself for several days and then start again in the same manner with a larger quantity of the medicine. Do this, adding or subtracting each time a greater or lesser quantity of the medicine, until you come to the first perception of the first degree. (In my view this can be more easily determined for the second degree, and the others, than for the first.) Now [that you have established the proper dose for one medicine of a given degree], note the quantity of this medicine together with its nature and other circumstances. . . . Determine two things, its *pondus* and its *dimensio per vasculum vel per filium* . . . , and with these measures of *dimensio* and *pondus* known for one medicine it will be possible to compare another medicine of the same degree in these respects . . . and by estimation to arrive at the dose [for it] that you believe called for by this rule.

Or again: choose a compound medicine from among those listed by Avicenna in the third or fifth [book of the *Canon*]; look at its dose and take into consideration its quality and degree as well as you can, as well as the purpose for which it was compounded. Then consider the qualities of other medicines and how distant they are from the mean . . . , the illness and organ at which they are directed, and how often they are to be taken, with respect to a temperate body and to their application to a temperate body. Then you can use this compound medicine [and its dosage] to regulate your doses of simple medicines intended for illnesses of the members or for other purposes. I present these ideas not as demonstrations, rather as *inductiones* to make medical practice better.[40]

I have set out Jordanus's account in full because it reveals so vividly his understanding of the practical niceties involved in making an experimental measurement or, indeed, in adapting the measurements of other authorities to one's own particular circumstances.[41] And in the end, as Jordanus is also well aware, the correctness of even your best results is not guaranteed. His conclusion – "istas non assero demonstrationes, sed potius inductiones ad bene operandum" – captures perfectly the practitioner's recognition of priorities.

Admittedly, we can see this statement, if we wish, as consciously derived from the scholastic's medical epistemology, from the view that medicine-as-*scientia* can only guide *operatio* and cannot determine it. But we can equally well see it as a practitioner's spontaneous acknowledgement that medical theory must always be controlled by a physician's sensitivity to the realities of nature, illness, and therapy. I have already mentioned that it was a commonplace of academic medicine that the limits of *scientia* were defined by the particular realities of individual cases, but Jordanus goes well beyond mere parroting of commonplaces on the primacy of experience – "signum verorum sermonum est ut concordent rebus sensatis" (fol. 57rb, quoting Averroës' commentary on the *Physics*) – to explain over and over the need to base medical judgments on sense-experience.

It is this conviction that allows him to attack the ideas of Arnald of Villanova. While Arnald was teaching at Montpellier in the 1290s, he began to develop heterodox, somewhat Joachimite ideas on the nature and future of the Christian community, and it was an increasingly vehement defense and propagation of these ideas that took him in the last decade of his life so much away from medicine and into oneiromancy and theology.[42] Already in 1303 there is some suggestion that Arnald's fellow masters at Montpellier thought of his religious opinions as wild ravings, but there is no sign that either these opinions or his growing fascination with dream-interpretation were lowering his colleagues' estimation of his purely medical teachings.[43] Two decades later, however, Jordanus extended this criticism of Arnald's thought from his theology to his medicine, missing few opportunities to hint that his famous predecessor had drawn many of his ideas out of the air. "These," says Jordanus of his own teachings concerning *mete*, teachings in conflict with Arnald's, "were not *in somno reperta sed perscrutativa consideratione inventa*" (fol. 59rb); and earlier, offering an un-Arnaldian definition of *gradus*, he was even more sarcastic about such an approach: "The present definition seems to be more complete than any, and is my own formulation, the best I am capable of; nor have I chosen to assert it arbitrarily, as things are asserted to man on the basis of reason or divination or prophecy or divine revelation, whether by God or by the separated substances, – or by a dream, as has recently been the case."[44]

In keeping with such criticism, Jordanus's detailed refutation of the theories of dosage set out by Arnald in the *Speculum* is based to a remarkable extent on arguments drawn from experience, on thought experiments, even though he acknowledges their limitations. How can

Arnald assert that a mere one dram of any medicine hot in the fourth degree, such as pepper, can corrupt the whole body? Not even a dram of fire itself can do this.

> For example: imagine a statue of some sort, of reeds, for example, to be heated to the heat of a living man; it is obvious that one dram of fire in a local inflammation will not create a great heat like that of the fourth degree; wherefore, etc. Or imagine a hollow man created out of leather, and let it be heated to the heat of a temperate body and suspend within it one dram of fire or charcoal with a good draft; it is obvious that even two [drams] could not heat this man to a strong heat – whence this would appear not to be true. Granted that these examples do not fit the case exactly (*plenarie similitudinem non habeant*), they still in their way contradict the aforesaid [Arnaldian] position to the shrewd and sensible student.[45]

Again, how can Arnald suppose that an ounce of food will alter the body's complexion as much as will a dram of medicine of the same degree? "This too appears to contradict experience. We see that lettuce, purslane, and cucumber are eaten in summer, and cucumber in autumn, to a quantity of half a pound at a time, and yet they do not have the effect described. Truly I believe that in his reasoning and . . . in his conclusions he was equally distant from the truth."[46]

Jordanus proceeds to link Arnald's *Speculum* implicitly with Francis of Piedmont's writings as having accomplished nothing "nisi quod pergamenum album scriptum scriptura vel litteris denigratur et animus legentis piger et lentus in posteriora reperitur" (fol. 61rb). A harsh assessment to make of an accepted authority, but one that Jordanus justifies by insisting on the faulty speculative basis of Arnald's medicine; he is clearly alluding to Arnald's *Speculum* when, in a pleasant play on words, he criticizes the defects of the "liber speculatoris."

We see that for an able practitioner like Jordanus medical experience should control theory at the same time that it is guided by it. Can it modify and help generate theory, too? Indeed it can, and did for Jordanus. We must remember, however, that experience itself can never be free of prior theory; the form in which Jordanus was led to cast his experience is not always what we today would expect. This is particularly true of his medical-clinical experience, for this was shaped by preexisting conceptions of health and sickness, which have a peculiarly subjective character in every age and culture. Consequently, most of the cases where Jordanus uses clinical experience to enlarge theory seem to us to have no solid experiential foundation at all.

Thus, for example, Jordanus extended his pharmaceutical theory by applying to it the universally recognized stages distinguishable by sense in every illness, stages that medieval physicians had learned to recognize from their reading of Galen's *De crisi* (I.ii): *principium, augmentum, status* or *crisis*, and *declinatio*.

> In any illness . . . there are four stages: onset, development, crisis, and recovery. Suppose an illness to be hot and wet in the second degree; then suppose four parts hot and wet to be assigned to its onset, five to development, ten to crisis, and six to recovery. A medicine will therefore be needed dry and cold in the second degree; necessarily, if the physician wants a perfectly curative procedure, he should divide the cold and dry medicine into four *mete*, of which the first will contain four [parts] cold and dry, the second five, the third ten, and the last six. By proportioning the first *meta* of the cold and dry medicine in this way, it is suited to the onset of the illness, which has four parts; the second is suited to development, the third to crisis, the fourth and last to recovery. For unless the curative increase or decrease in the numbers is proportional to the remedy needed by that stage, [the treatment] will be that of another stage, which is undesirable.[47]

This argument is meant to disprove Arnald of Villanova's aprioristic claim that every degree contains three *mete*, defended on the grounds that degree has breadth (*latitudo*) and hence, like all breadth, must have three parts: two extremes and a middle. Jordanus's counterclaim that there are four *mete*, not three, is based instead on what he understands to be its utility to the physician: four *mete*, he insists, best allow him to attain his ends, the restoration of the patient's health. In an analogous manner, Jordanus argues that the preservation of health will also be easier using this approach because of the correspondence of the four *mete* to the four ages of man. We today would be hesitant to quantify four stages of illness in this way, and furthermore cannot readily see why Jordanus felt it legitimate (as he evidently did) to transfer the meaning of *meta* from a subdivision of a degree to a subdivision of a dose. Even so, we must recognize that he has introduced into theory an innovation he has derived from his own operative understanding of the course of illness or of a healthy life.

Where natural-philosophical rather than medical understanding shaped Jordanus's experience, it is much easier for us to acknowledge that experience could, in turn, enhance theory. The "magna controversia" at Montpellier over whether increasing a drug's quantity increased its degree had grown out of a tension between natural philosophy and medical – indeed, everyday – experience. Aristotle had placed quantity and quality in different categories, and this seemed to make philosophically inadmissible the measurement of the intensity of a quality by the extension of its subject; moreover, as medical sophists had pointed out, in that case a medicine of first degree could have the effect of any degree at all, given a large enough dose. On the other hand, experience did seem to show the relation between magnitude and qualitative intensity in a variety of cases: Heavier bodies fall faster to earth, burning twigs heat less than do piles of coals, and medicines do require definite quantities, doses, to produce their characteristic effects.

Jordanus resolved this tension – with what has been called "the first theoretically neat solution to the problem"[48] – when he explained how increasing a medicine's quantity can increase its effect while leaving its intensity unchanged:

> It would seem that we should say that degree does not increase with increase or decrease of quantity *per se* but *per accidens*. The reason for this is that the heat produced by a reflection of hot or cold, . . . whether rectilinear or angular, is a hotness or coldness that heats or cools not *per se* but *per accidens*; but this sort of hotness is produced by multiplication and diminution of quantity; wherefore, et cetera. For when the entire body is heated up to the extremities or to the heart by a precise, fixed quantity of medicine, [because the heat from a larger quantity] cannot extend further, it is reflected back here or there and the hotness or coldness is increased; and thus the effect of another degree of increased or reflected quantity is brought about as well as that of the [first and] non-reflected quantity – as is the case with the heat of fire in a darkened room. For when the quantity is so small that it cannot heat the entire body by multiplication of its *virtus*, its further multiplication in the extension of the subject depresses and remits its *virtus*, since the greater the subject in which some quality is multiplied, the weaker is its effect, and the lesser, the more powerful it is, as is evident with the heat of fire and similar actions. Hence one ought to acknowledge a definite quantity [of a medicine] whose hotness, diffused through its subject in proper proportion, will heat (or chill) the whole body, neither more nor less. This is the true nature of *dosis*, for if [it is] less, it loses the strength of its proportion, and if more, brings about a reflection and produces another reflected quality as well as the direct [non-reflected quality], and of a higher degree.[49]

Jordanus is using here the model of optics and of the multiplication of species – the model that the Merton school would soon use to the same end of quantifying intensities – to formulate a new theory of drug action that will conform to experience. It was to make this argument possible that he began his discussion by distinguishing Aristotle's extended quantity from another kind, a "quantity of some form (*rei formalis*) capable of intension and remission . . . ; and such a quantity can be called a 'quantitas virtualis.'" Because the *quantitas virtualis* necessary to bring about a given effect is linked to extended quantity – that is, dosage – in a way that varies from drug to drug, Jordanus is approaching what would have been difficult for a strict Aristotelian: a "basis for considering different distributions of the same or equal quantities of quality into greater or lesser extensions." It was Nicole Oresme a generation later, and in the Middle Ages only Oresme, who was able consciously to develop this modern-seeming approach to "quantity of quality."[50]

I am suggesting, then, that whatever we may choose to imagine about the sterility or uselessness of scholastic medicine, it cannot be taken to show (1) that scholastic physicians did not involve themselves in medical practice; nor (2) that, while they may have practiced, they kept theory and practice well separated from each other; nor even (3) that they allowed theory a certain role in shaping practice but not vice versa. As I have tried to show here, an able practitioner-master like Jordanus used his medical experience to control and enlarge his scientific understanding, and even if he did not conceive of testing his own theories, he could appeal to practical experience in order to disprove the ideas of an opponent. It should be emphasized, too, that the scholastic's scientific thinking was not necessarily narrowly medical, kept tightly insulated from other intellectual realms. In Jordanus's case, at least, the teachings of the trivium and the quadrivium were fully integrated into his medical thought and, like that thought, were subject to shaping by experience. Like other academics, Jordanus acknowledged the two faces of his activity, as concerned both with *scientia* and with *operatio*, and it is often convenient for us to maintain the same categories; but we must not forget that ultimately they were coexisting, interdependent aspects of his identity as a physician.

Notes

1 A recent statement of the more traditional view is John M. Riddle, "Theory and Practice in Medieval Medicine," *Viator* 5 (1974):157–84. The case for the practical activities of academic physicians is made by Luke Demaitre, "Theory and Practice in Medical Education at the University of Montpellier in the Thirteenth and Fourteenth Centuries," *Journal of the History of Medicine* 30 (1975):103–23, and by Nancy G. Siraisi, *Taddeo Alderotti and his Pupils* (Princeton, N.J.: Princeton University Press, 1981).

2 Consider, for example, the revealing story of Geralda de Codines, a healer of Subirats (near Barcelona). In 1304 the bishop of Barcelona interrogated

her about her practices and was told that she diagnosed illnesses through uroscopy; her diagnostic terminology makes clear her rough familiarity with academic medical language, which in a later interrogation she revealed had been picked up thirty years before from a physician passing through town. Josep Perarnau i Espelt, "Activitats i fórmules supersticioses de guarició a Catalunya en la primera meitat del segle XIV," *Arxiu de Textos Catalans Antics* 1 (1982):67–70.

3 It is not unlikely that Jordanus de Turre was Italian by origin: his given name (Giordano) is commoner in Italy than in France, and he shows a knowledge of contemporary Italian medical authors in his writings and alludes to Italian domestic practices (e.g., "fungi . . . multum nutriunt, sicut volunt italici, qui comedunt in multa et magna quantitate" [*De adinventione graduum* (see below, note 14), fol. 60 va]). On Jordanus, see Ernest Wickersheimer, *Dictionnaire biographique des médecins en France au moyen âge* (Paris: E. Droz, 1936), vol. 2, 513–14; and its *Supplément*, ed. Danielle Jacquart (Geneva: Droz, 1979), 195.

4 Alexandre Germain, *Cartulaire de l'Université de Montpellier*, I, 119 (Montpellier: Ricard, 1890); the document (#30) is dated July 16, 1313.

5 The text, beginning "Et primo quoad quassaturas . . . " is found in MSS Oxford, All Souls College 80, fols. 47–52; Bamberg L.III.37, fols. 186–89; Prague lat. 1534, fols. 57–65; London, British Library, Sloane 2527, fols. 241v–50v; and Paris, BN nal 1391, fol. 70r–v (where it is incomplete).

6 The king named Jordanus "clericum et fisicum ac familiarem nostrum" on December 28, 1318 (Archivo de la Corona de Aragón (ACA), Canc. Reg. 216, fol. 106) and gave him a mule worth 1000 sueldos (Canc. Reg. 259, fol. 167v); at about the same time, Jordanus was paid a further 200 sueldos "pro aliquibus missionibus que facere habebat eo quia nos ipsum vocabamus pro infirmitate nostra" (Canc. Reg. 259, fol. 197v).

7 Germain, *Cartulaire*, I, 249–50; document #39, dated July 30, 1320. Jean XXII, *Lettres communes* (ed. G. Mollat; Paris, 1905), III, no. 11840.

8 The king gave "magistro Jordano fisico Montispessulani" 100 gold florins, worth 1450 sueldos, late in the summer of 1335 (ACA, Canc. Reg. 502, fol. 177r); in September and again on October 5 and 26, he was given 200 sueldos *per provisio*, and a final payment of 149 sueldos was made on November 2 (ACA, Real Patriomonio Reg. 307, fols. 45v, 48v, 57v, 58r). Jordanus seems to have returned to Montpellier before the king's death, which occurred on January 24, 1336.

A statute of the medical faculty from Montpellier dated November 2, 1335 (Germain, *Cartulaire*, I, 292–93; document #61) includes "Johannes de Turre" among the thirteen masters listed. P. Pansier, "Les maîtres de la faculté de médecine de Montpellier au moyen âge," *Janus* 10 (1905):3, interpreted this as an error for "Jordanus" de Turre; however, as Jacquart (*Supplément*, 188), has pointed out, it could perfectly well refer to Jordanus's son, Johannes, who had been in practice for nearly twenty years, and this is now established conclusively, because the documents just cited show that Jordanus himself was in Barcelona on that very day.

9 The *Regimen ad curandum epilepticum* is attributed to Jordanus in MS Leipzig, Univ. 1183, fols. 95–96v, but to Arnald of Villanova in Wolfenbüttel, 89. Gud. Lat. 2°, fols. 235v–37v, and, more indirectly, in Sevilla, Biblioteca Colombina 5-1-45, fols. 156v–58v. The Leipzig manuscript, however, is earlier than the other two and in other respects reveals itself as a particularly close witness to the doings of fourteenth-century Montpellier.

10 The texts are the *Signa leprosorum* (inc., "Cognoscuntur leprosi a quinque signis . . ."), usually assigned to Arnald of Villanova but attributed to Jordanus in MS Erfurt, Amplon. Q 320, fols. 212v–17, and *De lepra nota* in Basel D.I.11, fols. 97v–99 (inc., "Quoniam omnis operatio medicinarum . . ."). I have not been able to decide which (if either) of these two texts Guy had in mind when he referred to Jordanus' views on leprosy (*Chirurgia magna* VI.i.2; ed. L. Joubert [Munich, 1585; rpt. Darmstadt: Wissenschaftliche Buchgesellschaft, 1976], 252). Another reference to Jordanus in the *Chirurgia magna* (VI.ii.7; ed. Joubert, 349) is a further testimony to his continuing reputation for practical skill. I owe my knowledge of Guy's references to Jordanus to the kindness of Dr. Margaret S. Ogden.

11 Michael McVaugh, "Two Montpellier Recipe Collections," *Manuscripta* 20 (1976):175–80.

12 MS Leipzig, Univ. 1183, fol. 89ra.

13 Karl Sudhoff, "Eine Diätregel für einen Bischof, aufgestellt von vier Professoren von Montpellier in der Mitte des 14. Jahrhunderts," *Archiv für Geschichte der Medizin* 14 (1923):184–86.

14 This manuscript, to which all my subsequent citations of *De adinventione* refer, is described by Lynn Thorndike, "Some Medieval and Renaissance Manuscripts on Physics," *Proceedings of the American Philosophical Society* 104 (1960):195–200.

15 Precise though this date appears to be, it is still equivocal, since we cannot tell whether Jordanus dated the beginning of the year from the Nativity or the Circumcision, in which case the work was finished on February 13, 1325, or from the Incarnation or Easter, in which case the work was completed on January 29, 1326; the latter usage seems to have been the more common in Languedoc.

16 For the views of classical antiquity, see Georg Harig, *Bestimmung der Intensität im Medizinischen System Galens* (Berlin: Akademie Verlag, 1974).

17 The background and development of Arnald's mathematical pharmacy are treated in *Opera Medica Omnia Arnaldi de Villanova* (*OMOAV*), II: *Aphorismi de gradibus* (Barcelona: Universitat de Barcelona, 1976). A brief description of Montpellier's response to Arnald's pharmaceutical theory is given in Michael R. McVaugh, "Quantified Medical Theory and Practice at Fourteenth-Century Montpellier," *Bulletin of the History of Medicine* 43 (1969):397–413.

18 Jordanus mentions here (fol. 58ra) that some masters at Montpellier were arguing that the four degrees correspond to four *mutationes* (*dissolutionem; liquefactionem; cursibilitatem*; and *resolutionem*) in the *humidum radicale*.

19 For the impact of the new curriculum of 1309, see *OMOAV*, XV: *Commentum supra tractatum Galieni de malicia complexionis diverse . . .* (Barcelona: Universitat de Barcelona, 1985), 16–37.

20 Cf. Marshall Clagett, *The Science of Mechanics in the Middle Ages* (Madison: University of Wisconsin Press, 1959), 333–35.

21 *De adinventione*, fols. 59ra and 65rb, reflecting *Physics* VI. The continuing importance of quadrivial instruction in the arts curriculum is shown by Pearl Kibre, "The Boethian *De institutione Arithmetica* and the *Quadrivium* in the Thirteenth Century University Milieu at Paris," in M. Masi, ed., *Boethius and the Liberal Arts* (Berne: Peter Lang, 1981), 67–80.

22 Germain, *Cartulaire*, I, 219–21; document #25 (dated September 8, 1309).

Luke Demaitre, "Bernard of Gordon (ca. 1258–ca. 1315): A Representative of the Montpellier Academic Tradition" (Ph.D. dissertation, City University of New York, 1973), 58–67, discusses the relation of arts to medicine at Montpellier. He suggests that an interest in the liberal arts was strong enough in some medical masters of the 1330s to lead the faculty to more sharply "separate the study of medicine from that of liberal arts."

23 Pearl Kibre, "Arts and Medicine in the Universities of the Later Middle Ages," in Josef Ijsewijn and Jacques Paquet, eds., *The Universities in the Late Middle Ages* (Leuven: Leuven University Press, 1978), 213–27, argues that in fourteenth-century Paris (and schools founded on its model), training in arts had become a prerequisite for medical studies, and discusses the possible bearing of the several liberal arts on medical education.

24 Cf. *Ethics* VI.4, or, for a medieval restatement, Thomas Aquinas, *Summa theologiae* I.II.57.3: "ars nihil aliud est quam ratio recta aliquorum operum faciendorum." See also Galen, *De sectis* [text 1]: "Generalis omnium artium diffinitio est collectio ex comprehensis premeditata ad quandam utilitatem vita aptissimam" (*Opera Galieni*, 1490, fol. 2rb).

25 See in particular *Ethics* VI.3–4 and *Metaphysics* I.1 for these ideas. Much of my thinking on this subject has been shaped by the unpublished paper of Nancy G. Siraisi, "Views on the Certitude of Medical Science Among Late Medieval Medical Writers."

26 "Omnis ergo medicine aut theorica est aut practica. Theorica est perfecta notitia rerum solo intellectu capiendarum subiecta memorie rerum operandarum; practica est subiectam theoricam demonstrare in propatulo sensum et operatione manuum secundum preeuntia theorice intellectum." Haly Abbas, *Liber pantegni*, theorice I.3 (*Omnia Opera Ysaac*; Lyons, 1515, II, fol. 1v).

27 "Potest autem aliquis dicere quod medicina dividitur in theoricam et practicam, sed tu totam ipsam posuisti theoricam cum dixisti quod est scientia. Nos autem respondebimus: . . . Cum ergo de medicina dixerimus quod eius est theorica et ex ea est practica, non est existimandum quod velimus dicere quod una divisionum medicine est scire et altera operari, quemadmodum multi hunc locum perscrutantes existimant. Sed debes scire . . . quod nulla divisionum medicine est nisi scientia; sed una earum est ad sciendum principia et altera ad sciendum operandi qualitatem. . . . Cumque duarum divisionem qualitatem sciveris iam scientiam scientialem et scientiam operativam adeptus eris etsi nunquam operatus fueris." Avicenna, *Liber Canonis* I.i.1.i (Venice, 1507; rpt. Hildesheim: Georg Olms, 1964), fol. 1r.

28 Nancy G. Siraisi, *Taddeo Alderotti*, 121–22, shows Taddeo (d. 1295) insisting on just this point.

29 A stimulating analysis of the interrelationship of theoretical and practical elements in scholastic medical instruction is provided by Jole Agrimi and Chiara Crisciani, " 'Doctus et expertus': la formazione del medico tra due e trecento," *Quaderni della Fondazione Giangiacomo Feltrinelli* 23 (1983):149–171.

30 I am preparing an edition of *De intentione medicorum*, of which the most generally satisfactory manuscript appears to be Oxford, Merton College 230, fols. 56r–60r. The references that follow are to this manuscript.

31 "Unde iuxta hos duos labores operandi duplicem oportuit esse partem illius doctrine vel partis medicinalis que operativa vel practica dicitur, scilicet conservativam et curativam. . . . Propter hoc alia pars doctrine fuit

necessaria in arte medicine, in qua sanitas primo et per consequens morbus cum suis causis doceretur convenienter cognosci, et illa pars doctrine cognitiva vel theorica." *De intentione medicorum*, tr. I; fol. 56r-v.

32 "Omnia enim in arte sunt propter finem eius; quoniam igitur medicus cognitionem ad operationem ordinat, in omni doctrina cognitiva vel cognitionis hiis sermonibus utitur quibus facilius et paucioribus impedimentis ad recte operandum dirigitur. . . . Cum namque opus circa singulare existat, plus in opere proficit notitia singularis et proprii quam speculatio universalis et communis, et ideo in doctrina sua ad propria descendit quantum potest, et obmittit contraria quantum potest similiter."Ibid., fol. 57r.

33 Galen, *De sectis* [text 2]; *Opera*, 1490, fol. 3rb.

34 This point is emphasized by Nancy G. Siraisi, "Some Recent Work on Western European Medical Learning, Ca. 1200–Ca. 1500," *History of Universities* 2 (1982):233 n.4.

35 Earlier, in "Quantified Medical Theory" (above, note 17), I attributed this failure to apply pharmaceutical theory to practice to Jordanus's lack of interest in doing so, which I now think too strong a judgment [cf. below, note 50].

36 Guy Beaujouan, "Réflexions sur les rapports entre théorie et pratique au moyen âge," in J. E. Murdoch and E. D. Sylla, eds., *The Cultural Context of Medieval Learning* (Dordrecht: D. Reidel, 1975), 437–84.

37 See J. Ernesto Martinez Ferrando, *Jaime II de Aragón: Su vida familiar*, I (Barcelona: Consejo Superior de Investigaciones Cientificas, 1948), 255–60, for an account of the king's illness.

38 ACA, Canc. Reg. 281, fol. 31v, dated November 15, 1318. The archives of the Real Patrimonio (Reg. 282, fol. 85r) record a payment in November 1318 of 154s 6d "a Maestre G. Jorda Especier de Casa del SR . . . per l letovari que Maestre Johan Amell e Maestre Marti faeren per a obs del SR." This may or may not be a reference to the same electuary; if it is, the responsibility for its prescription was taken not by Jordanus but by the king's personal physicians, Joan Amell and Martí de Calçaroja.

39 The best-known drug embodying these two materials was probably *aurea alexandrina*, which the *Antidotarium Nicolai* claimed "valet ad omne capitis vitium ex frigiditate maxime et ad omnem reumaticam passionem que a capite ad oculos et aures et gingivas descendit, et ad gravedinem omnium membrorum que fit de eodem humore" (W. S. Van den Berg, *Eene middelnederlandsche vertaling van het Antidotarium Nicolai* [Leiden: E. J. Brill, 1917], 5); it goes on to explain that "detur in modum avellane febricitantibus cum acqua calida."

40 *De adinventione*, fol. 61rb-va:

> Cum ratio ad hoc non inveniatur sufficiens et experimenta preterita, quid tunc operandum erit? . . . Duplex est via. Una est quod medicus in regione sua et in loco temperato magis quam ibidem reperire poterit debet eligere sibi similem temperatum ut melius poterit in regione illa; deinde eligere medicinam calidam in primo gradu, meliorem in omnibus suis dispositionibus in illo gradu, et exhibere ei, eo sano existente, non repleto vel famelico, non subdolente vel infrigidato vel calido existente, tempore die hora temperato; et ei administrare aliquam medicinam in primo gradu, ut puta aliquam quantitatem medie 3, gratia exempli; et notetur quid fecerit impressionis, et consideretur urina pulsus corporis color et calor corporis et alia pertinentia signa caliditatis. Investigentur

an senserit caliditatem vel ne; si notabilem senserit, iam quantitas excessiva fuit, cum prima obviacione hoc facere non debuerit. Si vero non, iterum, condicionibus servatis, exhibeatur eidem et percipiatur quid operatum fuerit; si quatuor vel .v. vicibus fuerit exhibita, et nichil perceptum fuerit mutationis in sensu, sibi ipsi relinquatur per multos dies, et servato quod prius redeatur ad quantitatem eius maiorem. Et hoc operaberis addendo vel subtrahendo in eadem vel maiorem vel minorem quantitatem usquequo perveneris ad inventionem operationem prime experiencie primi gradus. (Sed ego estimo quod facilius et melius reperietur hoc in secundo gradu et in aliis quam in primo). Et nota hanc quantitatem huius medicine in natura sua in condicionibus suis. . . . Notentur ergo dua, pondus eius et dimensio per vasculum vel per filium, et cum ista mensura dimensionum et ponderum medicine experte, poterit eis aliquid comparare eiusdem gradus addendo vel subtrahendo habitudine ad illam et existimatione accipere dosim quam credes cum hac regula esse detentam.

Vel aliter: accipe unam medicinam compositam earum que ponuntur ab Avicenna in tertio vel in quinto, et vide dosim eius et considera cuius gradus sit et cuius qualitatis, ut melius poteris, et ad quam ponitur intentionem in composita medicina; et considera etiam qualitates omnium medicinarum et quantum recedunt a medio vel ab intentione alia; et iterum considera morbum ad quem ordinantur et membra, cum quotlibet accipiatur in corpore temperato et respectu istorum positorum in corpore temperato. Tunc dico quod respectu passionis membrorum vel intentionem aliarum per illam compositam in simplici medicina te poteris in dosibus regulare. Istas non assero demonstrationes, sed potius inductiones ad bene operandum.

41 For other examples of such a sensitivity in other Montpellier academics, see McVaugh, "Quantified Medical Theory," 403; and *OMOAV*, III: *De dosi medicinarum tyriacalium* (Barcelona: Universitat de Barcelona, 1985), 67–68.

42 A good introduction to this period in Arnald's life is provided by Joaquín Carreras Artau, *Relaciones de Arnau de Vilanova con los reyes de la casa de Aragón* (Barcelona: Real Academia de Buenas Letras, 1955), 31–59; and see, too, Paul Diepgen, "Arnalds Stellung zur Magie, Astrologie und Oneiromantie," *Archiv für Geschichte der Medizin* 5 (1912):88–115; reprinted in Paul Diepgen, *Medizin und Kultur* (Stuttgart: Ferdinand Enke, 1938), 150–72.

43 Michael McVaugh, "Nota sobre las relaciones entre dos maestros de Montpellier: Arnau de Vilanova y Bernardo Gordon," *Asclepio* 25 (1973):331–36.

44 "Ista autem diffinitio videtur esse completa inter omnes, et hec est nostre proprie inventionis ut melius potuimus. Nec ponere volumus manifeste vel inmanifeste absolute, cum quedam sint homini manifesta ratione, divinatione, prophetia, vel revelatione divina – vel a deo vel a substantiis separatis vel sompno, sicut nuper apparuit." *De adinventione*, fols. 55vb–56ra.

45 *De adinventione*, fol. 60rb:
Verbi gratia, accipiatur statua cuiuscumque rei calefacta ad calorem hominis vivi, ut puta arundinum; certum enim est quod una [3] ignis circumscripta inflammatione non faciet calefactionem magnam, scilicet 4 gradus; ergo et cetera. P. Supponamus hominem factum de corio

concavum, et calefiat calefactionem corporis temperati et suspendatur in medio [3] l ignis vel carbonum cum bona flabellacione; certum est etiam si essent due quod non calefacerent hunc hominem ad fortem caliditatem tangentem, quare hic non videtur habere veritatem. Et licet hec plenarie similitudinem non habeant, tamen suo modo contrarium ostendunt predicte positionis perspicaciter et sensibiliter intuenti.

46 "Item hec apparet contra experientiam: videmus comedi lactucas, portulacas, cucumeres in estate, cucurbite in autumpno ad quantitatem medie libri, et tamen non faciunt operationem dictam. Et in veritate credo quod in hac cogitatione sua et . . . in solutione equaliter fuit distans a veritate." *De adinventione*, fol. 60va.

47 *De adinventione*, fol. 58vb:

In quolibet enim morbo materiali et salubri sunt quatuor tempora: principium, augmentum, status et declinatio. Ponatur morbus in secundo grado et humidus; ponantur in principio quatuor partes caliditatis et humiditatis, in augmento .v., in statu .x., in declinatione 6. Accipitur ergo medicina et sicca et frigida in secundo gradu; necessario ergo, si velit habere perfectum ingenium curativum medicus, oportet habere et diversificare medicinam frigidam et siccam in quatuor methis, quod prima contineat quatuor frigiditatis et siccitatis, secunda contineat .v., tertia .x., et ultima 6. Modo sic proportionando prima metha medicine frigide et sicce, competit principio morbi, que est .iv. partium; secunda competit augmento, tertia statui, quarta et ultima declinationi. Nisi enim esset elevatio numerorum vel depressio curativa proportionalis cure unius temporis, esset cura alterius temporis, quod est inconveniens.

48 For a full discussion of Jordanus's contribution, see Edith Sylla, "Medieval Quantifications of Qualities: The 'Merton School,'" *Archive for History of Exact Sciences* 8 (1971):22–23.

49 *De adinventione*, fol. 62ra. The text is furnished by Sylla, "Medieval Quantifications, " 22–23 n. 39.

50 In my article "Quantified Medical Theory," my brief remarks about Jordanus's achievement ("links to medical experience have become irrelevancies [for him]") may seem to contradict the argument of this paper by suggesting a divorce in his mind between theory and practice. Those remarks were meant to distinguish Jordanus from his colleagues at Montpellier, who give some evidence of having put these mathematical theories of pharmacy into practice. There is indeed no firm evidence of a link from pharmaceutical theory to practice in Jordanus's work, as we have seen; but that does not mean that a link from experience to theory cannot exist – which is, after all, one of the points of this chapter.

PUBLICATIONS OF MARSHALL CLAGETT

Books

Giovanni Marliani and Late Medieval Physics. New York: Columbia University Press: P. S. King and Son, 1941; Reprinted AMS Press: New York, 1967.

Chapters in Western Civilization. Vol. 1. Selected and edited by the Contemporary Civilization Staff of Columbia College, Columbia University. New York: Columbia University Press, 1948, pp. 3–122 (chap. 1: "The Medieval Heritage: Political and Economic"; chap. 2: "The Medieval Heritage: Religious, Philosophic, Scientific"). 2d ed., 1954, pp. 3–123 (chap. 1: "The Medieval Heritage: Economy, Society, Polity"; chap. 2: "The Medieval Heritage: The Classical Influence"; chap. 3: "The Medieval Heritage: The Christian Conception of Life").

Medieval Science of Weights. Madison: University of Wisconsin Press, 1952. Co-edited with Ernest A. Moody. Reprinted 1960 (with corrections).

Greek Science in Antiquity. New York: Abelard-Schuman, 1955. Reprinted London: Abelard-Schuman, 1957. Revised edition New York: Collier Books, 1963.

The Science of Mechanics in the Middle Ages. Madison: University of Wisconsin Press, 1959. Reprinted 1961, 1979 (with corrections). Italian translation: *La scienza della meccanica nel medioevo*. Milano: Feltrinelli Editore, 1972.

(Editor) *Critical Problems in the History of Science*. Proceedings of the Institute for the History of Science at the University of Wisconsin, September 1–11, 1957. Madison: University of Wisconsin Press, 1959.

Twelfth-Century Europe and the Foundations of Modern Society. Proceedings of a symposium sponsored by the Division of Humanities of the University of Wisconsin and the Wisconsin Institute for Medieval and Renaissance Studies. Madison: University of Wisconsin Press, 1961. Co-edited with Gaines Post and Robert Reynolds. Reprinted 1966, 1980.

Archimedes in the Middle Ages. 5 vols. in 10 tomes. Vol. 1: Madison: University of Wisconsin Press, 1964; Vols. 2–5: Philadelphia: American Philosophical Society, 1967–84.

Nicole Oresme and the Medieval Geometry of Qualities and Motions. A Treatise on the Uniformity and Difformity of Intensities Known as *Tractatus de configurationibus qualitatum et motuum.* Edited with an Introduction, English Translation, and Commentary. Madison: University of Wisconsin Press, 1968.

Studies in Medieval Physics and Mathematics. London: Variorum Reprints, 1979.

Articles

"Some general aspects of physics in the Middle Ages." *Isis* 39 (1948):29–44.

"Richard Swineshead and late medieval physics." *Osiris* 9 (1950):131–61.

"Check list of microfilm reproductions in the history of late medieval physics." *Progress of Medieval and Renaissance Studies in the United States and Canada,* Bulletin No. 21, August, 1951, pp. 36–51.

"Archimedes in the Middle Ages: *The De mensura circuli.*" *Osiris* 10 (1952):587–618.

"A medieval fragment of the *De sphaera et cylindro* of Archimedes." *Isis* 43 (1952):36–38.

"Use of the Moerbeke translations of Archimedes in the works of Johannes de Muris." *Isis* 43 (1952):236–42.

"The medieval Latin translations from the Arabic of the *Elements* of Euclid, with special emphasis on the versions of Adelard of Bath." *Isis* 44 (1953):16–42.

"Medieval mathematics and physics: a check list of microfilm reproductions." *Isis* 44 (1953):371–81.

"The *De curvis superficiebus Archimenidis*: a medieval commentary of Johannes de Tinemue on book I of the *De sphaera et cylindro* of Archimedes." *Osiris* 11 (1954):295–358.

"King Alfred and the Elements of Euclid." *Isis* 45 (1954):269–77.

"A medieval Latin translation of a short Arabic tract on the hyperbola." *Osiris* 11 (1954):359–85.

"The *Quadratura per lunulas*: a thirteenth-century fragment of Simplicius' commentary on the *Physics* of Aristotle." In *Essays in Medieval Life and Thought Presented in Honor of Austin Patterson Evans.* New York: Columbia University Press, 1955, pp. 99–108.

"The *Liber de motu* of Gerard of Brussels." *Osiris* 12 (1956):73–175.

"Impact of Archimedes on medieval science." *Isis* 50 (1959):419–29.

"A medieval treatment of Hero's theorem on the area of a triangle in terms of its sides." In *Didascaliae, Studies in Honor of Anselm M. Albareda,* ed. Sesto Prete. New York: Bernard M. Rosenthal, 1961, pp. 79–95.

"The use of points in medieval natural philosophy and most particularly in the *Questiones de spera* of Nicole Oresme." In *Actes du Symposium International R. J. Bošković 1961.* Belgrade, 1962, pp. 215–21.

"Nicole Oresme and medieval scientific thought." *Proceedings of the American Philosophical Society* 108, no. 4 (August 1964):298–309.

"Archimedes and scholastic geometry." In *L'aventure de la science. Mélanges Alexandre Koyré.* 2 vols. (*Histoire de la Pensée,* vols. 12, 13). Paris: Hermann, 1964, vol. 2, pp. 40–60.

"A medieval Archimedean-type proof of the law of the lever." *Divinitas* 11 (1967):805–820.

"The Pre-Galilean configuration doctrine: 'the good treatise on uniform and difform [surface].'" In *Pubblicazioni del Comitato nazionale per le manifestazioni celebrative del IV centenario della nascita di Galileo Galilei. Saggi su Galileo Galilei,* ed. Carlo Maccagni. Florence: G. Barbèra Editore, 1967, vol. 1, pp. 1–24.

"Some novel trends in the science of the fourteenth century." In *Art, Science, and History in the Renaissance,* ed. Charles S. Singleton. Baltimore: Johns Hopkins University Press, 1967, pp. 275–303.

"Johannes de Muris and the problem of proportional means." In *Medicine, Science and Culture: Historical Essays in Honor of Owsei Temkin,* ed. Lloyd G. Stevenson and Robert P. Multhauf. Baltimore: Johns Hopkins University Press, 1968, pp. 35–49.

"Leonardo da Vinci and the Medieval Archimedes." *Physis* 11 (1969):100–151.

"Prosdocimus de Beldomandis and Nicole Oresme's proof of the Merton Rule of uniformly difform." *Isis* 60 (1969):223–25.

"The quadrature by lune in the later Middle Ages." In *Philosophy, Science, and Method: Essays in Honor of Ernest Nagel,* ed. Signey Morgenbesser, Patrick Suppes, and Morton White. New York: St. Martin's, 1969, pp. 508–22.

"Adelard of Bath." In *Dictionary of Scientific Biography,* ed. Charles C. Gillispie, 16 vols. New York: Scribner, 1970–80, vol. 1 (1970), pp. 61–63.

"Archimedes." In *Dictionary of Scientific Biography.* vol. 1 (1970), pp. 213–31.

"Archimedes in the Late Middle Ages." In *Perspectives in the History of Science and Technology,* ed. Duane H. D. Roller. Norman: University of Oklahoma Press, 1971, pp. 239–59.

"Gerard of Brussels." In *Dictionary of Scientific Biography,* vol. 5 (1972), p. 360.

"Leonardo da Vinci: Mechanics." In *Dictionary of Scientific Biography,* vol. 8 (1973), pp. 215–34.

"Nicole Oresme." In *Dictionary of Scientific Biography,* vol. 10 (1974), pp. 223–30.

"The Works of Francesco Maurolico." *Physis* 16 (1974):149–98.

"The life and works of Giovanni Fontana." *Annali dell'Istituto e Museo di Storia della Scienza di Firenze,* Anno I (1976):5–28.

"Francesco of Ferrara's 'Questio de proportionibus motuum.'" *Annali dell'Istituto e Museo di Storia della Scienza di Firenze,* Anno III (1978):3–63.

"Francesco Maurolico's use of medieval Archimedean texts: the 'De sphaera et cylindro.'" In *Science and History: Studies in Honor of Edward Rosen*, ed. Erna Hilfstein, Pawel Czartoryski, and Frank D. Grande. Wroclaw: Ossolineum, 1978, pp. 37–52.

"Conic sections in the fourteenth century." In *Studi sul xiv secolo in memoria di Anneliese Maier*, ed. A Maierù and A. Paravicini Bagliani. Rome: Edizioni di Storia e Letteratura, 1981, pp. 179–217.

"William of Moerbeke: translator of Archimedes." *Proceedings of the American Philosophical Society* 126 (1982):356–66.

INDEX